Magnetic Resonance in Colloid and Interface Science

Magnetic Resonance in Colloid and Interface Science

Henry A. Resing, EDITOR
Naval Research Laboratory

Charles G. Wade, EDITOR
University of California at Berkeley

A symposium sponsored by the
Divisions of Colloid and Surface
Chemistry, Physical Chemistry,
and Petroleum Chemistry at the
172nd Meeting of the
American Chemical Society,
San Francisco, Calif.,
Aug. 29–Sept. 2, 1976.

ACS SYMPOSIUM SERIES **34**

AMERICAN CHEMICAL SOCIETY

WASHINGTON, D. C.   1976

Library of Congress CIP Data

Magnetic resonance in colloid and interface science.
(ACS symposium series; 34 ISSN 0097-6156)

Includes bibliographical references and index.

1. Colloids—Congresses. 2. Surface Chemistry—Con-
gresses. 3. Magnetic resonance—Congresses.
I. Resing, Henry Anton, 1933-   . II. Wade,
Charles Gordon, 1937-   . III. American Chemical
Society. Division of Colloid and Surface Chemistry.
IV. American Chemical Society. Division of Physical
Chemistry. V. American Chemical Society. Division of
Petroleum Chemistry. VI. Series: American Chemical
Society. ACS symposium series; 34.

QD549.M19   541'.345   76-44442
ISBN 0-8412-0342-3

ACS Symposium Series

Robert F. Gould, *Editor*

FOREWORD

The ACS SYMPOSIUM SERIES was founded in 1974 to provide
a medium for publishing symposia quickly in book form. The
format of the SERIES parallels that of its predecessor, ADVANCES
IN CHEMISTRY SERIES, except that in order to save time the
papers are not typeset but are reproduced as they are submitted by the authors in camera-ready form. As a further
means of saving time, the papers are not edited or reviewed
except by the symposium chairman, who becomes editor of
the book. Papers published in the ACS SYMPOSIUM SERIES
are original contributions not published elsewhere in whole or
major part and include reports of research as well as reviews
since symposia may embrace both types of presentation.

CONTENTS

PREFACE

Magnetic resonance, although a mature tool at this date, continues to be exploited in novel ways. The purpose of the symposium on "Magnetic Resonance in Colloid and Interface Science" is to present a current statement for this area of research. The scope is broadly encompassing; reports of nuclear, quadrupole, electron paramagnetic resonance, and Mössbauer studies are included, with a proper mix of theory and experiment. Our aim is to include all experimental systems in which molecules or ions are bound to interfacial structures or within colloidal systems. We span the range from the biological to the petrological with a sharp emphasis on those systems in which catalytic processes occur. Thus, the first papers focus on anisotropy, gradually yielding to gas–solid adsorption systems and surface kinetics. Aqueous macromolecular systems are then discussed; work with living systems appears finally. A truly international and interdisciplinary roster of participants has made certain that our aims and desires were achieved. That we were able to bring such a large contingent from abroad to participate in this symposium results from not only our desire to fittingly celebrate the Bicentennial of the United States of America, the Centennial of the American Chemical Society, and the Semicentennial of the Division of Colloid and Surface Chemistry, but also from the generosity of our financial supporters, namely: The Petroleum Research Fund of the American Chemical Society, the Experimental Program Development Fund of the American Chemical Society, Office of Naval Research, Marathon Oil Co., Nicolet Instrument Co., Wilmad Glass Co., Inc., Bruker Instruments, Inc., Perkin Elmer, and Union Carbide.

The organizers heartily thank these donors for their substantial part in the success of the symposium. We acknowledge also the encouragement and support of W. H. Wade, Chairman of the Division of Colloid and Surface Chemistry, Noel H. Turner, Secretary of the Division of Colloid and Surface Chemistry, the able secretarial assistance of Mrs. Brenda Russell, and the encouragements of our respective spouses, Kay and Kim.

HENRY A. RESING　　　　　　　　　　　　CHARLES G. WADE
Washington, D. C.　　　　　　　　　　　　Berkeley, Calif.
July 1976

ESR Studies of Spin Probes in Anisotropic Media

JACK H. FREED

Department of Chemistry, Cornell University, Ithaca, N. Y. 14853

We wish to summarize some of our recent studies of spin probes in liquid crystalline media. We also wish to indicate some theoretical aspects of spin-dependent phenomena on surfaces and interfaces.

Order Parameters and Equation of State

One of the most useful ways in which magnetic resonance may be applied to anisotropic media is to the determination of the molecular order parameter:

$$S = \frac{1}{2} \langle 3\cos^2\beta - 1 \rangle \qquad 1$$

where β is the angle between the preferred spatial direction in the medium, usually referred to as the director and specified by unit vector $\hat{n}$, and the symmetry axis of the molecule. Actually, this parameter is sufficient only if the molecule is ordered with cylindrical symmetry (either prolate with positive S or oblate with negative S). In general, there is an ordering tensor, which is completely specified once the principal axes of ordering of the molecule relative to the director are known, and the values of S and the "asymmetry parameter" are given.

$$\delta = \frac{\sqrt{6}}{2} \langle \sin^2\beta \, \cos 2\alpha \rangle \qquad 2$$

In eq.2 α is the azimuthal angle for the projection of the direc-tor in the molecular x-y plane. One finds, in the case of nitro-xides, that if the principal axes can be chosen on the basis of molecular symmetry, it is then possible to determine S and δ from ESR measurements of the effect of the ordering on hyperfine and g-shifts ([1,2]). It is, of course, first necessary to accura-tely measure the hyperfine and g-tensors from rigid-limit spectra preferably in the solvent of interest, since these tensors are generally somewhat solvent-dependent ([1,2,3]). It was found that, by a proper choice of the molecular z-axis, it is usually possible to keep δ small relative to S. The ordering tensor may be related to $P_0(\Omega) = P_0(\beta,\alpha)$ the equilibrium distribution by such expressions as:

$$S = \int d\Omega P_0(\Omega) \frac{1}{2} (3\cos^2\beta - 1) \qquad\qquad 3$$

where

$$P_0(\Omega) = \exp(-U(\Omega)/kT) \; / \; \int d\Omega \exp(-U(\Omega)/kT) \qquad 4$$

with $U(\Omega)$, the mean restoring potential of the probe in the field of the molecules of the anisotropic solvent. Since the mean torque on the probe is obtained by taking the appropriate gradient of $U(\Omega)$ (1,2,4), one has that it is statistically the potential of mean torque for the spin probe. This potential $U(\Omega)$ can be expanded in a series of spherical harmonics of even rank, since, in liquid crystals $\hat{n}$ and $-\hat{n}$ are equivalent. The leading terms in the expansion are then spherical harmonics of rank 2. So, if we keep only these lowest order terms, and employ the principal axis system of the ordering, then

$$U(\beta,\alpha)/kT \simeq -\lambda\cos^2\beta - \rho\sin^2\beta\cos2\alpha. \qquad\qquad 5$$

A non-zero δ then implies a non-zero ρ. Once these potential parameters λ and ρ are measured, then they may be related to theories for the equation of state of the liquid crystal. In the "mean-field" theory of the Maier-Saupe type, one assumes that, for example, $\lambda^{(s)} = S^{(s)} a^{ss}/V^{\gamma_s}T$ (where the superscript implies solvent molecules), where the coefficient a^{ss} which is independent of T and V refers to the strength of the anisotropic potential and $\lambda^{(s)}$ is taken to be proportional to the mean ordering of the surrounding solvent molecules given by $S^{(s)}$. This expression also allows for dependence of the interaction on molar volume V. A simple point of view would be to let $V^{\gamma_s} \propto r^{3\gamma_s}$, so a measurement of γ_s would indicate the nature of the intermolecular forces leading to the orienting potential. The case of $\gamma_s=2$, implying van der Waals attractions, would correspond to Maier-Saupe Theory. More generally, 3γ would represent some mean radial dependence averaged over the different kinds of interaction forces. In the case of a dilute solution of probe molecules, which are ordered by their interaction with solvent, one should write: $\lambda^{(p)} = S^{(s)} a^{sp}/V^{\gamma_p}T$. Thus, studies of the pressure and temperature variation of the ordering should be useful in testing theories of the equation of state. Interesting thermodynamic derivatives one may hope to measure are $(\partial\ln S^{(i)}/\partial\ln T)_V$ and $(\partial\ln T/\partial\ln V)_{S(i)}$. Actually one obtains the analogous expressions with V replaced by P from pressure-dependent studies. Then one uses P-V-T data to obtain these derivatives . Using the simple mean field theory outlined above, but with $\rho=0$, one finds:

$$\left(\frac{\partial \ln S^{(s)}}{\partial \ln T}\right)_V = \frac{R^s}{1+R^s}. \qquad\qquad 6$$

where $\qquad R^s \equiv (\Delta S^{(s)})^2 \, a^{ss} / V^{\gamma_s} T \qquad\qquad 6a$

with $(\Delta S^{(s)})^2$ the mean square fluctuation in order parameter, and R^s is a quantity that is uniquely predicted by Maier–Saupe theory for a given value of $S^{(s)}$ or $\lambda^{(s)}$. Also:

$$\left(\frac{\partial \ln S^{(p)}}{\partial \ln T}\right)_V = \frac{S^{(s)}}{S^{(p)}}\frac{R^{(p)}}{1+R^{(s)}} \qquad\qquad 7$$

with $\qquad R^{(p)} \equiv (\Delta S^{(p)})^2 \, a^{sp} / V^{\gamma_p} T . \qquad\qquad 7a$

Also,

$$\left(\frac{\partial \ln T}{\partial \ln V}\right)_{S^{(s)}} = -\gamma_s \qquad\qquad 8$$

and $\qquad \left(\dfrac{\partial \ln T}{\partial \ln V}\right)_{S^{(p)}} \equiv -\gamma'_p = -\gamma_p\left[1+R^{(s)}\left(1-\dfrac{\gamma_s}{\gamma_p}\right)\right]. \qquad 9$

Thus we see that experimental results on $(\partial \ln S^{(i)}/\partial \ln T)_V$ enable tests of the mean-field theories, while those on $(\partial \ln T/\partial \ln V)_{S^{(i)}}$ provide information on the nature of the orienting potentials. The experimental results on $(\ln S^{(i)}/\ln T)_V$ both for i=s from NMR studies on the nematic PAA (5), and for i=p from ESR studies on the weakly-ordered perdeuterated nitroxide spin probe PD-Tempone in the nematic solvent Phase V, are in qualitative agreement with eqs. 6 and 7 respectively, but the predictions are typically about 50% larger in magnitude compared to the experimental results. Since $R^{(s)}$ is predicted to range from about -3/4 to -1 in the nematic phase, this is not surprising. Thus, small contributions from other terms in the expansion of $U(\Omega)$ (e.g. the fourth rank spherical harmonics), can readily "explain" such effects (4). The main point is that the thermodynamic measurements of ordering, whether for the pure solvent or the spin probe, compare reasonably well with the single parameter (λ) mean-field theory. Thus, it is reasonable to attach some physical significance to eqs. 8 and 9. The NMR result on PAA gives $\gamma_s = 4$, while recent more approximate results from P-V-T measurements at the isotropic-nematic phase transition give $\gamma_s \sim 3.3$-4 for a range of solvents including MBBA, which is very similar in properties to Phase V (6). Now eq.9, after allowing for the small corrections in magnitude in $R^{(s)}$ as discussed above, should be moderately well approximated by $\gamma'_p \sim \frac{1}{2}(\gamma_p + \gamma_s)$. The experimental results on PD-Tempone in Phase V have yielded $\gamma'_p \sim 2$. This suggests

that γ_p itself is rather insensitive to changes in the molar volume. This may not be so surprising if we have (1) the weakly
aligned spin probe located in a cavity of the liquid crystal;
(2) a specific short-range attractive interaction between probe
and solvent molecules dominant in $\lambda^{(p)}$; and (3) the size of the
cavity is not rendered small relative to the molecular size of
PD-Tempone at the pressures achieved. This model is supported and
amplified by the spin-relaxation results below. Clearly, it would
be of considerable interest to determine γ_p as a function of the
size and shape of different probe molecules.

Relaxation and Rotational Reorientation

We now discuss detailed results on ESR relaxation and linewidths: how they are to be analyzed and the microscopic dynamical
models they suggest ($\underline{1}$-$\underline{4}$,$\underline{7}$). As long as the concentration of spin
probes is low, the line shapes are determined by rotational
modulation of the anisotropic interactions in the spin Hamiltonian,
in particular the hyperfine and g-tensors; (the nuclear quadrupole
interaction exists for nitroxides, but is too small to be of
importance except in ENDOR experiments). The important feature of
anisotropic media that is different from isotropic media is the
existence of the restoring potential $U(\Omega)$, to which the molecular
reorientation is subject. The simplest model for molecular
reorientation is then Brownian rotational diffusion including
$U(\Omega)$ in a rotational Smoluchowski equation. This may be most
compactly written in terms of $P(\Omega,t)$, the probability the
molecular orientation (relative to a fixed lab frame) is specified
by Euler angles Ω at time t, and $\vec{M}$ the vector operator which
generates an infinitesimal rotation, by ($\underline{1}$,$\underline{2}$,$\underline{8}$):

$$\frac{\partial P(\Omega,t)}{\partial t} = -\vec{M} \cdot \overleftrightarrow{R} \cdot \left[\vec{M} + \left[\vec{M}U(\Omega) \right]/kT \right] P(\Omega,t) \quad 10$$

The dyadic $\overleftrightarrow{R}$ represents the rotational diffusion tensor. One may
solve this diffusion equation by various methods and then obtain
various expressions for the ESR line widths in the motional
narrowing region ($\underline{2}$,$\underline{9}$,$\underline{10}$). The main points to note are that
1) the restricted rotation due to the restoring potential means
that the rotational modulation of the spin Hamiltonian $\mathcal{H}_1(\Omega)$ is
reduced in magnitude, and 2) the rotational relaxation represented by τ_R for an isotropic liquid, must now be replaced by a
collection of correlation times associated with the different
orientational components to be relaxed in the spin Hamiltonian.
That is, we first expand $\mathcal{H}_1(\Omega)$ in the usual fashion in terms of
the generalized spherical Harmonics $\mathcal{D}_{KM}^L(\Omega)$:

$$\mathcal{H}_1(\Omega) = \sum_{L,K,M,i} (-1)^K \mathcal{D}_{-KM}^L(\Omega) \; F_i'^{(L,K)} \; A_i^{(L,M)} \qquad 11$$

with $F_i'^{(L,K)}$ and $A_i^{(L,K)}$ irreducible tensor components of rank L, where $^1F'$ is in molecule-fixed co-ordinates while A is a spin operator in the lab axes. Then the correlation functions:

$$C(KK';M,M';\tau) \equiv \langle \mathcal{D}^{*(2)}_{K,M}(t)\, \mathcal{D}^{(2)}_{K'',M'}(t+\tau) \rangle$$

$$- \langle \mathcal{D}^{*(2)}_{KM}(t) \rangle \langle \mathcal{D}^{*(2)}_{K',M'}(t+\tau) \rangle \qquad\qquad 12$$

will exhibit different time dependences for each set of values of the "quantum numbers" K, K', M, and M'. (Note that $S=\langle \mathcal{D}^{(2)}_{00}(\Omega) \rangle$.) Such a theoretical analysis has been found, in various studies ($\underline{12}$), to give reasonable semi-quantitative agreement with experiment. However, liquid crystalline solvents tend to be somewhat viscous, and large spin probes tend to reorient slowly. Consequently, motional narrowing theory need not apply ($\underline{1},\underline{2},\underline{4}$). This fact has not been adequately appreciated in past work, since the effect of the ordering in reducing the modulation of $\mathcal{H}_1(\Omega)$ yields narrower lines. Nevertheless, the slow motional features manifest themselves in significant shifts of the resonance lines and in lineshape aysmmetries. This former effect can lead to significant errors in determination of S as outlined above, if the slow-tumbling contributions are not corrected for. The line-width asymmetry starts to be appreciable at somewhat shorter τ_R's, so it can be used as an indicator of slow tumbling in anisotropic media. The slow-tumbling theory of ESR lineshapes in anisotropic media is based upon a generalization of eq. 10 known as the stochastic Liouville equation in which one introduces a probability density matrix that includes simultaneously the spin and orientational degrees of freedom ($\underline{1},\underline{8}$).

Recent careful studies have now shown the importance of slow-tumbling corrections, but, even more importantly, have demonstrated that the rotational reorientation is not so simply described as given by eq. 10 for Brownian reorientation subject to the potential of mean torque ($\underline{1}-\underline{4}$, $\underline{7}$, $\underline{13}$, $\underline{14}$). This matter has been analyzed theoretically from several points of view ($\underline{2},\underline{3},\underline{8}$). The main point to be made is that molecular probes, which are comparable in size to the solvent molecules, are subject to a range of intermolecular torques affecting the reorientation and these torques are characterized by a wide spectrum of "relaxation times". Brownian motion is the limiting model appropriate when τ_R, the rotational reorientation time of the B particle, is much slower than the "relaxation times" for the intermolecular torques experienced by it, such that the latter are well approximated as relaxing instantaneously. Instead, for particles of molecular dimensions, one must distinguish between forces that are relaxing much faster than, comparable to, and slower than τ_R, since they will tend to play a different role in the reorientational motion. These considerations are particularly important for ordered media such as liquid crystals.

According to our approximate theoretical analysis, fluctuating torques relaxing at a rate that is not slower than τ_R will
contribute to a frequency-dependent-diffusion coefficient, e.g.
for a simple model wherein the fluctuating torques decay with a
single exponential decay constant τ_M, one has:

$$R(\omega) = R_0(1-i\omega\tau_M) \ . \qquad\qquad 13$$

where R_0 is the zero-frequency diffusion coefficient. Such a model
might approximately account for the reorientation of a probe
molecule of comparable size to the surrounding solvent molecules
wherein the reorientation of the latter at rates comparable to τ_R,
provide the fluctuating torques causing the reorientation of the
probe molecule. This is called the "fluctuating torque" model
(2,3,4). In the case of an isotropic fluid (i.e. $U(\Omega)=0$), eq.13
leads to a spectral density:

$$j(\omega) \equiv Re\int^{\infty} e^{-i\omega t}\ dt\ C(K;M;t) = \tau_R/\left[1+\epsilon\omega^2\tau_R^2\right] \ . \qquad 14$$

with $\epsilon \equiv \left[1+\tau_M/\tau_R\right]^2$. Recent observations of anomalous behavior of
the non-secular spectral densities (i.e. $\omega=\omega_e$ the electron spin
Larmour fequency) for some nitroxide spin probes peroxylamine
disulfonate (PADS) and PD-Tempone are amenable to such an
interpretation with $\tau_M \sim \tau_R$ (3,15,16). There is also some indication that even for pseudo-secular spectral densities (i.e. ω_n,
the nuclear-spin Larmour frequency) such an analysis is appropriate (3). Also, it has been noted (3) that such a model may be
appropriate for explaining anomalies in slow-tumbling spectra
that had previously (15,16) been attributed to rotational reorientation by jumps of finite angle. The basic point is that the
spectral density of eq.14 is similar, but not identical, to
typical spectral densities predicted for models of jump diffusion
(8).

For anisotropic media, the "fluctuating torque" model with
eq.13 may also be applied to eq.10 (or more precisely its
Fourier-Laplace transform). Because $U(\Omega) \neq 0$, the spectral
densities are more complex than given by eq.14. In fact, it becomes
necessary to recognize the tensorial properties of $\overleftrightarrow{R}(o)$ as well
as τ_M^{-1}. In an "anisotropic diffusion" model, the principal axes of
diffusion are fixed in the molecular co-ordinate frame (2,4).
This is a model which is also appropriate for isotropic media. In
an "anisotropic viscosity" model the principal axes of $\overleftrightarrow{R}$ (as well
as τ_M^{-1}) would be fixed in a lab frame, e.g. with respect to the
macroscopic director (2,4). One can, furthermore, introduce a
"localized anisotropic viscosity" model in which there is ordering
on a microscopic scale (i.e. a local director), but it is randomly distributed macroscopically.

Experimental results on the weakly ordered PD-Tempone spin
probe have been found to exhibit rather large anomalies in the
incipient slow-tumbling region ($\tau_R \gtrsim 10^{-9}$ sec.), where the ESR

lines are beginning to lose their Lorentzian shape. These anomalies could be successfully accounted for by a fluctuating torque model in which τ_M is highly anisotropic when referred to the lab frame (i.e. the torques inducing reorientation about the axis parallel to the director are very slow). But, for a variety of aspects (not the least of which is the fact that ε would have to be very large (ca. 20) implying $\tau_M > \tau_R$), which are inconsistent with the basic "fluctuating torque" model, this mechanism does not appear to be an adequate explanation. (A simple "anisotropic viscosity" model may successfully be applied to predict the observations, but it too would lead to physically untenable predictions).

Combined studies of the pressure dependence as well as the temperature dependence of the ESR lineshapes have been useful in demonstrating that this anomaly (in which the two outer lines of the ^{14}N hyperfine triplet are significantly broader than predicted from simple theory for $\tau_R \sim 10^{-9}$ sec,) is mainly dependent on $\tau_R(T,P)$ and nearly independent of the particular combination of T and P. This is good evidence that the anomaly is associated with the viscous modes, as is τ_R itself (2,4).

We now turn to the models appropriate when there are important fluctuating components of the anisotropic interactions which relax significantly slower than τ_R. In this case, it is necessary to augment the potential of mean torque $U(\Omega)$ by an additional component $U'(\Omega,t)$ which is slowly varying in time, but with a time average $<U'(\Omega)> = 0$. In this "slowly relaxing local structure" (SRLS) model, each spin probe sees a net local potential given by $U(\Omega)+U'(\Omega,t)$ which remains essentially constant during the timescale, τ_R, required for it to reorient. Then, over a longer timescale, the local reorienting potential $U'(\Omega,t)$ relaxes. An approximate analysis of ESR lineshapes using this model does appear to have the correct features for explaining the observed anomalies, but a full treatment would be very complex (2,4). This analysis yields rough estimates of the order parameter relative to the local structure of $S_1^2 \sim 1/16$ and $\tau_X/\tau_R \sim 10$, where χ is the relaxation-time of the local structure. (Again, one can bring in anisotropies in τ_X^{-1}). This SRLS mechanism is one that may also be applied to macroscopically isotropic media, which however, on a microscopic level display considerable structure. A very approximate analysis (2,4) for the isotropic spectral densities analogous to eq.14 yields

$$j_0(\omega) = \left[D_0(\omega) + S_1^2 \ \tau_X / \ (1+\omega^2\tau_X^2) \right] \qquad 15a$$

$$j_2(\omega) \simeq D_2(\omega) \qquad 15b$$

where $j_0(\omega)$ is the spectral density for the K=0 (cf. eqs.11-12) terms and $j_2(\omega)$ for the K=2 terms. For $\omega^2\tau_R^2 <<1$, one may approximate $D_0(\omega)$ and $D_2(\omega)$ by:

$$D_o(\omega=0) \simeq \left[1 + 0.27S_1 - 2.87S_1^2 + 1.522S_1^3\right]\tau_R \qquad 16a$$

$$D_2(\omega=0) \simeq \left[1 + 0.052S_1 + 0.264S_1^2 + 0.177S_1^3\right]\tau_R \qquad 16b$$

provided S_1 is positive and $\lesssim 0.8$; while

$$D_o(\omega=0) \simeq \left[1 - 0.180S_1 - 3.11S_1^2 - 6.34S_1^3\right]\tau_R \qquad 17a$$

$$D_2(\omega=0) \simeq \left[1 - 0.134S_1 - 0.601S_1^2 - 2.654S_1^3\right]\tau_R \qquad 17b$$

provided S_1 is negative and $\lesssim -0.4$. One effect of the results expressed by eqs. 15 is to cause the $J_o(\omega)$ and $J_2(\omega)$ spectral densities to be different, just as though there were anisotropic diffusion. It is the more complex frequency dependence, as well as $\tau_X/\tau_R \gg 1$, that enables eqs.15 to have features which are consistent with the anomalous experimental observations.

It is interesting to note that the Pincus-deGennes model of spin relaxation by director fluctuations (17,18) can be applied to ESR relaxation (2) just as it has been to NMR relaxation (19). It is also, in a sense, a slowly relaxing structure mechanism. The important differences with the above SRLS model are the Pincus-deGennes model (1) is a hydrodynamic one based upon long-range ordering and not a local structure, and (2) includes only small fluctuations of the director orientation about its mean orientation. Feature 2) results in the prediction (2) that its ESR contribution is opposite in sign (as well as much too small in magnitude for a weakly ordered probe) to "explain" our observed anomalies.

Recent analyses on a highly ordered cholestane nitroxide spin-probe have exhibited the importance of applying a slow tumbling analysis (1,13,14). Also, a spin-relaxation theory which recognizes the statistical interdependence of rotational reorientation and director fluctuations (or of the former and SRLS) has recently been developed (29).

It should be clear, then, that interesting microscopic features of molecular motions in anisotropic media manifest themselves in ESR spectra, and it is still a challenge to unravel the intricate details of observed spectral anomalies.

Many of the general features applied to anisotropic fluids, as discussed above, would be applicable for studying rotational reorientation of molecules on surfaces and at interfaces. That is, one could determine an ordering tensor as given by eqs.1 and 2 with the director usually normal to the surface. However, the director will no longer obey the symmetry relation $\hat{n} = -\hat{n}$. Thus, the expansion of $U(\Omega)$ in spherical harmonics would have to include odd rank terms. In particular, one should add to the L=2 terms, such as given by eq.5, the L=1 term proportional to $\cos\beta$. Rotational diffusion would still be described by eq.10;

the correlation functions defined by eq.12 would still be appropriate for motional narrowing, although their detailed form is
altered; and the SLE equation for slow tumbling would have to
include the new $U(\Omega)$. It remains to be seen what particular features of the microscopic motions will manifest themselves in
careful ESR studies.

Phase Transitions

In the pressure-dependent study of PD-Tempone in Phase V
solvent, the nematic-solid phase transition was also studied (4).
It was found that in frozen solution, PD-Tempone undergoes rapid
isotropic rotational reorientation with values of τ_R about the
same on either side of the phase transition. This suggests that
the probe is located in a cavity, and its structure is very
similar, whether above or below the phase transition. However,
as the pressure is increased, τ_R actually becomes shorter! This
is taken to imply that increasing the pressure freezes out
residual movement of solvent molecules into the cavity, and the
motion of the spin probe becomes less hindered in the cavity.
Also, we may note that a large relaxation anomaly, of the type
discussed above, is observed in the solid phase, and it is
suppressed as the pressure is increased. This is consistent with
our model of the probe in a cavity in the solid and with the SRLS
mechanism.

The isotropic-nematic phase transition of nematic solvent
MBBA with PD-Tempone probe has been studied with careful temperature control (7). The ESR linewidths are fitted to the usual
expression: $\delta = A+BM+CM^2$, where M is the ^{14}N nuclear-spin z-com-
ponent quantum number. The parameters B and C are observed to
behave anomalously as the phase transition at temperature T_c is
approached from either side. In fact they appear to diverge.
The non-anomalous, or background contributions to B and C (i.e.
B_0 and C_0) may be analyzed in the usual manner to show that τ_R
is again nearly the same on either side of the phase transition.
The anomalous contributions to B and C (i.e. ΔB and ΔC) are found
to be fit by the form:

Isotropic Phase:

$$\Delta B = (0.034)(T-T^*)^{-0.6\pm0.2} \qquad\qquad 18a$$
$$\Delta C = (0.041)(T-T^*)^{-0.5\pm0.1} \qquad\qquad 18b$$

Nematic Phase

$$\Delta B = (0.051)(T^+-T)^{-0.5\pm0.1} \qquad\qquad 19a$$
$$\Delta C = (0.077)(T^+-T)^{-0.5\pm0.1} \qquad\qquad 19b$$

These results have been successfully analyzed in terms of Landau-
deGennes mean-field theory (21,22) for the weak first order

transition, as applied to ESR relaxation ($\underline{7}$). In this theory, the free energy of orientation (F') of the liquid crystal is expanded in a power series in the nematic order parameter Q; i.e. $F' = \frac{1}{2} \tilde{A} Q^2 - \frac{1}{3} \tilde{B} Q^3 + \frac{1}{4} \tilde{C} Q^4$. One then minimizes the free energy to obtain the values of Q in the nematic phase (Q_N) as well as the location of T_c. One then allows for small fluctuations of Q about the mean values of Q=0 in the isotropic phase and Q_N in the nematic phase by including in the free energy a term: $\int L(\vec{\nabla}Q)^2 d^3r$, where L is a force constant for distortions and $\vec{\nabla}Q(\vec{r})$ is the gradient of Q. One then studies small fluctuations in F' (or $\Delta F'$) by Fourier analyzing $Q(\vec{r})$, and keeping only lowest order terms in Fourier components Q_q in the isotropic phase (and ΔQ_q the FT of $\Delta Q \equiv Q - Q_N$ in the nematic phase). Thus we have:

$$\Delta F' = \frac{1}{2}V \sum_q (\tilde{A} + Lq^2)|Q_q|^2 \qquad \text{isotropic phase} \qquad 20a$$

$$\Delta F' = \frac{1}{2}V \sum_q (\overline{A} + L_N q^2)|\Delta Q_q|^2 \qquad \text{nematic phase} \qquad 20b$$

where $\overline{A} = \tilde{A} - 2BQ_N + 3CQ_N^2$ and V is the sample volume. It is then assumed that $\tilde{A} \simeq a(T-T^*)$. One then determines that the relaxation times, τ_q for the qth mode are given by:

$$\tau_q^{-1} = L(\xi^{-2} + q^2)/\nu \qquad \text{isotropic} \qquad 21a$$

$$\tau_q^{-1} = L_N(\overline{\xi}^{-2} + q^2)/\nu_N \qquad \text{nematic} \qquad 21b$$

where $\xi^2 \equiv L/a(T-T^*)$ and $\overline{\xi}^2 \equiv L_N/A \simeq L_N/3a(T^+-T)$ (with $T^+ \equiv T_c + \frac{1}{2}(T_c-T^*)$ and the approximation applies only near T_c). The quantities ξ and $\overline{\xi}$ are the coherence lengths of the order fluctuations in the isotropic and nematic phases respectively, and ν and ν_N are the respective viscosities. One then obtains for the ESR linewidth contributions for the weakly ordered spin probe:

$$\Delta B \simeq 5B_o \tau_R^{-1} K_{o,o}(0) \qquad\qquad\qquad 22$$

$$\Delta C \simeq C_o \tau_R^{-1}[8K_{o,o}(0) - 3K_{o,1}(\omega_n)] \qquad 23$$

where

$$K_{o,M}(\omega) = \frac{kTv\xi(S^{(p)})^2}{\sqrt{2}\ 4\pi\ L^2(S^{(s)})^2} (1+[1+(\omega/\omega_\xi)^2]^{\frac{1}{2}})^{-\frac{1}{2}} \quad 24$$

and $\omega_\xi \equiv L/\nu\xi^2$. Below T_c, one replaces ξ everywhere by $\overline{\xi}$ in eq.24. Eq.24 predicts $K_{o,M}(0) \propto \xi$ or $\overline{\xi}$, and we have $\xi \propto (T-T^*)^{-\frac{1}{2}}$ and $\overline{\xi} \propto (T^+-T)^{-\frac{1}{2}}$ from their derivations. This is in exact agreement with our observations of eqs.18 and 19. We also find experimentally that $T_c-T^* \simeq 1^\circ$ and $T^+-T_c \simeq 0.5^\circ$ also in agreement with prediction. The magnitude of ΔB (and ΔC) in the isotropic phase

can be predicted from eqs. 22-24 and values of the experimental
parameters measured by Stinson and Litster (23) in the isotropic
phase. The agreement with our experimental results is again very
good.(Stinson and Litster measured a,L and ν by light scattering).

Thus the ESR observations about the isotropic-nematic phase
transition display a symmetry for spin-relaxation due to critical
fluctuations, and the characteristic features are predicted
rather well by simple mean-field theory combined with a motional-
narrowing relaxation theory. It is significant to note that light
scattering and NMR studies only successfully deal with order
fluctuations above T_c and could not demonstrate this symmetry
about T_c. The two above examples should clearly demonstrate the
utility of ESR studies of phase transitions in the anisotropic
media.

Spin Exchange and Chemically-Induced Spin Polarization in Two Dimensions

Translational diffusion may be monitored by spin probes as
a function of the concentration of the probes. This has been
extensively studied in isotropic liquids (24,25) and some work
has been performed in liquid crystals (13,26). In these works,
one monitors the translational diffusion by the concentration-
dependent line broadening due to the Heisenberg spin-exchange
of colliding spin probes in solution. It is interesting to
speculate on the nature of such phenomena if radicals are confined
to translate on a two dimensional surface. From the point of view
of radical-radical interactions on surfaces, one may ask the
related questions about spin-dependent reaction kinetics on surfaces and the associated magnetic resonance phenomena of CIDNP and
CIDEP which have been amply studied in three dimensions (27,28).
Deutch (29) has pointed out that such phenomena should be qualitatively different in two vs. three dimensions, mainly because
the re-encounter probability of two radicals which have initially
separated is always unity for two dimensions, but less than unity
in three dimensions. A recent study (30) focuses on the two
dimensional aspects from the point of view of the Pedersen-Freed
theory (28). In this theory, the stochastic Liouville equation
(SLE), which simultaneously includes the spin and diffusive
dynamics, is solved by finite difference methods subject to an
initial condition (usually that the radicals are initially in
contact or slightly separated). The results then show the accumulated spin-dependent effects after the radical pair have had the
opportunity to re-encounter many times (and possibly react). This
is referred to as the "complete collision". One then obtains
(1) the probability of spin exchange per collision (ΔP), (2) the
fractional probability of reaction per collision if there are no
spin-dependent selection rules (Λ), (3) the probability per
collision of conversion of non-reactive triplets to singlets which
immediately react ($\mathcal{F}^*$), and (4) the polarization of radical A

from the radical-pair mechanism per collision (P_a). However, in
two dimensions, because the re-encounter probability is always
unity, the "collision" is never complete as t $\to \infty$, unless other
processes act to interfere (e.g. radical scavenging, radical T_1,
or radicals leaving the surface). Thus a finite time scale is
brought on by these other processes. It has been pointed out that
a finite time scale can be replaced by a finite outer radial
collecting (or absorbing) wall boundary at r_N in these problems
(28). It is convenient to solve for finite r_N and then transform
to the equivalent time representation. We summarize approximate
expressions obtained from the numerical solutions.

The re-encounter probability (t_f) for finite r_N is given by:

$$t_f = 1 - Ln(\frac{r_I}{d}) \, / \, Ln(\frac{r_N}{d}) \qquad\qquad 25$$

where r_I is the initial radical separation and d the encounter
distance. (The next set of results are quoted for r_I=d). The
quantity Λ obeys :

$$\Lambda = k\tau_1 \, Ln(\frac{r_N}{d}) \, / \, \left[1 + k\tau_1 \, Ln(\frac{r_N}{d}) \right] \qquad\qquad 26$$

where k is the first order rate constant for irreversible
disappearance of singlet radical pairs when in contact with a
"sphere of influence" from d to Δr_k, and $\tau_1 = d\Delta r_k/D$ is a
characteristic lifetime of the encounter pair with D the relative
diffusion coefficient.(One recovers the 3D result by replacing
$Ln(\frac{r_N}{d})$ in eq.26 by unity). Thus as $r_N \to \infty$, $\Lambda \to 1$ as it should,
since there will be many re-encounters to guarantee reaction for
finite k.

Then $\mathcal{F}*$ approximately obeys:

$$\mathcal{F}* \simeq \frac{\frac{1}{40} (\frac{r_N}{d})^4 (\frac{Qd^2}{D})^2}{1+\frac{1}{40}(\frac{r_N}{d})^4 \frac{Qd^2}{D}^{1.8} \left[Ln(\frac{r_N}{d})\right]^{-1} \left[1+Ln(\frac{r_N}{d})(\frac{Qd}{D})^{0.2}\right]} \qquad 27$$

where 2Q is the difference in Larmour frequencies of the two
interacting radicals. As r_N gets large (and/or for large $\frac{Qd^2}{D}$)
eq.27 goes as

$$\mathcal{F}* \simeq (\frac{Qd^2}{D})^{0.2} Ln(\frac{r_N}{d}) \left/ \left[1+Ln(\frac{r_N}{d})(\frac{Qd}{D})^{0.2}\right] \right. \qquad\qquad 28$$

which increases with $Ln r_N$ until $\mathcal{F}* = 1$, its maximum value.
(Eqs. 27 and 28 neglect the relatively small effects of the
exchange interactions (30)). For typical values of the parameters
eq. 28 is usually appropriate. Note that CIDNP polarizations are
typically closely related to the product $\Lambda \mathcal{F}*$ (28) . Eq.27 is
different in its Q dependence from the Q^2 power-law typically

found in 3D. This Q dependence suggests that in 2D the mechanism does not require the time period between successive encounters for the spin-dependent evolution due to having $Q \neq 0$. Instead the relevant spin-dependent evolution occurs while the radicals are in contact during each encounter.

For CIDEP, a contact exchange model yields:

$$P_a \simeq \frac{\frac{1}{2}(\frac{r_N}{d})^2 (\frac{Qd^2}{D})^1}{1 + b(\frac{r_N}{d})^2 (\frac{Qd^2}{D})^\varepsilon} \left(\frac{2J_o\tau_1}{1+\frac{13}{4} Ln(\frac{r_N}{d})(2J_o\tau_1)^2} \right) \qquad 29$$

with $\varepsilon \simeq 1.2$ and $b \simeq 3/2$ for $2J_o\tau_1 \ll 1$ and $\varepsilon \simeq 0.85$ and $b \simeq 1/5$ for $2J_o\tau_1 > 1$. Here J_o is the magnitude of the exchange interaction of infinitesimal range Δr_J and $\tau_1 = d\Delta r_J/2D$. Again, the Q-dependence indicates the importance of the spin dependent evolution during each encounter. For finite range of exchange interaction (i.e. $J(r) = J_o e^{-\lambda(r-d)}$) one has $\tau_1(\lambda)\frac{d}{D}\left(1+\frac{1}{2d\lambda}\right)$ with a similar, but not identical, form to eq.29. In particular, for $2J_o\tau_1(\lambda) > 1$ (and/or large r_N),

$$P_a \simeq \frac{\frac{3}{8}(\frac{r_N}{10d})^{3.3}(\frac{Qd^2}{D})^1 (2J_o\tau_1(\lambda))^{-0.02}}{1+ \frac{1}{8}\left[Ln(\frac{r_N}{d})\right]^{2.75} (\frac{r_N}{10d})^{3.3} (\frac{Qd^2}{D})^{0.85} (\frac{d^2}{\tau_1(\lambda)D})^{\frac{1}{2}}} \qquad 30$$

For Heisenberg-spin exchange, a contact exchange model yields:

$$\Delta P(d_t) \simeq (Ln(\frac{r_N}{d}))^2 (2J_o\tau_1)^2 \Big/ \left[1+(Ln(\frac{r_N}{d})^2 4(J_o^2+Q^2)\tau_1^2\right] \qquad 31$$

This is like the 3D result, which is recovered by letting $Ln(\frac{r_N}{d}) \to 1$. For finite exchange range, one has the possibility of values of $\Delta P(d_t)$ greater than unity as in 3D (28):

$$\Delta P(d_t) \simeq 1 + \left[Ln(1+J_o d^2/D)\right]3 \Big/ \left[Ln(\frac{r_N}{d})\right]^2 \lambda d$$

$$\text{for } J_o^2\tau_1^2(\lambda ln\frac{R_N}{d})^2 \gtrsim .001 \qquad 32$$

which becomes the 3D result if $\frac{1}{3}\left[Ln(\frac{r_N}{d})\right]^2 \to 1$.

Lastly, we can replace the dependence upon r_N in the above expressions with that on s, the Laplace transform of time by using:

$$s^{-1} \approx \left[\text{Ln}(\frac{r_N}{d}) - \frac{1}{2} \right] (r_N)^2 / 4D. \qquad 33$$

Then, these results are rigorously for radical pairs which are scavenged or "collected" by a (pseudo) first-order rate process with rate constant s. More approximately then, they correspond to the values at time $t \sim s^{-1}$. The generally weak dependence of these equations on r_N (e.g. the validity of eq. 28, etc.) hence on s, suggests that the approximate point of view is entirely satisfactory. Then we can approximately replace s by $k_r + T_1^{-1}$ in eq.33 (where k_r would take care of radical scavenving, rate of leaving the surface, etc.) for use in the previous expressions. When desired, however, the SLE equation may be solved, explicitly including the disruptive effects.

In some applications, in particular that of Heisenberg spin exchange, the radicals are initially randomly distributed. Thus, one must first determine the steady-state rate for new bimolecular collisions of radical-pairs, $(2k_2, \text{cf. } \underline{28})$. While in 3D this is a well-defined quantity, it is not quite true for 2D, since the rate always has some time-dependence $(\underline{31})$. Thus, in this case the actual rate (for unit concentrations) of spin exchange at time t (or times $t \rightarrow \infty$) would be obtained from the convolution $\int_0^t \Delta P(t-\tau) k(\tau) d\tau$ or alternatively from its Laplace transform: $\Delta P(s)k(s)$ using eq.33 and eq.31 or 32 for ΔP. Similar comments apply for CIDEP $(P_a(s))$ and CIDNP $(\approx \Lambda(s) \mathcal{F}^*(s))$ due to random initial encounters.

In addition , we note that for magnetic resonance in 2D, spin relaxation by modulation of intermolecular dipolar interactions is an important concentration-dependent mechanism $(\underline{32})$. It is expected to be sensitive to the structure (i.e. the pair correlation function) appropriate for the fluid or surface $(\underline{33})$, just as are the other mechanisms we have discussed $(\underline{28})$.

Literature Cited

1) Freed,J.H. in "Spin Labeling Theory and Applications", L.J. Berliner, Ed. Academic Press, New York, 1973. Ch.3.
2) Polnaszek, C.F. and Freed, J.H., J. Phys. Chem. (1975) 79, 2283.
3) Hwang, J.S., Mason, R.P., Hwang, L.P. and Freed, J.H., J. Phys. Chem. (1975) 79, 489.
4) Hwang, J.S., Rao, K.V.S. and Freed, J.H., J. Phys. Chem. (1976) 80, 1490.
5) McColl, J.R. and Shih, C.S., Phys. Rev. Letts. (1972), 29, 85.
6) Lin W.J. (private communication).
7) Rao, K.V.S., Hwang, J.S. and Freed, J.H., (submitted for publication).

8) Hwang, L.P. and Freed, J.H., J. Chem. Phys. (1975) 63, 118.

9) Nordio, P.L. and Busolin, P., J. Chem. Phys. (1971), 55, 5485.

10) Nordio, P.L., Rigatti, G. and Segre, V., J. Chem. Phys. (1972) 56, 2117.

11) Polnaszek, C.F., Bruno, G.V. and Freed, J.H., J. Chem. Phys. (1973), 58, 3185.

12) See references cited in 2.

13) Polnaszek, C.F., Ph.D. Thesis, Cornell University, 1975.

14) Rao, K.V.S., Polnaszek, C.F. and Freed, J.H. (to be published)

15) Goldman, S.A., Bruno, G.V., Polnaszek, C.F. and Freed, J.H., J. Chem. Phys. (1972), 56,716.

16) Goldman, S.A., Bruno, G.V. and Freed, J.H., J. Chem. Phys. (1973), 59, 3071.

17) deGennes, P.G., Mol. Cryst. Liq. Cryst. (1969), 7, 325.

18) Pincus, P. Solid State Commun. (1969), 7, 415.

19) Doane, J.W., Tarr, C.E. and Nickerson, M.A., Phys. Rev. Lett. (1974), 33, 620.

20) Freed, J.H. (to be published).

21) Landau, L.D., Lifshitz, E.M., Statistical Physics (Addison-Wesley, Reading, Mass. 1958), Ch. 14.

22) deGennes, P.G., Mol. Cryst. Liq. Cryst. (1971), 12, 193.

23) Stinson, T.W., and Litster, J.D., Phys. Rev. Letts. (1973), 30, 688 and references given.

24) Eastman, M.P., Kooser, R.G., Das, M.R. and Freed, J.H., J. Chem. Phys. (1969), 51, 2690 and references cited.

25) Lang, J.C. Jr. and Freed, J.H., J. Chem. Phys. (1972), 56, 4103.

26) Bales, B.L., Swenson, J.A., Schwartz, R.N., Mol. Cryst. Liq. Cryst. (1974), 38, 143.

27) Lepley, A.R., and Closs, G.L., eds. "Chemically Induced Magnetic Polarization". Wiley, New York, 1973.

28) Freed, J.H. and Pedersen, J.B., Adv. Mag. Res. (1976), 8, 1.

29) Deutch, J.M., J. Chem. Phys. (1972), 56, 6076.

30) Zientara, G. and Freed, J.H. (to be published).

31) Naqvi, K.R., Chem. Phys. Letts. (1974), 28, 280.

32) Brulet, P. and McConnell, H.M., Proc. Nat. Acad. Sci. (1975), 72, 1451.

33) Hwang, L.P. and Freed, J.H., J. Chem. Phys. (1975), 63, 4017.

2

Measurements of Self-Diffusion in Colloidal Systems by Magnetic-Field-Gradient, Spin-Echo Methods

JOHN E. TANNER

Naval Weapons Support Center, Applied Sciences Department, Crane, Ind. 47522

A very useful application of NMR methods to colloidal systems is the measurement of translational self-diffusion coefficients in the fluid (usually liquid) components. Rates of diffusion indicate the restrictions to mobility due to adsorption or due to physical barriers. In many cases, one may also obtain information on the dimensions of the barriers or other inhomogeneities.

Types of systems which have been or might be studied by these methods include biological cells and tissues, porous rocks and minerals, emulsions, liquids adsorbed on fibers, polymer solutions, and the synthetic and natural adsorbent materials used for various types of chromatography.

NMR Methods of Measuring Diffusion

There are two distinct NMR methods used for measuring translational diffusion. In one method[1] the nuclear spin-spin (T_2) and spin-lattice (T_1) relaxation times due to random motions over distances of a few Angstroms are measured by cw or pulse methods. By repeating the measurements over a range of temperatures and/or resonance frequencies, and/or degrees of isotopic substitution, and by the use of other information about the system it is frequently possible to analyze these motions in terms of molecular rotations and translations. The simpler the experimental system, the easier the analysis, and the less ambiguous are the results. With certain metals[2,3] and plastic crystals[4] the measurement of spin relaxation is the preferred method of determining translational self diffusion.

A somewhat different application of spin-relaxation measurements to determine diffusion coefficients involves the determination of exchange times between spatial regions which have different spin relaxation rates. Exchange times may be determined if the spin relaxation function can be resolved into two (or more) components,[5] and if independent knowledge of the relative spin populations or of the relaxation rates in the isolated regions is available. If the dimensions of the regions are also known, then the diffusion coefficients or the barrier permeabilities may be calculated from the dimensions and the exchange times. This method has recently been applied to measure intracrystalline diffusion of butane in zeolite[6] and the permeability of human red cells to water.[7]

The other NMR method measures the loss of phase coherence of a set of precessing nuclear spins due to their random motion over a gradient in the magnetic field.[8-10] This method (MFG-NMR) is the subject of the present paper. Numerous applications of this method to colloidal systems have been made in the past decade. A part of this has been reviewed by Stejskal.[11] The emphasis here will be on more recent experimental results and developments of methods.

The MFG-NMR Experiment

The magnetic-field-gradient, spin-echo experiment has been well described in many places (see e.g. ref. 8-10); only an outline will be given here. Briefly, the sample is placed in a high magnetic field which creates a small net magnetization among the nuclear and electronic spins present. A small oscillating field of appropriate duration and magnitude ("90°") at the resonance frequency of the nuclear spins of interest (usually in the radio frequency, rf, range) rotates this magnetization into a plane perpendicular to the static field, about which it then precesses, producing a detectable signal. This signal decays as the spins in their precession lose phase coherence due to various energy transfer processes (time constant T_2) and due to inhomogeneities in the magnetic field. Magnetization reappears in the direction of the static field. The time constant for this, T_1, gives the upper limit of T_2.

To the extent that the spins remain in place, the loss of phase coherence due to field inhomogeneities can be reversed by applying a second rf pulse of appropriate magnitude ("180°"), which causes the spins to briefly regain phase coherence forming an increased signal, the "spin echo". Any displacement of the nuclear spins after the 180° pulse from their position before the 180° pulse lessens this signal recovery, so that the spin echo is attenuated. This attenuation of the echo can be amplified by artificially increasing the magnetic field inhomogeneities already present, applying a (linear) field gradient.

In the earlier experiments, a steady field gradient was applied.[8,9] More recently it has become common to apply the gradient in two pulses, one before and one after the 180° rf pulse.[10] For experimental reasons, a much larger gradient can be applied in this manner, and thus lower diffusion coefficients can be measured. If the pulses are short compared to their separation, the diffusion time of the experiment is well defined and is the pulse separation interval. This experiment is illustrated schematically in Figure 1(b).

For the shortest diffusion times, it is necessary to multiply the effect by repeating the gradient pulses a number of times before observing the echo. One means of doing this is to apply trains of gradient pulses of alternating sign.[12]

For the longest diffusion times, in cases where $T_1 >> T_2$, which is typical for colloidal systems, a three-rf pulse sequence, with observation of the "stimulated echo" is advantageous.[8,13] These variations are also illustrated in Figure 1.

Special methods have been developed to help overcome problems associated with the short T_2 of certain types of solid systems. Pines and Shattuck[14] have demonstrated the long T_2 of naturally abundant ^{13}C in a plastic crystal, using a special rf sequence designed to impart a high degree of spin magnetization to these nuclei. As they imply, it should be possible to use long enough gradient pulses so that the diffusion coefficient can be measured.

Blinc and co-workers[15,16] have performed experiments in which a Waugh-type rf sequence was combined with pulsed gradients, with the object of measuring self diffusion of liquid crystals in nematic and smectic phases. This method is useful where $T_{1\rho} >> T_2$, where $T_{1\rho}$ is the rotating frame T_1.

Where nuclei of the same type but belonging to different molecular species contribute to the observed signal, the individual signals may in principal be resolved by means of differences in the T_1 or T_2 spin relaxation times, by differences in the chemical shifts (using Fourier transformation of the echo),[17] or by differences in the diffusion coefficients themselves. Naturally a higher initial signal strength is required so that each component will be measurable, and to compensate for losses in the resolution process.

Diffusion Time and Distance Ranges

With the best of presently developed magnetic-field-gradient techniques, and with simple systems where only one component contributes to the signal, it should be possible to measure diffusion coefficients with at least 20% precision (and usually better) over a range of diffusion times, $10^{-4} < t_D < 1$ seconds, corresponding to root mean square (rms) diffusion distances between a few tenths of a micron and tens of microns.

The upper limit of diffusion times is determined primarily by the T_1 and the concentration of the spins of interest, and by the apparatus sensitivity, which determines the amount of relaxational signal decay that can be tolerated. The sensitivity can usually be enhanced by time averaging. With high concentrations of protons or fluorine nuclei, it is readily possible to make a measurement at $t_D = 1.5\ T_1$.

The lower limit of attainable diffusion times and distances depends on the spin concentration, the diffusion coefficient, and the T_2 relaxation time of the spins of interest, and on the sensitivity of rf detection and amplification, and on the magnitude, rise/fall times, and stability of the applied field gradient. At the lowest attainable diffusion times with a given apparatus, signal averaging generally can not be used. The inevitable instability of the field gradient pulses at high strength and close spacing requires monitoring of the individual experiments.

We make a quantitative estimate of the lower limits as follows: The diffusional echo attenuation, R, is given by

$$\ln R = -\gamma^2 DNg^2\delta^2 (\Delta -\delta /3),$$

where γ is the nuclear gyromagnetic ratio, D is the diffusion coefficient, N is the number of gradient pulse pairs, and g, δ, and Δ are the magnitude, duration, and separation of the individual gradient pulses as illustrated in Figure 1. Idealizing the experiment, we imagine that the gradient at its maximum value has negligible rise/fall times, and sufficient stability to allow a reproducible echo. We fill the time between the 90° rf pulse and the echo maximum (interval 2τ) with a train of gradient pulses of alternating signs, ignoring the usual need for a millisecond or so of dead time before the 180° rf pulse and before the echo. Specifically, we set $\delta = \Delta = \tau /N$. We allow the largest value of τ which will permit accurate measurement of a signal attenuation of say 30%. This is usually $\tau \lesssim T_2$. Substituting in the above equation, the minimum possible diffusion time is then

$$t_D \equiv \Delta -\delta /3 = 3.1 \times 10^{-5}(-\ln R/D\tau)^{1/4};$$

and the corresponding minimum diffusion distance is

$$x_{rms} = 0.0078(-D \ln R/g^2\tau)^{1/4}.$$

We present calculated values of t_D and x_{rms} for a few typical values of g, D, and τ in Table I. As can be seen, both from the formulas and from the calculations, the lower limit of the method is at about half a micron. Extensions below this by increasing the available field gradient or by increasing apparatus sensitivity, and thus the permissible value of τ, are small, and are obtained with considerable difficulty. Nevertheless, for the study of systems whose dimensions are near the limit, the improvements could be well worth the trouble.

TABLE I. Minimum values of the diffusion time and rms displacement for various diffusion coefficients and maximum gradient and
maximum rf pulse spacing.

D cm^2/sec	g G/cm	τ msec	t_D msec	x_{rms} μm
10^{-5}	500	20	0.12	0.48
10^{-5}	100	20	0.58	1.07
10^{-8}	500	20	0.65	0.09
10^{-5}	500	5	0.23	0.68

Evaluation of Colloidal Parameters from Diffusion Information

        The magnetic-field-gradient method obtains the self-
diffusion coefficient of the observed species directly and in as
few as two observations. It has the additional advantage that
the experimental diffusion distances are of the order of
colloidal dimensions, so that a change in the apparent diffusion
coefficient may be detected as the diffusion time is varied.[18]
Figure 2 shows some typical geometries of colloidal systems.
One or more phases are divided into compartments by means of
various boundaries which may be permeable or impermeable to the
substance of interest. At short enough diffusion times most of
the molecules will not have contacted the boundaries, they
experience only a homogeneous medium, and their diffusion
coefficient will represent the "true" diffusion coefficient, D_O,
for the pure phase in the absence of boundaries. At longer
diffusion times more of the molecules will have contacted the
boundaries. If these boundaries or the adjoining compartments
are not completely permeable, the measured diffusion coefficient
will appear to decrease. At long enough diffusion times the
apparent diffusion coefficient, D_∞, will be a function of the
diffusion coefficients of the observed species in all of the
phases it penetrates, and of the permeability and geometry of
the barriers.
        By making measurements over all or most of the range between
D_O and D_∞ and fitting the data to mathematical functions
appropriate to the known or suspected geometry of the system, it
is possible to extract quantitative information about the
dimensions and permeabilities of the barriers.
        In practice, this is made difficult by the fact that mathe-
matical expressions for the diffusion coefficient as measured by
magnetic-field-gradient NMR, and valid over the entire relevant
range of diffusion times have been derived for only a few simple
cases, all involving impermeable boundaries, where the diffusion
coefficient D approaches zero as the diffusion time becomes
long. To the best of my knowledge, these cases are the
following: (1) parallel planar boundaries, an approximate

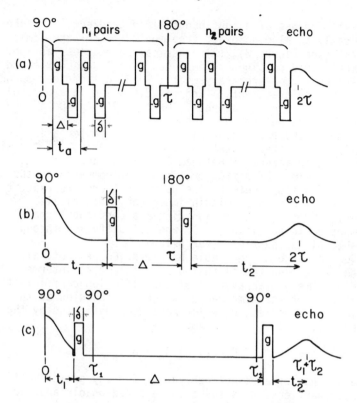

Figure 1.   *Radio frequency (90°, 180°) and field gradient (g, —g) pulse sequences suitable for diffusion measurements at (a) short, (b) intermediate, and (c) long diffusion times, respectively*

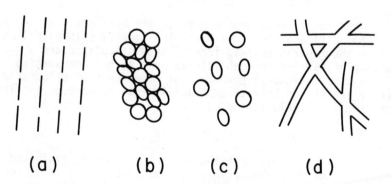

Figure 2.   *Geometries for restricted diffusion as might be found in (a) a layered mineral, (b) and (c) cells and emulsions, and (d) fibers*

treatment for use of a continuous field gradient,[19,20] checked
numerically,[21] also an exact treatment[22,23] for the limit
$\delta << D\Delta/\underline{a}^2$ (see Figure 3); (2) spherical boundaries, an approximate
treatment for a continuous field gradient,[20] also extended to
gradient pulses of arbitrary width;[24] (3) cylindrical boundaries,
an approximate treatment,[20] and (4) diffusion in a parabolic
potential, an exact treatment[25] for the limit $\delta << D\Delta/\underline{a}^2$. Here $\underline{a}$
is the distance between barriers.

Some of these relationships have been tested using artificially constructed systems of simple geometry,[21-23,26] with
generally good agreement between experiment and theory.

In the limit of diffusion times long compared to the time
required to diffuse between the barriers there are exact treatments for (1) diffusion within impermeable spherical
boundaries,[22,23] (2) diffusion within a continuous phase containing impermeable particles,[27,28] and (3) diffusion through
permeable planar barriers (see below).

For the case of permeable barriers, Cooper, et. al.,[29]
propose expressing the diffusion coefficient at intermediate
diffusion times as an average of the diffusion coefficients at
the extremes of the time range, weighted according to the
fraction of molecules, p, which have been deflected by the
barriers, i.e.,

$$D = (1-p)\ D_0 + p\ D_\infty.$$

In the absence of an exact derivation for any geometry of
permeable barriers, this seems to be a reasonable approach.
They use the diffusion time, t_a, at the point where

$$D = (D_0 + D_\infty)/2$$

to calculate a "characteristic length"

$$\underline{a} = (2D_0 t_a)^{1/2},$$

which represents the distance between the barriers. It can be
seen that p(t) and hence the diffusion time at this point, as
well as the "characteristic length" thus calculated are a
function of barrier permeability as well as barrier spacing.
This is unsatisfactory; however, the dependence of calculated
barrier spacing on barrier permeability can be removed to first
order simply by calculating

$$\underline{a} = [2(D_0 - D_\infty)t_a]^{1/2}.$$

It seems probable that the barrier parameters evaluated
from diffusion data are not highly sensitive to subtle differences in the geometry assumed for the system. For instance,
mathematical relations pertaining to parallel planar barriers

have been used to analyze diffusion data taken on biological cells;[22,23] and the theory pertaining to diffusion near an attractive center has been used to analyze data taken on diffusion in pores of hydrated vermiculite.[30,31] The calculated cell or pore sizes were in semiquantitative agreement with microscopic observations.

However, when a large distribution in barrier distances exists it should be approximated with a reasonable functional form. In a study of diffusion within emulsion droplets Packer and Rees[32] obtained a much better fit to their data by assuming a log-normal distribution of droplet sizes than by the use of a cruder distribution[22,23] containing just as many adjustable parameters.

Examples

Now we show some examples of what has been done with the MFG-NMR method and give a few comments as to what else might have been learned from these or other systems using presently available experimental and mathematical methods. The list of examples is not all-inclusive, but is intended to sample the major areas where work has been done.

The first demonstration of a dependence of the diffusion coefficient on diffusion time in a colloidal system was by D. E. Woessner.[18] He studied diffusion of water in a porous sandstone of known porosity and pore size, diffusion of water in an aqueous suspension of silica spheres of known size, and diffusion of benzene in two rubber samples. For the sandstone and the silica spheres, the diffusion coefficients extrapolated to nearly the value for bulk water at zero diffusion time. A dependence of the diffusion coefficient on the diffusion time was observed for one of the rubber samples. For both rubber samples the diffusion coefficient extrapolated to a value well below that of bulk benzene. Woessner did not attempt to analyze the form of the dependence of diffusion coefficient on diffusion time to confirm the known barrier parameters.

Recently Lauffer[33] has measured diffusion of water and n-undecane in porous tetrafluoroethylene and in sandstone samples. He assumed a log-normal distribution of pore sizes. Using appropriate mathematical expressions to analyze the diffusion-time dependence of the diffusion coefficient he obtained pore sizes in semiquantitative agreement with results of optical observations.

A considerable number of diffusion measurements by MFG-NMR have been made on biological cells in recent years. In materials such as commercial cake yeast,[22,23] apple flesh,[22,23,34] tobacco stem pith tissue,[22,23] red blood cells,[29,35] and rat brain tissue,[36] a marked dependence of the diffusion coefficient on diffusion time has been demonstrated. In a few cases (apple, tobacco pith) the diffusion coefficient seems to extrapolate to

the value for bulk water at zero diffusion time, while in the
other cases it extrapolates to a value which is lower by a
factor of 2-4. In cases where the functional relation between D
and t_D was analyzed,[22,23,29,35] the calculated cell diameters
were in rough agreement with the known values.

As an example, recent results of the author and colleagues[35]
on red cells are presented in Figure 4. The range of t_D clearly
extends to the constant limiting values of the diffusion
coefficients (D_O and D_∞) at short and long diffusion times (the
upward trend for $t_D > 1$ sec is probably a systematic error due to
very low signal strength). The analysis in terms of membrane
permeability and cell dimensions is made difficult by the
complicated sample geometry--toroidal, disc-shaped cells
(2 μm x 9 μm) oriented at random.

As an attempt to get by with available theory, we use a
relation which applies to diffusion between parallel planes.
The midpoint of the decrease in diffusion coefficient (Figure 4)
is at approximately 6 msec. The midpoint in the upper curve of
Figure 3 is at $\log(\pi^2 D t/a^2) \approx 0.1$. Substituting, we obtain
a = 4 μm, where D is interpreted as referring to D_O-D_∞. This
value is reasonable, since a value larger than a simple geometric
average is expected. Using the formula $1/D_\infty = 1/D_O + 1/aP$, we
obtain a value of P = 0.011 cm/sec for the outer membrane
permeability. This agrees reasonably well with P = 0.005, which
may be calculated from the data of Shporer and Civan,[7] and the
values P = 0.005, 0.012 for tracer and osmotic measurements,
respectively, tabulated by Whittam.[37]

By contrast, diffusion coefficients of water in muscle
tissue generally show only a weak dependence on the diffusion
time.[22,29,35,36,38-40] The outer cell membranes are widely
spaced and there are apparently no other significant barriers
within the cells, spaced at intervals within the accessible
range of experimental diffusion distances.

Diffusion coefficients of liquids and gases absorbed in a
number of minerals have been measured by MFG-NMR.

Kaerger and his colleagues have made a number of studies of
water[41] and several alkanes[42-44] absorbed on several types of
zeolites. By measuring the apparent diffusion coefficient over
a wide range of temperature and with several different average
crystalite sizes, they have been able to distinguish diffusion
within the crystals from diffusion outside. In some cases,
where the apparent diffusion coefficient had a temperature
independent region they were able to verify the known crystal
size. They have produced evidence that the external diffusion
takes place on the surfaces of the crystals rather than in the
space between them; and they have deduced the energy of transfer
of molecules from within the crystals to the surface.

They have further shown that in the upper temperature range
where intercrystalline diffusion dominates, the diffusion
coefficient increases with increasing degree of adsorption (pore

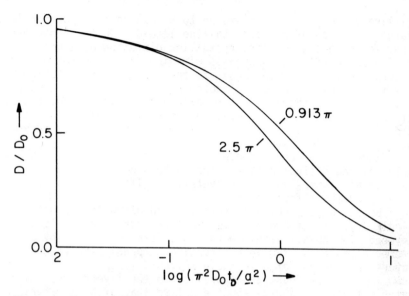

Figure 3.   Impermeable, parallel planes.   Theoretically calculated apparent diffusion coefficient vs. reduced diffusion times, under the condition $\delta \ll D\delta/a^2$ (Refs. 22, 23), for two values of $\gamma g\delta a$.

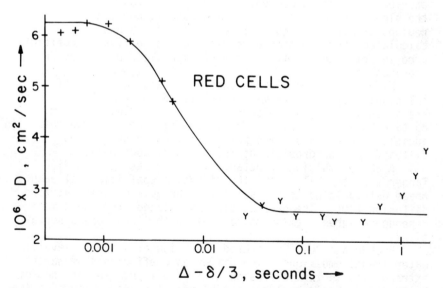

Figure 4.   Measured diffusion coefficients of water in a sample of packed human red cells at various values of the diffusion time, $\Delta - \delta/3$.   The symbols $+$ and Y indicate the use of pulse sequence (a) and (c) respectively of Figure 1.

filling factor). This is stated to be due to a proportionately larger fraction of extracrystalline liquid. On the other hand, they found that when intracrystalline diffusion dominates, the diffusion coefficient <u>decreases</u> with increasing degree of adsorption. This is analogous to the decrease in diffusion coefficient with increasing pressure in a gas. It is also an indication that there is not a large variability in the degree of binding of alkane molecules on the various internal sites in the zeolite crystal, in contrast to the situation with many other adsorption systems.

Each of the above studies on zeolites was made at a single value of the diffusion time. The authors did not attempt to demonstrate inter- and intracrystalline diffusion simultaneously by varying the diffusion time.

Boss and Stejskal[30,31] have studied diffusion of water in single crystals of hydrated vermiculite. Vermiculite is a layered alumino-silicate mineral which can be swollen by a factor as much as x100 in a direction perpendicular to the layers, by the use of a suitable solvent.

The authors found that parallel to the layers, most of the water diffused at a rate about equal to that of bulk water. A small amount of water diffusing more slowly was resolvable, but could not be ascribed to any specific structural features.

Diffusion perpendicular to the layers was resolvable into a fast, time-independent component and a slow time-dependent component. On the basis of known structural information[45] they ascribed the fast component to water in macroscopic cracks, and the slow component to water in microscopic pores consisting of regions of buckling and dislocation of the layers. Pore sizes calculated from the time dependence of the diffusion coefficients were in semiquantitative agreement with sizes estimated from previous work.[45]

Liquid-liquid emulsions constitute another class of colloidal systems which may be studied by MFG-NMR. Here there are a continuous and a discontinuous phase. Each phase may contain molecules whose diffusion coefficient is measurable. In addition, there exists the possibility of measuring the diffusion of the droplets of the discontinuous phase.

A 50/50 emulsion of octanol in water has been studied by Tanner and Stejskal.[22,23] The diffusion coefficient of each phase extrapolated to the value for the bulk material at zero diffusion time. The droplet size calculated from the time dependence agreed semiquantitatively with microscope observations.

Packer and Rees[32] measured diffusion of water in several water-in-oil emulsions. The diffusion coefficients of water extrapolated to zero diffusion time were within 20% of that of bulk water. Their analysis showed that the assumption of a log-normal distribution of droplet sizes gave a much better fit to the data than the assumption of a single average size or of a three-size distribution.

Moll and Baldeschwieler[46] measured the self-diffusion coefficient of micelles of dimethyldodecylamine oxide, obtaining a value $D \approx 4 \times 10^{-7}$ cm^2/sec. They were unable to observe diffusion within the micelles. However, for dilute solutions of droplets of ~0.5 μm there exists a possibility of measuring diffusion of the droplets and diffusion within the droplets at the same time.

Metals must be dispersed into colloidal-size particles for NMR studies so as to allow penetration of the rf field. Emulsions of liquid ^{23}Na,[47] and of ^{6}Li and ^{7}Li,[24,48,49] in oil, have been studied by the pulsed gradient NMR technique. The measured apparent diffusion coefficients were corrected for the restricting effects of the droplet boundaries by means of appropriate formulas. The ratio of self diffusion coefficients for the two isotopes of lithium was found to be smaller than the ratio of their mutual diffusion coefficients, which had been measured by another method.[50]

Polymeric systems would seem to offer the possibility of measuring inhomogeneous diffusion by NMR since polymer molecules are of colloidal size. In a dilute polymer solution, the solvent molecules might experience a reduced diffusion coefficient within the polymer coil compared to outside it. However, polymer segment densities are usually quite low even at the center of polymer coils, and the decrease in segment density occurs gradually, progressing outward from the center. This would smear out any difference in diffusion coefficients which occurs. Furthermore, reductions of solvent diffusion coefficients by moderate concentrations of dissolved polymers are not large. For example, measurements of solvent diffusion in a 20% solution of high molecular weight polydimethylsiloxane showed a reduction of the self-diffusion coefficient of only a factor of two compared to pure solvent.[51] In the much lower segment densities inside a polymer coil in a dilute solution, it would not be possible to resolve the difference of diffusion coefficients inside and outside the polymer coils.

In more concentrated solutions, there would be even less spatial variation of the polymer segment density, due to interpenetration of the polymer coils.

Another type of inhomogeneous diffusion would be that of the polymer chain segments themselves. Over short distances they move rapidly, but over longer distances their motion is considerably slower due to their attachment to the rest of the polymer molecule.

Polymers of molecular weight 10^6 have dimensions of a few microns. Flexible polymers such as polyethyleneoxide or polydimethylsiloxane have T_2 long enough to permit measurements of the diffusion coefficients of the entire molecule.[52] It is probable that in favorable cases, the experimental diffusion distance could be decreased below the dimensions of the polymer coil, where an increase in the apparent diffusion coefficient of the polymer spins would be observed. However, it is doubtful whether quantitative information of value could be resolved.

Vapors adsorbed on fibers present another type of colloidal system where diffusion may be studied by MFG-NMR.[53] Oriented samples should show an anisotropy in the diffusion coefficient of the adsorbed material, depending on the closeness of contact of the fibers. The possibility of observing time-dependent diffusion coefficients exists if the fiber diameters or the spacing of bends in the fibers falls within the range of the method as given above.

This work was supported by ONR Contract NO001476WR60015. Thanks are due to the Chemistry Department of Indiana University for the use of equipment to perform the measurements presented by the author,[35] and to Mr. Arthur Clouse and Mr. Robert Adleman for assistance in setting it up.

Literature Cited

1. Bloembergen, N., Purcell, E. M., Pound, R. V., Phys. Rev. (1948) 73, 679.
2. Norberg, R. E. and Slichter, C. P., Phys. Rev. (1951) 83, 1074.
3. Ailion, D. and Slichter, C. P., Phys. Rev. Letters (1964) 12, 168.
4. Andrew, E. R. and Allen, P. S., J. Chim. Phys. (1966) 63, 85.
5. Zimmerman, J. R. and Brittin, W. E., J. Phys. Chem. (1957) 61, 1328.
6. Kaerger, J. and Renner, E., Z. Phys. Chemie, Leipzig (1974) 255, 357.
7. Shporer, M. and Civan, M. H., Biochim. Biophys. Acta (1975) 335, 81.
8. Hahn, E. L., Phys. Rev. (1950) 80, 580.
9. Hahn, E. L., Phys. Today (1953) 6 (11), 4.
10. Stejskal, E. O. and Tanner, J. E., J. Chem. Phys. (1965) 42, 288.
11. Stejskal, E. O., Adv. Mol. Rel. Processes (1972) 3, 27.
12. Gross, B. and Kosfeld, R., Messtechnik (1969) 7/8, 171.
13. Tanner, J. E., J. Chem. Phys. (1970) 52, 2523.
14. Pines, A. and Shattuck, T. W., Chem. Phys. Letters (1973) 23, 614.
15. Blinc, R., Pirs, J., and Zupancic, I., Phys. Rev. Letters (1973) 30, 546.
16. Blinc. R., Burgar, M., Luzar, M., Pirs, J., Zupancic, I., and Zumer, S., Phys. Rev. Letters (1974) 33, 1192.
17. James, T. L. and McDonald, G. G., J. Mag. Res. (1973) 11, 58.
18. Woessner, D. E., J. Phys. Chem. (1963) 67, 1365.
19. Robertson, B., Phys. Rev. (1966) 151, 273.

20. Neuman, C. H., J. Chem. Phys. (1974) 60, 4508.
21. Wayne, R. C. and Cotts, R. M., Phys. Rev. (1966) 151, 264.
22. Tanner, J. E., Ph.D. Thesis, University of Wisconsin, 1966.
23. Tanner, J. E. and Stejskal, E. O., J. Chem. Phys. (1968) 49, 1768.
24. Murday, J. S. and Cotts, R. M., J. Chem. Phys. (1968) 48, 4938.
25. Stejskal, E. O., J. Chem. Phys. (1965) 43, 3597.
26. Lauffer, D. E., Phys. Rev. (1974) A9, 2792.
27. Wang, J. H., J. Am. Chem. Soc. (1954) 76, 4755.
28. Crank, J. and Park, G. S., "Diffusion in Polymers", Chapter 6, Academic Press, London, 1968.
29. Cooper, R. L., Chang, D. B., Young, A. C., Martin, C. J., Ancker-Johnson, B., Biophys. J. (1974) 14, 161.
30. Boss, B. D. and Stejskal, E. O., J. Chem. Phys. (1965) 43, 1068.
31. Boss, B. D. and Stejskal, E. O., J. Colloid Interface Sci. (1968) 26, 271.
32. Packer, K. J. and Rees, C., J. Colloid Interface Sci. (1972) 40, 206.
33. Lauffer, D. E., Poster Session Abstracts, 17th Experimental NMR Conference, 25-29 April 1976, Pittsburgh, Pennsylvania.
34. Packer, K. J., Rees, C., and Tomlinson, D. J. in "Diffusion Processes" Vol. 1, ed. Sherwood, J. N., Chadwick, A. V., Muir, W. M., and Swinton, F. L., Gordon and Breach, London, 1971.
35. Tanner, J. E. and Strickholm, A., unpublished measurements.
36. Hansen, J. R., Biochim. Biophys. Acta (1971) 230, 482.
37. Whittam, R., "Transport and Diffusion in Red Blood Cells", p. 58, Edward Arnold, London, 1964.
38. Finch, E. D., Harmon, J. F., and Muller, B. H., Arch. Biochem. Biophys. (1971) 147, 299.
39. Abetsedarskaya, L. A., Miftakhutdinova, F. G., and Fedotov, V. D., Biofizika (1968) 13, 630 (Biophys. (1968) 13, 750).
40. Pearson, R. T., Duff, I. D., Derbyshire, W., and Blanshard, J. M. V., Biochim. Biophys. Acta (1974) 362, 188.
41. Riedel, E., Kaerger, J., and Winkler, H., Z. Phys. Chemie, Leipzig (1973) 3/4, 161.
42. Kaerger, J. and Walter, A., Z. Phys. Chemie, Leipzig (1974) 1, 142.
43. Kaerger, J., Shdanov, S. P., and Walter, A., Z. Phys. Chemie, Leipzig (1975) 256 319.
44. Kaerger, J., Walter, A., and Riedel, E., Zhur, Fiz. Khim. (1974) 48, 663 (Russ. J. Phys. Chem. (1974) 48, 382).
45. Venkata Raman, K. V. and Jackson, M. L., Clays Clay Minerals Proc. Natl. Conf. Clays Clay Minerals (1963) 12, 423 (1964).
46. Moll, R. E. and Baldeschwieler, J. D., paper presented at the 8th Experimental NMR Conference, 2-4 March 1967, Pittsburgh, Pennsylvania.

47. Murday, J. S. and Cotts, R. M., J. Chem. Phys. (1970) 53, 4724.
48. Murday, J. S. and Cotts, R. M., Z. Naturforsch. (1971) 26a, 85.
49. Krueger, G. J. and Mueller-Warmuth, W., Z. Naturforsch. (1971) 26a, 94.
50. Loewenberg, L. and Lodding, A., Z. Naturforsch. (1967) 22a, 2077.
51. Tanner, J. E., Macromolecules (1971) 4, 748.
52. Tanner, J. E., Liu, Kang-Jen, and Anderson, J. E., Macromolecules (1971) 4, 586.
53. Davis, S. M., M. S. Thesis, University of Wisconsin, 1966.

Measurement of Diffusional Water Permeability of Cell Membranes Using NMR without Addition of Paramagnetic Ions*

D. G. STOUT†
Department of Floriculture and Ornamental Horticulture, Cornell University, Ithaca, N. Y. 14853

R. M. COTTS
Laboratory of Atomic and Solid State Physics, Cornell University, Ithaca, N. Y. 14853

Conlon and Outhred[1] have described a means for determining the mean residence time and the diffusional water permeability for water in a cell surrounded by a permeable membrane. They measure the value of an NMR (nuclear magnetic resonance) spin relaxation time of the mobile water as a function of concentration of paramagnetic impurities outside the cell. They used Mn^{++} ions. If the relaxation time of the extracellular water can be made sufficiently small the Conlon-Outhred experiment can be intrepeted with little difficulty, but, especially in biological systems, there is, as Conlon and Outhred point out, the concern that the presence of Mn^{++} concentrations in non-physiological amount might perturb the cell membrane enough to alter its water permeability and the mean residence time for intracellular water molecules.

The experiment described below eliminates the need for addition of paramagnetic salts in systems having small cells. The extracellular water which must have relatively unrestricted motion can have its transverse relaxation rate enhanced by application of a pair of pulsed magnetic field gradients as is done in NMR measurements of the self diffusion coefficient. The intracellular water has, in small cell systems restricted diffusion[2] and responds with a relaxation rate that is less enhanced. The amplitude of this "slow" component of the signal decays with a rate partially determined by the mean residence time.

Let p_a and p_b equal the fractions of mobile intracellular and extracellular water, respectively, with intrinsic transverse relaxation rates (T_{2a}^{-1}) and (T_{2b}^{-1}). We assume that $T_{2a} \gg \tau_D$ and $T_{2b} \gg \tau_D$ where $\tau_D = (R^2/D_a)$, R is the measure of the cell size, and D_a is the self diffusion coefficient of p_a. We also assume that gradients in the applied D.C. magnetic field are negligible. Let τ_r equal the mean residence time for water within the cell. If $\tau_r \geq \tau_D$ then τ_r is determined by diffusional water permeability, P_d, of the cell membrane and $\tau_r \propto R/P_d$ with a proportionality constant the order of two and determined by cell shape[3].

The two pulse spin echo experiment is done with a field gradient $G = (dH_o/dZ)$ turned on for a time δ immediately after the $\pi/2$ r.f. pulse and similarly after the π r.f. pulse a time τ later. For the water p_b, the echo amplitude[4]

$$M_b \propto p_b \exp[-\gamma^2 G^2 \delta^2 D_b(\tau - \delta/3)]$$

where τ equals the $\pi/2 - \pi$ rf pulse spacing and γ is the nuclear spin gyromagnetic ratio. For sufficiently large values of $G\delta$ this component of the signal is much attenuated. For the water fraction p_a, the signal at $\tau \gg \tau_D$ (and for usual operating conditions $\tau \gg \delta$), approaches a limiting form[2] which, due to restricted motion of p_a, is independent of D_a.

$$M_a \propto \exp[-\gamma^2 G^2 \delta^2 R^2/5].$$

The strength of this component of the signal is proportional to the fraction of the water p_a which was inside the cell at the time of the $\pi/2$ rf pulse and which has neither relaxed due to T_{2a} processes (negligible) or escaped the cell due to P_d. This proportionality should be valid in the limit of large values of $G\delta$ which "destroy" phase coherence of water in the b region.

Then $M_a \approx p_a M_o \exp[-(2\tau/\tau_r) - \gamma^2 G^2 \delta^2 R^2/5].$

The experiment is done on a sample of cells in water with cells sufficiently loosely packed that there are ample channels for diffusion of extracellular water. The echo amplitude is measured as a function of gradient strength $G\delta$ at various values of fixed τ. Since $\tau \gg \tau_D$ then $D_a(\tau-\delta/3) \gg (R^2/5)$ and at sufficiently large values of $G\delta$ only the signal from M_a remains. For each value of τ, a straight line is fit to the plot of $\ell n(M_a)$ vs. $G^2\delta^2$. The intercept of this line at $(G\delta) = 0$ is equated to $-(2\tau/\tau_r) + \ell n(p_a M_o)$. From the set of intercepts the value of τ_r is calculated.

The experiment was performed using Chlorella vulgaris var. "virides" plant cells. These cells are spherical with radii ranging from about 0.8 to 2.4 x 10^{-4} cm.

The value of T_{2a} exceeded 55 msec and since $\tau_D = R^2/D_a \approx$ 1 msec (using $D_a \approx 1$ x 10^{-5} cm^2/sec), the condition $T_{2a} \gg$ τ_D was met. The growth solution had $T_{2b} = 700$ msec and the NMR experiment was done in a packed cell sample of cells in their growth solution.

The maximum values of G and δ used were 173 G/cm and 1.7 msec. Semilog graphs of echo height vs values of $I^2\delta^2$ (where $G = cI$) and $c = 9.7$ (Gauss/cm.amp.) are shown in Fig. 1. Extrapolations of each curve of Fig. 1 to $(I\delta) = 0$ were fit to exp $-(2\tau/\tau_r)$ and a value $\tau_r = 25 \pm 2$ msec is obtained if only data for $\tau \geq 10$ msec (where $\tau \gg \tau_D$) is utilized, see Fig. 2. Since $\tau_r \gg \tau_D$, a valid measure of P_d can be calculated. For spherical cells[3], $P_d = (R/3\tau_r)$ and using a mean radius of 1.6 x 10^{-4} cm, we find $P_d = 2.1 \pm 1$x10^{-3} cm/sec with the large uncertainty due to the distribution of cell sizes.

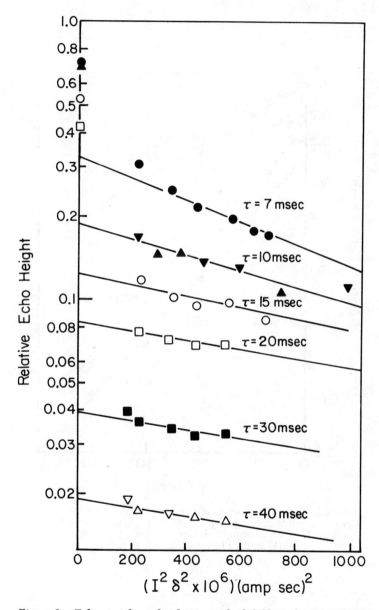

Figure 1. Echo signal amplitude in a pulsed field gradient spin echo experiment on chlorella *vs.* $(I\delta)^2$ *for different values of* τ *in a centrifuged sample of cells in their own growth solution. Except for data points at* $I\delta = 0$, *data at low values of* $(I\delta)$ *are not plotted. The solid curves are straight line fits at large* $(I\delta)$ *and which for* $\tau \geq 20$ *msec have had slopes set equal to* $(\gamma^2 C^2 R^2/5)$ *where* $\overline{C} = (G/I)$.

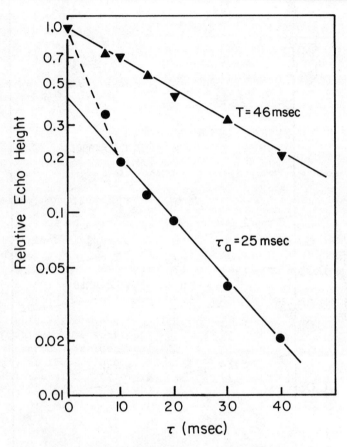

Figure 2.   Echo signal amplitude vs. τ for chlorella: ▼ total echo
signal amplitude for p_a and p_b observed with zero applied field
gradient; ● signal amplitude obtained from the (Iδ) = 0 intercept
of the straight lines fit to data in Figure 1

The decay time of the signal obtained by extrapolating back
to $G\delta = 0$ differs from the value actually measured at $G = 0$. The
echo amplitude at $G = 0$ depends upon all the characteristics of
the two water populations p_a, p_b, τ_r, T_{2a}, and T_{2b} in a manner
already described in the literature [5]. The value extrapolated
to $G\delta = 0$ depends upon only the characteristics of intracellular
water. Fig. 2 shows semi-logarithmic plots of the $G = 0$ echo
amplitude data for Chlorella and the data obtained by extrapolating data for large values of $(G\delta)$ to $G\delta = 0$. The $G=0$ data gives a
relaxation time of 46 msec which is considerably longer than the
relaxation time of the extrapolated data, 25 msec. The fact
that the $G=0$ data does not have a longer relaxation time is due
to a weak concentration of paramagnetic ions from the growth
solution on the outside of the Chlorella cells. These ions
reduce extracellular water relaxation times when cells are
packed together by centrifuging as they were in this experiment.
A detailed report on T_1, T_2, and some diffusion constant measurements in samples of Chlorella cells will be published elsewhere.

* Supported by the National Science Foundation, Grant GH-37259.

† Predoctoral Fellow, National Research Council of Canada;
Current address: Crop Science Department, University of
Saskatchewan, Saskatoon, S7N 0W0, Canada.

1. T. Conlon and R. Outhred, Biochim. Biophys. Acta, 288, 354-61
   (1972).
2. C. H. Neuman, J. Chem. Phys. 60, 4508-11 (1974).
3. Crank, J., The Mathematics of Diffusion, University Press,
   Oxford (1956), Pgs. 56, 57, 73, and 91. While these sections
   deal with "surface evaporation", the mathematics is identical
   to the problem of diffusional permeability at the surface.
   Similar boundary conditions appear in heat transfer problems.
4. E. D. Stejskal and J. E. Tanner, J. Chem. Phys. 42, 288 (1965).
5. H. D. McConnell, J. Chem. Phys. 28, 430-31 (1958) and
   D. C. Chang, C. F. Hazlewood, B. C. Nichols, and H. E.
   Rohrschach, Biophys. J. 14, 583-606 (1974).

4

Adsorption Phenomena in Zeolites as Studied by Nuclear Magnetic Resonance

H. PFEIFER

Sektion Physik der Kark-Marx-Universität, DDR 701 Leipzig, Linnéstrasse 5

Introduction

Zeolites are porous crystals with a well-defined structure and a high specific surface area, widely used in industry as molecular sieves, as catalysts, and as ion exchangers. Their common formula is

$$(Al\,O_2)^- \; M^+ \; (Si\,O_2)_n \qquad (1)$$

where M^+ denotes an exchangeable cation which can be also replaced by $1/2\ M^{2+}$ or $1/3\ M^{3+}$, and n is a number greater or equal 1. The adsorption phenomena to be discussed in this report are concerned with molecules in the pores of a special "family" of zeolites (faujasite group) to which belong the so-called zeolites NaX and NaY generally denoted as NaF. These synthetic zeolites, available as small crystallites with a mean diameter between 1 and 100 um, have the same structure and differ only in the value of n. Their pore system accessible to hydrocarbons, is a three-dimensional network of nearly spherical cavities, commonly referred to as supercages, each with a mean free diameter of 11.6 Å and connected tetrahedrally through windows of 8 - 9 Å diameter. Because each silicon or aluminium ion of the zeolite lattice is surrounded by four larger oxygens, the internal surfaces of ideal crystals consist of oxygen, the only other elements exposed to adsorbates being the exchangeable cations which are sodium ions in the case of NaF zeolites. Some of these cations are not accessible to hydrocarbons since they are localized at sites SI and SI' outside of the supercages, some are localized at the walls of the

36

supercages opposite to the windows (sites SII)
and some could not be localized by X-ray measurements,
they are assumed, to be loosely bound (sites S3)
to the walls of the supercages between sites SII.
As was mentioned already it is possible to replace
Na^+ by other cations, and since di- and especially
trivalent cations occupy first of all sites SI and
SI', the number of sodium ions per supercage can
be varied not only by n, but also through an ion
exchange. Examples are given in Table I.

Table 1
Number of exchangeable cations in dehydrated
zeolites of faujasite type per 1/8 unit cell
(corresponding to one supercage)

Site	maximum number	NaX n=1.37	NaY n=2.6	NaCeY70[x)] n = 2.6
SI	2	1.2	0.7	⎫ 1.5 Na^+
SI'	4	1.7	2.3	⎬ +1.5 Ce^{3+}
SII	4	3.2	3.3	0.5 Na^+
S3 (non-loc.)	6	4.0	0.4	–

[x)] NaCeY70 is a NaY zeolite where 70 % of Na^+
has been replaced by Ce^{3+}.

The physical state of adsorbed molecules depends on the ratio of intermolecular interaction
energy to the energy of interaction between a molecule and the surface. According to Kiselev (1)
molecules and surfaces may be classified with regard
to their capacity for nonspecific and specific interaction. Nonspecific interaction is caused mainly by
dispersion forces and is present therefore in all
cases. Specific interaction may appear additionally
whenever electron density (e.g. π bonds of molecules)
or a positive charge (Na^+ in the case of NaF
zeolites) is localized on the periphery between
the adsorbed molecules and the surface.
    We have studied the physical state of cyclic
hydrocarbons, adsorbed in NaF zeolites where the
degree of specific interaction has been systematical-
ly varied through the number of π electrons per
molecule (cyclohexane, cyclohexene, cyclohexadiene,
benzene and toluene) and through the location and

number of sodium ions at the walls of the supercages.

Adsorbed cyclohexane

Using various mixtures of cyclohexane and
deuterated cyclohexane it could be shown (2)
that the magnetic interaction with paramagnetic
impurities of the zeolite (called proton electron
interaction) is controlling the proton relaxation
times (T_1 and T_2) of C_6H_{12} adsorbed in a commercial
NaY zeolite (VEB Chemiekombinat Bitterfeld, n =2.6)
These paramagnetic impurities are Fe^{3+} ions located
at Al^{3+} sites of the zeolite skeleton, although an
even greater number is localized at sites SI or as
an iron oxide-like phase on the crystals (3).
The total concentration, determined optically corresponds to about 500 ppm Fe_2O_3 for this zeolite.
The temperature dependence of T_1 and T_2 has been
measured at 16 MHz between − 140°C and +90°C for
pore filling factors Θ = 0.16, 0.4, 0.7, and 0.88.
This quantity Θ is the ratio of the adsorbed amount
of cyclohexane to its maximum value which corresponds to about four molecules per supercage.
In all cases the minimum of T_1 occured at the same
temperature (−100°C) while the ratio of T_1 to T_2 at
this point increased from about 2 to 4, and T_1
versus reciprocal absolute temperature becomes
flatter with increasing coverage. Using Torrey's
model for translational motion (4) the following
formulae can be derived for this case (5)
where $\omega_s \tau_c \gg 1$ has been assumed

$$\frac{1}{T_1} = k \frac{\tau_c}{d^2} f_t (\alpha, \omega_I \tau_c) \qquad (2)$$

$$\frac{1}{T_2} = \frac{2}{3} k \frac{\tau_c}{d^2} \{1 + f_t (\alpha, \omega_I \tau_c)\}. \qquad (3)$$

Here ω_I and ω_s are the Larmor frequencies of the
proton and paramagnetic ion respectively, K is a
constant proportional to the concentration
of paramagnetic ions, d denotes the minimum
distance between proton and ion, and τ_c is the
correlation time for the translational motion of the
adsorbed molecule:

$$\tau_c = \frac{\tau}{10\alpha} + \frac{\tau}{2} \qquad (4)$$

where

$$\alpha = \frac{\langle r^2 \rangle}{12\,d^2} \quad . \tag{5}$$

The mean quadratic jump length $\langle r^2 \rangle$, the average time τ between two jumps and the diffusion constant D are related by Einstein's equation

$$\langle r^2 \rangle = 6\,D\,\tau . \tag{6}$$

The function f_t (α, $\omega_I \tau_c$) is defined as

$$f_t(\alpha, \omega_I \tau_c) = 3\left(\frac{1}{5}+\alpha\right)\left\{ v\left(1 - \frac{1}{u^2+v^2}\right) \right.$$
$$+ \left[v\left(1 + \frac{1}{u^2+v^2}\right)+2\right] e^{-2v} \cos 2u$$
$$\left. + u\left(1 - \frac{1}{u^2+v^2}\right) e^{-2v} \sin 2u \right\} (\omega_I \tau_c)^{-2} \tag{7}$$

where

$$\begin{matrix} u \\ (v) \end{matrix} = \frac{1}{2\sqrt{\alpha}}\sqrt{q\,(1 \mp q\,)} \tag{8}$$

$$q = \left[1 + \left(1 + \frac{1}{5\alpha}\right)^2 (\omega_I \tau_c)^{-2}\right]^{-1/2} . \tag{9}$$

A fitting procedure (6) yields $\alpha \sim 1/5$ for $\theta = 0.88$, and since for this value of α, at the minimum of T_1 it is $\omega_I \tau_c = 1.58$, we have

$$\tau_c = \tau = \frac{1.58}{\omega_I} . \tag{10}$$

In contrast, for the lowest coverage ($\theta = 0.16$) it follows from the fitting procedure $\alpha \to \infty$. Because in this case the minimum of T_1 occurs if $\omega_I \tau_c = 1$, so that we have

$$\tau_c = \frac{\tau}{2} = \frac{1}{\omega_I} . \tag{11}$$

Therefore τ is about 16 ... 20 ns at -100°C and does not depend on coverage within these limits. The mean quadratic jump length of translational motion however, decreases with increasing coverage which is a direct hint for nonlocalized adsorption of cyclohexane. In accordance with this, measurements of proton relaxation in zeolites with different

number of sodium ions ($n = 1.37$, ([7]) and $n = 1.2$,([8]))
revealed about the same correlation times
as in the above system ($n = 2.6$). Since these
zeolites are of high purity (only 2 ... 3 ppm
Fe_2O_3 in the case of $n = 1.37$) and the crystallites
of big diameter (15 ... 25 μm in the case of $n = 1.2$)
it was possible to determine the intracrystalline
self-diffusion coefficient D by the pulsed field
gradient technique. The results are shown in
Table 2.

Table 2
Self-diffusion coefficient of C_6H_{12} in NaX zeolites
at room temperature (D_{20}) and activation energy (E_D)

$n = 1.37$ $\theta = 0.3$ ([7])	$D_{20} = 7 \cdot 10^{-6}$ cm^2/s $E_D=(3.0\pm0.2)\frac{kcal}{mole}$					
$n = 1.2$ θ	0.25	0.27	0.45	0.66	0.69	$\gtrsim$0.75
$D_{20}\left[cm^2/s\right]$ ([8])	$5.5 \cdot 10^{-6}$	$6 \cdot 10^{-6}$	$3.8 \cdot 10^{-6}$	$9 \cdot 10^{-7}$	$9 \cdot 10^{-7}$	$<10^{-7}$

From equs. (10) and (11) and the observed activation
energy $E_c = (3.0 \pm 0.3)$ kcal/mole, the value of τ
at room temperature comes out to be ~ 0.5 ns, so
that the rms jump length $\sqrt{<r^2>}$ (cf. equ.(6)) and
Table 2) is about 13 Å, and less than 2 Å for $\theta = 0.2$
and 0.8 respectively. Summarized, cyclohexane
adsorbed in zeolites of faujasite type behaves like
an intracrystalline liquid: the molecules do not
jump between sites which are fixed in space.
The mean jump length during translational motion
decreases with increasing coverage as a consequence
of mutual hindrance, while the average time between
jumps does not change significantly.

Adsorbed nonsaturated cyclic hydrocarbons

In Table 3 the ratios of proton relaxation
times T_1 and T_2 measured at the minima of T_1 for
various cyclic hydrocarbons adsorbed in the same
commercial NaY zeolite ($n=2.6$) as above, are listed
([9])

Table 3
T_1/T_2 at the minimum of T_1, activation energy E_c and correlation time τ_c for various cyclic hydrocarbons adsorbed in a commercial NaY zeolite (n = 2.6)

		C_6H_6	C_6H_8	C_6H_{10}	C_6H_{12}
T_1/T_2	θ =0.2	1.8	1.8	1.8	2.1
	θ =0.8	1.8	1.9	1.9	4
$E_c \left[\dfrac{kcal}{mole}\right]$	θ =0.2	\} 3.4 ± 0.3			3 ± 0.3
	θ =0.8				
τ_c [ns] at 20°C	θ =0.2	13	7.7	2.8 ±10%	$\tau_c = \tau/2 \sim 0.25$
	θ =0.8	50	18	9	$\tau_c = \tau \sim 0.5$

As can be seen, these ratios are close to the theoretical value (1.83) for dominating proton electron interaction and thermal motion characterized by a single correlation time τ_c (10). Therefore we have

$$T_1 \propto \frac{1 + (\omega_I \tau_c)^2}{\tau_c} \qquad (12)$$

and τ_c can be readily determined as a function of T, since T_1 has been measured between about $-100°C$ and $+200°C$ (9). The logarithm of τ_c plotted versus $1/T$ gives straight lines, so that an activation energy E_c can be defined which is listed in Table 2 together with absolute values for τ_c at room temperature. From these results the following conclusions concerning non saturated cyclic hydrocarbons adsorbed in NaY can be drawn:
(i) Even at high pore filling factors (θ=0.8) there is no distribution of correlation times: The molecules are adsorbed at sites fixed in space (localized adsorption). These adsorption centres will be the sodium ions at sites SII as we shall see later.
(ii) As a consequence of localized adsorption Eyring's equation can be applied :

$$\tau_c = \frac{h}{kT} \frac{{}_1f}{{}_2f^{\ddagger}} \exp(E_c/RT) \qquad (13)$$

where the initial state (partition function ${}_1f$)
corresponds to the molecule adsorbed on Na^+ and
the activated state (partition function ${}_2f^{\ddagger}$)
corresponds to the molecule during its jump from
one Na^+ to another. In accordance with this model
the energy of activation E_c is less than the heat of
adsorption (Q =18 kcal/mole for benzene). It does
not depend on the number of π electrons. The observed increase of τ_c from C_6H_{10} to C_6H_6 must be
explained by a change of the preexponential factor
and especially by a decrease of ${}_1f^{\ddagger}$ since a
stronger interaction always leads to a decrease in
the partition function. Correspondingly, we have an
increasing restriction of molecular mobility in
the activated state with increasing number of
π electrons.
(iii) The increase of τ_c with increasing coverage is
      also in accordance with equ.(13) because the
      preexponential factor is proportional to $1/u$,
      where u denotes the number of unoccupied
      adsorption sites which can be reached from
      the activated state: We can assume $u \sim 3$ and
      $u \leqslant 1$ for $\theta = 0.2$ and 0.8 respectively, since
      there are about 4 adsorption centres per super-
      cage, and $\theta = 1$ corresponds roughly to the
      same number of molecules.
     For a further check of this model we have measured
the proton relaxation times of toluene adsorbed on
the same zeolite. Although benzene and toluene
differ in their intermolecular interactions, as can
be seen from their quite different melting points
(+ 5.5°C and -95°C respectively), molecular motion
in the adsorbed state should be simular, since it
is determined by the interaction between the ring
of the molecules and the sodium ions at sites SII.
This indeed has been observed: There is no
difference in proton relaxation times between
benzene and toluene from about -20°C to +180°C (11).
In addition,this result states that the rotation
of the benzene molecule around its sixfold axis of
symmetry does not influence proton relaxation in
this temperature interval.

If benzene is adsorbed in a commercial NaX
zeolite (VEB Chemiekombinat Bitterfeld, n = 1.8,
~500 ppm Fe_2O_3) the minimum of T_1 shifts to much
lower temperatures (Fig. 1) and the ratio T_1/T_2
increases slightly from 1.8 to about 2.4. Therefore the mobility of benzene molecules adsorbed in
NaX is much higher (at room temperature $\tau_c \sim 10$ ns
in comparison to 50 ns in NaY) and a slight distribution of correlation times seems probable.
The much higher mobility of benzene molecules
adsorbed in NaX zeolite should be discussed in
connection with results for water where the
mobility does n o t depend on silicon to aluminium
ratio n of zeolites (12). For a pore filling
factor θ = 0.8 there are about 24 water molecules
and only about 4 benzene molecules per supercage.
Therefore in NaY (n = 2.6) all aromatic molecules
will be strongly bound by specific interaction
to the localized sodium ions which are fixed at
the wall of the supercage (sites SII, cf. Table 1).
This explains the low mobility of these molecules.
In NaX zeolites (n = 1.8) however, aromatic molecules will be bound to the nonlocalizable sodium
ions available here, since these are more exposed
than Na^+ at sites SII and act as primary adsorption
centres (13). So the high mobility in NaX can be
explained by translational and rotational motion of
aromatic molecule-sodium ion-complexes. In agreement
with this, it was found recently (14) that with increasing Si/Al-ratio (n) the mobility decreases at
first, but remains constant above n = 2.43 where
all sodium ions are localized. Another model has
been proposed before (15), according to which the
difference in mobility is caused by different
electric fields in NaX and NaY zeolites. But this
model would predict a monotonous decrease of
mobility with increasing values of n opposite to
the experimental results.
      In contrast to benzene, for water at higher
pore filling factors ($\theta \gtrsim 0.5$) the mobility of
adsorbed molecules does n o t depend on n, so
that the reduced mobility with respect to liquid
water cannot be caused by the sodium ions. This is
confirmed by two experimental findings: Firstly,
even in concentrated aqueous solutions of Na^+ the
mobility of water molecules is only slightly reduced and corresponds at room temperature to an
increase of correlation time from the free water
value 2.5 ps (16) to only about 5 ps (17) for a

water molecule in the first coordination sphere of
a sodium ion. Secondly the small width of the distribution function for τ_c (11), (12) is equivalent
to the statement that the correlation time of a
water molecule adjacent to a sodium ion cannot be
very different from the correlation time of water
molecules surrounded by oxygen atoms belonging to
the wall of the supercage or to other water molecules, because the experimentally determined half
width of the distribution function (for the distribution parameter ß as defined in (18) it has been
found ß $\leqslant$ 1) corresponds to a ratio of correlation
times of less than about 5.

This only small change in mobility of water
molecules caused by Na^+ is due to the strong
water-water interaction (19) and therefore the
reduced mobility in the adsorbed state (τ_c =200 ps
at room temperature) must be a collective effect of
the rigid oxygen lattice of the zeolite which
stabilizes the ensemble of water molecules in
front of it ("stabilization effect" (11)).

To show directly the influence of sodium ions
on the mobility of benzene, measurements of proton
relaxation times in NaCeY 70 zeolite have been
performed, where through an exchange of 70 % Na^+
by Ce^{3+} the number of sodium ions in the supercages has been drastically reduced (cf. Table 1).
Although the minimum of T_1 is shifted to extreme
low temperatures (to less than -150°C (19))
this cannot be taken as a proof for a higher
mobility of the adsorbed benzene molecules, since
it is an artifact of nuclear magnetic relaxation
method. The values of the relaxation times are
significantly g r e a t e r than in NaY in
contrast to the usual effect if paramagnetic ions
(Ce^{3+}) are introduced into a sample. The reason is
that Ce^{3+} reduces by magnetic dipole interaction
the electron relaxation time of the paramagnetic
impurities which control (Fe^{3+} ions located at Al^{3+}
sites of the zeolite skeleton) the proton relaxation
of adsorbed benzene to such a value that is becomes
shorter than the thermal correlation time (19).

Therefore a zeolite was prepared where the
very similar but diamagnetic La^{3+}, instead of
the paramagnetic Ce^{3+} was introduced. The result,
shown in Fig. 1, is in full agreement with the
above model, since one observes that τ_c at room
temperature decreases from 50 ns for NaY to 10 ns
or to ~ 5 ns, if nonlocalized sodium ions are

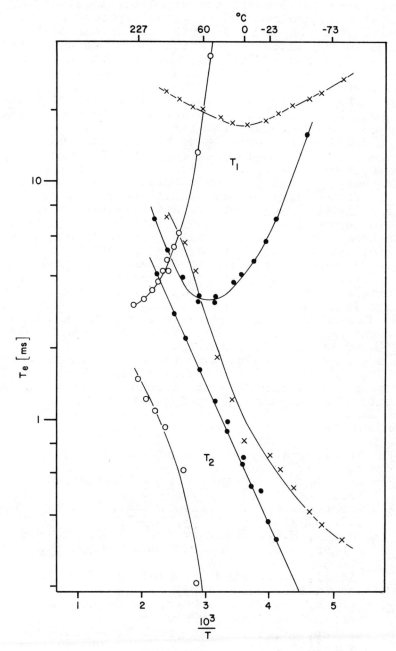

Figure 1.   *Benzene, ν = 30 MHz, θ = 0.8;* ○, *NaY, n = 2.6;* ●, *NaX, n = 1.8; X, NaLaY 74, n = 2.6*

introduced (NaX) or if these and the localized
sodium ions are removed from the supercages
(NaLaY 74). The broad distribution of correlation
times in the latter case which can be deduced from
the great value of T_1/T_2 at the minimum of T_1 may
be due to a few tightly-bound benzene molecules
(e.g. at the few Na^+ still occupying sites SII)
and partly due to nonlocalized adsorption of the
major part of adsorbate (small jump lengths during
the course of translational motion).

Finally it should be mentioned that the
benzene-sodium ion-complex suggested by these
measurements was studied also theoretically using
the semiempirical quantum chemical CNDO/2 method
(20). From energy minima of full potential curves
the optimum arrangement and stabilization energy
could be obtained. The most favourable configuration
has been found when the sodium ion is lying on the
sixfold axis of symmetry of the aromatic ring
(3.2 Å), because in this case there is a good
possibility for overlapping of the free electron
orbitals of Na^+ with the π orbitals of the adsorbed
molecules, and an electron transfer was found from
the π system toward the cation. This model of
benzene adsorption has been confirmed experimentally
(21) by nmr broad line technique, insofar, as the
independence of second moments of proton resonance
at 77 K on pore filling factors could be explained
only by a flat arrangement of benzene molecules in
front of sodium ions occupying sites SII.

References

(1) Kiselev, A.V., Disc.Farad.Soc. (1966) 40,
     205-218
(2) Michel, D., Thöring, J., Z.phys.Chemie
     (Leipzig) (1971) 247, 85-90
(3) Derouane, E.G., Mestdagh, M.M., Vielvoye, L.,
     J.Catalysis (1974), 33, 169-175
(4) Torrey, H.C., Phys.Rev. (1953) 92, 962-969
(5) Krüger, G.J., Z.Naturforsch. (1969), 24a,
     560-565
(6) Michel, D., Thesis (Prom.B) Karl-Marx-Uni-
     versität Leipzig (1973)
(7) Lorenz, P., Diplomarbeit, Karl-Marx-Uni-
     versität Leipzig (1974)
(8) Walter, A., Thesis in prep. Karl-Marx-Uni-
     versität Leipzig
(9) Nagel, M., Pfeifer, H., Winkler,H.,
     Z.phys.Chemie (Leipzig) (1974),255, 283-292

(10) Pfeifer, H., NMR - Basic Principles and Progress
      Vol.7, p.53-153, Springer, New York 1972
(11) Pfeifer, H., Surface Phenomena Investigated
      by Nuclear Magnetic Resonance, Phys.Reports
      (in press)
(12) Pfeifer, H., Gutsze, A., Shdanov, S.P.,
      Z.phys.Chemie (Leipzig) (in press)
(13) Dzhigit, O.M., Kiselev, A.V. et al.,
      Trans.Farad.Soc.(1971), 67, 458-467
(14) Lechert, H., Haupt, W., Kacirek, H.,
      Z.Naturforsch. (1975), 30a, 1207-1210
(15) Lechert, H., Hennig, H.J., Mirtsch, Sch.,
      Surface Sci. (1974), 43, 88-100
(16) Hindman, J.C., Zielen, A.J. et al.,
      J.Chem.Phys. (1971), 54, 621-634
(17) Hertz, H.G., Angew.Chemie (1970), 82, 91-106
(18) Nowick, A.S., Berry, B.S., JBM J.Res.
      Development (1961), 5, 297-320
(19) Geschke, D., Pfeifer, H., Z.phys.Chemie
      (Leipzig) (in press)
(20) Geschke, D., Hoffmann, W.-D., Deininger, D.,
      Surface Sci. (in press)
(21) Hoffmann, W.-D., Z.phys.Chemie (Leipzig)
      (in press)

5

DMR Study of Soap-Water Interfaces

J. CHARVOLIN and B. MÉLY

Physique des Solides, Université Paris Sud, 91405 Orsay, France

Soap-Water Interfaces

We shall deal with potassium soaps of formula $CH_3-(CH_2)_n - CO_2K$ with $n \sim 10$. Owing to their amphiphilic character such molecules agregate when put in presence of water and their polar heads determine interfaces separating aqueous and paraffinic media . Extensive X-ray works on lyotropic liquid crystals have shown the wide variety of possible interfaces (1). As illustrated on figure 1 a plane interface exists with either ordered or disordered paraffinic medium but the interface can also be cyclindrical or spherical in the second case.

The well documented structural description of those systems is being completed now by a description of their molecular dynamics through the use of magnetic resonance techniques. We shall present some recent information about molecular behaviors on each side of the interface, gained studying deuterons of deuterated molecules(DMR).

DMR Results

The search for well resolved spectra, not broadened by the dipolar interactions between protons of ordinary protonated molecules, led to the use of deuterated ones(2). Deuteron spectra of D_2O and perdeuterated potassium laurate (n = 10) in a lamellar L_α phase are shown in figures 2 and 3 respectively. No line appears at the centers of the spectra; i.e. the deuteron quadrupolar interactions are not totally averaged out. This shows the anisotropic influence of the interface.Clearly, in such structures, the interactions of water molecules with soap polar heads can be viewed as anisotropic states of adsorption and rotational isomerism of

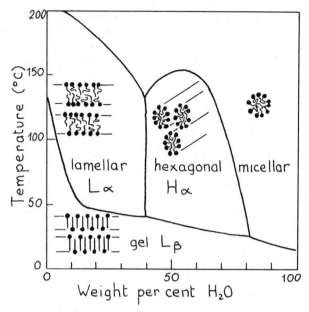

Figure 1.   *Schematic of a potassium soap–water lyo-*
tropic liquid crystal

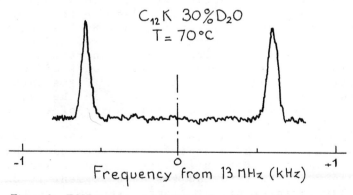

Figure 2.   *DMR spectrum of D₂O in a lamellar Lα phase of potas-*
sium laurate–water. The magnetic field is in the plane of the lamella.

the chains around their C-C bonds cannot lead to isotropic reorientations of their C-D bonds. Methylene and
methyl groups along a chain were identified assuming
that the smaller the splitting is, the further from the
polar head the group should be. This identification was
confirmed later on by DMR studies of selectively deuterated molecules (3)(4).

Anisotropic Behaviors

Water. No detailed microscopic description of
soap-water interactions has been presented yet. Many
variables can be expected to intervene : nature and
ionisation of the polar heads, density on the interface
chain behavior, temperature and water content.
    In the phases with disordered chains temperature
and water content variations lead to simultaneous variations of all other parameters   through the modification of the soap bilayer(1), systematic studies are
needed to treat every parameter as a unique variable(5).
    In the L_β phase with ordered chains the structure
of the soap bilayer does not change appreciably with
water content at constant temperature(6). The situation
is then simpler than in L_α ; chain behavior and density
do not vary, the state of the polar head can be assumed
constant in the studied hydration range, and the only
variable left is the water content. Variations of D_2O
splitting with water content in the L_β phase of potassium stearate (n = 16) is shown in figure 4. The splitting remains constant up to about 6 water molecules per
soap molecule, in the phase separation zone (L_β+crystal)
This shows that a minimum water content is needed for
the L_β phase to exist and the molecules are in anisotropic states of interaction with the soap polar heads.
When extra water molecules are added the splitting decreases as the inverse water content. Such a behavior
is typical of the averaging effect introduced by the
exchange of water molecules between constant interfacial anisotropic states and isotropic ones of increasing statistical weight in the center of the water layer. Thus the range of the anisotropy in the aqueous
medium appears limited to a small number of molecules.

    Chains. An interesting feature of the L_α laurate
spectrum in figure 3 is that 7 methylenes have about
the same behavior. For an isolated chain, fixed at one
end, this would be very odd for each segment is expected to have greater orientational freedom than the pre-
ceeding one. Obviously steric repulsions between neighboring chains in bilayers have to be considered. As a

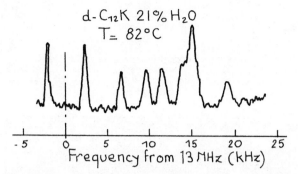

d-C₁₂K 21% H₂O
T = 82°C

Frequency from 13 MHz (kHz)

Figure 3. *DMR spectrum of deuterated soap molecules in a lamellar L_α phase of potassium laurate–water. The magnetic field is in the plane of the lamella. Only half the spectrum, which is symmetric around zero frequency, is shown.*

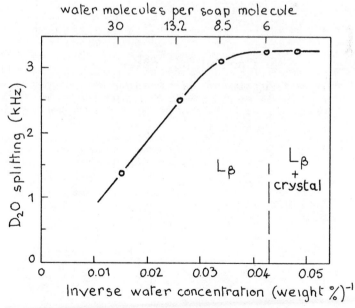

water molecules per soap molecule

Inverse water concentration (weight %)⁻¹

Figure 4. *Variations of D_2O splitting, as function of the inverse water content, in lamellar L_β phases of potassium stearate–water*

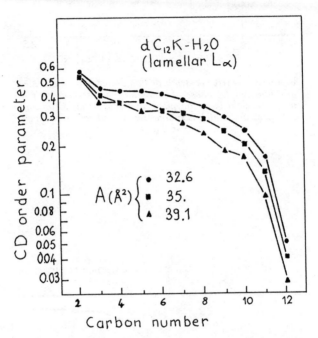

Figure 5. *Variations of the CD bond order parameter curves as the area A per polar head at the interface increases in a lamellar L_α phase of potassium laurate–water*

matter of fact the degree of orientational order depends upon the chain density at the interface, as shown
in figure 5. When the mean area A per chain at the interface increases(increasing temperature or water content) the order parameters $<3 \cos^2\theta-1>$ of the CD bonds
with respect to the bilayer normal decrease. The rate
of decrease is significantly smaller for the first
three methylene than for the others. This last result
reveals a stiffness effect in the vicinity of the interface. The proximity of water limit the conformations
accessible to the first C-C bonds. At this point it is
worthwhile to notice that the influence of the natures
of the polar heads on the anchoring conditions may be
seen when comparing those results with these of (3) for
a ternary system. Whereas the order parameter of the
first methylene is significantly higher than that of
others in the K laurate-water system, they are about
the same in the decanol-Na decanoate-water system.

These considerations suggest to consider three
regions in a bilayer : an interfacial one where the
first chain segments are stiffened, then a region where
the segments get more disordered but are nevertheless
sterically constrained by neighboring molecules, finally the disorder decreases rapidly in the central region
of the bilayer where the presence of chain ends removes
partially the steric constraints (4).

Conclusion

The very small variation of chain order in the
interfacial region, when important variations of the
mean area per polar head take place, let suppose that
water penetrates in between the polar heads and the
following three or four chain segments. Thus,all along
this discussion, the meaning of the term interface has
evoluated from that of a very limited zone where polar
heads and water interact to that of a more extended one
where soap and water intermingle. This is well known
in micellar studies where it has been shown that, in
the process of micelle formation, the hydration of the
methylene groups adjacent to the hydrophilic group is
retained (7).

Literature cited

(1) Luzzati V.,"Biological Membranes" Ed.by D.Chapman
    Academic Press (1968).
(2) Charvolin J.,Manneville P., Deloche B., Chem.Phys.
    Let., (1973) 23, 345.

(3) Seelig J., Niederberger W., Biochemistry(1974), 13, 1585.

(4) Mely B., Charvolin J., Chem.Phys. Lip., (1975), 15, 161.   °

(5) Johansson A, Lindman B., "Liquid Crystals and Plastic Crystals" Vol.1, Chap 8, Ed.by Gray G.W. and Winsor P.A., Ellis Horwood (1974).

(6) Mely B., Charvolin J., submitted to Chem.Phys.Lip.

(7) Corkill J.M., Goodman J.F., Walker T., Trans.Faraday Soc.(1967), 63, 768.

Chemical Aspects of the Hydrophobic/Hydrophilic Interface by Nuclear Magnetic Resonance Methods

L. W. REEVES

Instituto de Quimica, Universidade de São Paulo, Caixa Postal 20,780, São Paulo, Brazil

F. Y. FUJIWARA and M. SUZUKI

Chemistry Department, University of Waterloo, Waterloo, Ontario, Canada N2L 3G1

Introduction

Amphiphilic species such as detergent ions, lipids, soaps and certain polar compounds dissolve in water to form simple molecularly dispersed systems only at small concentrations, below what is termed the critical micelle concentration. Micelles are agglomerates of these amphiphilic compounds in modest number on the order 10^2 or less in which the solution structure seeks to minimize the hydrophobic/hydrophilic interactions by forming a hydrophobic inner core in an aqueous medium with an interface between them. At much more concentrated levels of ampiphilic substances the micellar structure becomes liquid crystalline in nature but the overall division into aqueous, and hydrophobic compartments with a structured interface between them remains (1). The nature of these liquid crystals, which can take several forms, has been attributed to superstructure arrays which have been studied in low angle x-ray scattering experiments (2). The concentration of interface area per unit volume is extremely high and furthermore some mesophases are oriented by magnetic fields (3). The completely oriented phases have advantages in magnetic resonance studies because the anisotropic parts of the parameters available become the dominant features of the spectrum which is generally of high resolution quality. With modern nuclear magnetic resonance (NMR) spectrometers almost all elements in the periodic table become accessible by the same sensitivity enhancement techniques available for more commonly studied low abundance nuclei such as C-13 and variation of chemical elements involved in lyotropic liquid crystals can be very wide, especially in the counter ions which reside in the electrical double half layer. It is evident therefore that the NMR technique can attack the interface problem on a broad front, which includes both averaged and dynamic information.

The anisotropic NMR parameters such as chemical shift, pseudo-dipolar coupling, dipole-dipole coupling and nuclear quadrupole interaction energies all depend on the angular factor $\frac{1}{2}(3\cos^2\theta - 1)$ where θ is the angle between some vector or principal

tensor component and the static magnetic field used in the NMR
spectrometer. This angular factor becomes partially averaged by
the anisotropic tumbling of the liquid crystalline component and
it is always non-zero in anisotropic fluids (4-6). The value

$$S = \tfrac{1}{2} \left\langle (3Cos^2\theta - 1) \right\rangle \qquad\qquad [1]$$

has been called the degree of orientation of the appropriate axis
with respect to the static magnetic field. It is an important
accessible quantity and gives a good indication of how highly
ordered an axis is and sometimes the direction of that order,
from the sign. ($S_{||}$ = + 1 and $S_{\perp}$ = -½ for perfect alignments).
The anisotropic tumbling of a molecule or ion is described by a
3 x 3 order matrix of 5 independent elements (4,5). Molecular or
ionic symmetry reduces the number of elements required to describe
the motion and with a three fold or more axis only one degree of
order parameter is required. The number of independent S axis
values required to describe the anisotropic motion of a rigid
molecule is thus related to molecular symmetry.

The new questions that may be posed concerning the interface
relate to processes at the molecular and ionic level. While it is
evident from previous studies that the interface region is the
most highly structured and that water plays an important role,
there has been no experimental work on the concentration distribution of ions in the electrical double layer (6). The distribution
of dissolved substances between hydrophobic and hydrophilic compartments is also indicated by NMR measurements (7). There is a great
deal of heterogeneity of order between the water, interface and
hydrocarbon chain regions and component molecules or ions reveal
their location from the micro degrees of order of axes in them.
The ionic head groups and counter ions influence the hydrocarbon
chain motions. The present talk will seek to illustrate these
and other questions on the general interface problem.

Micro-Degrees of Order of Species in the Hydrophobic/Hydrophilic Interface Systems

It is possible to prepare two types of lyotropic liquid
crystals, which orient in a magnetic field. We have designated
these phases as Type I and Type II, the first aligns with the
symmetry or director axis of the phase parallel to the magnetic
field and the second in a perpendicular manner (8). The degree of
orientation of phase components and small dissolved molecules has
been investigated in a wide range of chemical systems and representative results are listed in Table I (8-12). The degree of
orientation or order parameter S, is always referred to the
magnetic field direction in the ordered sample, but samples of
Type I and Type II differ in the direction of the phase director,
with respect to the applied field, as indicated above. For
comparison of the order parameters of ions, molecules or hydrocarbon chain segments the longest local symmetry axis in the

species is usually chosen in each case, and ranges of values are given in Table I with respect to the magnetic field direction. Since all S values are referred to a long axis in the species with respect to the magnetic field they may be compared in magnitude as an indication, more or less, of perfect local order. The water has clearly a low average local order, the experimental values being the exchange averaged value for interface structure water and the remotely located molecules in the aqueous compartment.

The hydrocarbon segments near the head group and the $-ND_3^+$ head group are the most highly ordered regions of these oriented lyotropic phases (13). The order parameter of long axes in ions varies a great deal from phase to phase, as also they do in polar molecules. The more mobile constituents of the electrical double half layer may or may not play an integral role in the highly structured interface region between the hydrophobic and hydrophilic compartments(6). In the case of the dimethylthallium ion, which is positively charged, the degree of orientation of the C_{∞} axis is very low in the cationic mesophase but almost two orders of magnitude higher in the anionic phase. In both instances the ion was added in about 2 weight % as the electrolyte dimethylthallium nitrate (9,10). In these cases it is clear that in the cationic phase the complex cation resides in a region relatively remote from the interface, but in the anionic phase it is sufficiently ordered to suggest a role in forming the interface. Such studies on one ion cannot give us a model of what is important among the factors influencing structure at the interface, but the studies must be extended as widely as possible, considering different chemical structures.

Aromatic organic ions and related molecules also vary a great deal in the degree of orientation of their long axes, though they appear to be less sensitive to the sign of the charge on the interface (9,10,14). Those aromatic organic ions, which are highly water soluble seem to be less ordered along their para axes, while those with a para substituent, which is hydrophobic, such as para chlorobenzoic acid become as highly ordered in a direction perpendicular to the interface as the first segment of the hydrophobic hydrocarbon chain. The results in Table I confirm the influence of charge and hydrophobicity in the examples investigated. Insertion of organic ions and molecules in the hydrophobic region influences their degree of orientation a great deal.

It is clear that such studies, as indicated in Table I, must be broadened to include as many structural types as possible within the regime that the lyotropic liquid crystalline regions can be oriented in a magnetic field; such wide chemical flexibility is evidently available to us.

Changes in the Phase Director with Respect to the Magnetic Field

The chemical nature of lyotropic liquid crystalline systems, which can be oriented in magnetic fields, has been varied in our

previous studies in several respects (7-14). The phases can be
binary mixtures of detergents and water, but often require the
addition of electrolyte in order to lower the viscosity and
facilitate the act of orientation (15). Mesophases of ternary
composition can include cationic, anionic and neutral amphiphiles
or mixtures of them with and without added electrolyte. The
experimenter has some control over the chemical nature of the head
group, which can be any of $-OH$, $-NH_2$, $-NH_3^+$, $-N(CH_3)_3^+$, $-SO_4^-$,
$-CO_2^-$ $-COOH$, etc. The charge density of the head groups per unit
area of interface is a useful experimental variable, but the
chemical nature of the head group and of counter ions is also
evidently extremely important, though often ignored in model
theories. The hydrocarbon chain length is usually C_8, C_{10} or C_{12}
in phases which can be oriented near room temperature, though some
phases can be oriented at higher temperatures and cooled slowly to
remain oriented at room temperature with other chain lengths. The
counter ion can be any soluble simple or complex ion stable in
water between $pH \sim 0.5$ and ~ 12. While some counter ions are
difficult to orient if they constitute 100% of the total counter
ions, they can often be introduced by substitution of less than 5%
of the total ionic detergent content, or as part of the added
electrolyte.

    One important physical result of these chemical variations
is the fact that two types of mesophase emerge, a Type I with
the phase director parallel to the applied magnetic field and
Type II with the director perpendicular to the magnetic field (8).
The direction of this director with respect to the magnetic field
can be determined in the following manner. A typical NMR sample
tube is shaken well to randomise the mesophase before entering
the magnetic field and a spectrum recorded immediately of the
D_2O, deuterium doublet, preferably, in the interests of speed of
execution, by giving a single pulse and transforming the free
induction decay to the frequency mode spectrum. A powder diagram
spectrum results. The sample is allowed to remain for a sufficient time in the magnetic field to become homogeneously oriented.
At this point in time another deuterium NMR spectrum is recorded
and the sharp doublet observed may have twice the powder diagram
doublet splitting (Type I phase) or the same splitting (Type II
phase) (16). An example of the general behaviour of mesophases
is given in figure 1, for the binary system decylammonium
chloride/water and the ternary system with added electrolyte.

    In binary mixtures between 35 and ~49 weight
per cent water there is a mesophase which gives a
powder doublet, the splitting decreasing with added
water. This phase is viscous, not oriented in detectable
times at 28°C by the magnet and is probably a lamellar
type. Between $\sim$ 49 and 52 weight per cent water two
phases separate, the upper one is clearly the lamellar
type phase, the powder spectrum having constant splitting
of the D_2O doublet. A second phase separates and orients

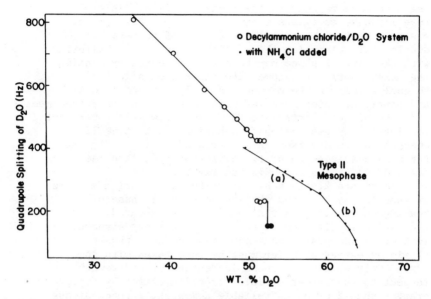

Figure 1. The mesophase behavior of the binary system decylammonium chloride/D₂O with and without added NH₄Cl electrolyte. The quadrupole splitting in Hertz of the deuterium doublet is used as a monitor of the numbers and types of mesophases. (a) and (b) refer to two possible Type II mesophases. The large open circles refer to the lamellar and Type I phases. The larger filled circles to the Type II phase which occurs without adding electrolyte. The small filled points refer to phases to which a few wt % of ammonium chloride has been added. These are Type II.

in the magnet to give a single crystal sharp doublet.
This second phase was shown to be type II with the
mesophase director perpendicular to the field. At
greater than $\sim$ 52 weight per cent water another two
phase system occurs. The upper layer is birefringent
and the lower layer an isotropic phase. The upper
phase orients rapidly in the magnetic field. This
type II phase co-exists with an isotropic phase giving
a single deuterium peak at the center of the type II
doublet. The lamellar phase which appears in equilibrium
with the type II phase may be partly oriented by heating
the sample until it becomes isotropic then allowing it
to cool slowly in the magnet. A well oriented type II
and partly oriented lamellar phase occur after this treatment.
Addition of the electrolyte NH_4Cl in a few weight per cent
completely changes the phase behavior. Only type II
mesophases are observed as illustrated in figure 1, but
there is a suggestion of two different type II phases
(a) and (b) as indicated in the figure.

In figure 2, the behavior of the equilibrium mixture
of lamellar type II and the second type II phase of
heating and cooling in the magnet is illustrated. There
are obviously different arrangements of super-structure
in type II mesophases and comparison between phases
of micro-degrees of order must be undertaken with some
care. Only when differences in degrees of order with
respect to the magnetic field direction differ by factors
greater than 2 or 3 can reliable comparative conclusions
be drawn. There is a lack of low angle x-ray diffraction
work on the oriented samples reported in our work.

Changes in Water Content of a Given Mesophase
Some mesophases which can be oriented sustain reasonably
large changes in water content without undergoing phase transitions. The effect of enlarging the volume of the water compartments of phases can be studied by always preserving the mole
ratios of amphiphilic molecules and merely using different water
contents. A good example is the system 37 parts decylammonium
chloride, 4 parts ammonium chloride and 54 to 69 parts by weight
D_2O. The deuterium magnetic resonance spectrum of specifically
deuterated chain segments was monitored as a function of D_2O
content and the doublet separation in oriented samples is shown
in figure 3A.

The composition of the amphiphilic compartment was also varied by studying the corresponding deuterium doublets in specifically deuterated chains for mesophases with compositions 0.197 gms
cholesterol, 1.11 gms d_7-decylammonium chloride, 0.12 gms ammonium chloride and 2.61 to 3.30 gms D_2O. These deuterium doublet
splittings versus water weight per cent are plotted in figure 3B.

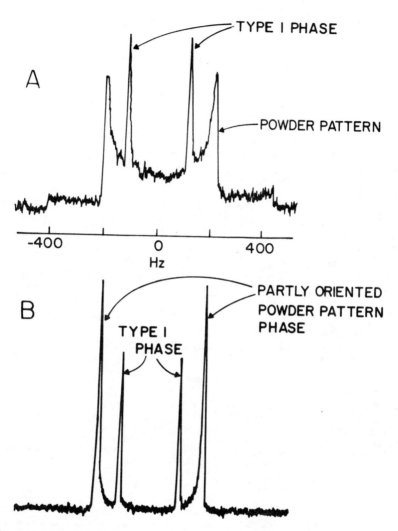

Figure 2. (A) *The deuterium magnetic resonance spectrum of the equilibrium mixture of lamellar and Type I mesophase from Figure 1 (49–52% D_2O in binary mixture) is shown. The superimposed powder pattern and single crystal spectrum of the Type I persist after 12 hr in the spectrometer at 28°C.* (B) *Heating to an isotropic medium and allowing it to cool slowly in the magnet causes both phases to appear more highly ordered at 28°C, but the powder pattern spectrum of the outer doublet partly persists. The phase leading initially to powder pattern spectra appears to be Type II.*

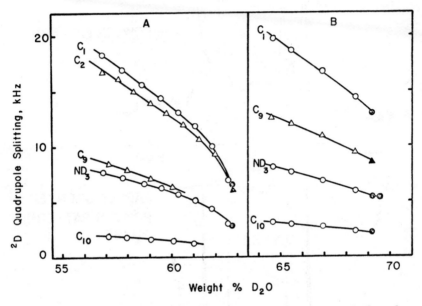

Figure 3. The deuterium quadrupole splittings for deuterium in hydrocarbon chain segments for mesophases (Type II) which differ only in water content. The phases were acidified slightly to prevent rapid proton exchange from the $-ND_3^+$ group. (A) Without cholesterol content in hydrophobic compartment. (B) With 9.2 mole ratio detergent/cholesterol. Compositions are as in the text. C_i indicates ith carbon from the head group.

The addition of water within one phase region leads to a decrease in the average degree of orientation of the deuterium principal electric field gradient for all components and segments [D_2O (figure 1) chain segments (figures 3A and 3B)]. The temperature of the experiments was $31.3\pm0.1°C$. The phases with added cholesterol require larger water contents in order to form the desired mesophase. The deuterium electric field gradient tensor has a principal axis approximately aligned with the C-D bond direction and it is generally considered to have a small asymmetry parameter η(0.15 or less) which is usually neglected to the order of this discussion. The deuterium quadrupole couplings are proportional to the order parameter along the C-D bonds. The effect of the cholesterol is to increase the degree of order considerably at all points in the hydrocarbon chains. This has been interpreted as increasing the population of all trans chains (13). The solid points in figure 3 correspond to a two phase region one of which is isotropic (higher density).

The characteristic quantity, which is independent of water content, provided counter ion and hydrophobic components are not changed, is the ratio of the quadrupole couplings to each other in partially oriented hydrocarbon segments. In figure 4 for a number of phases, in which the water content only is varied at constant temperature, the quadrupole splittings of the hydrocarbon chain segments are plotted against the deuterium doublet splittings of the $-ND_3^+$ head groups. A strict linearity is observed in all cases with zero intercept. The ratios are affected by adding cholesterol as in figure 3B. Comparisons of the ratios from results in 3A and 3B with and without cholesterol are; e.g. $(\Delta\nu_1/\Delta\nu_{ND}) = 2.325\pm0.015$ without and 2.42 ± 0.01 with cholesterol, $(\Delta\nu_9/\Delta\nu_{ND}) = 1.215\pm0.02$ without and 1.57 ± 0.02 with cholesterol (9.2 mole % of hydrophobic components). These ratios are only very slightly affected by temperature changes (19-50°C) at constant composition. An example of this behaviour are the changes $[\Delta\nu_9/\Delta\nu_{ND}] = 1.26\pm0.01$ at $21.7°C$ which linearly decreases to 1.10 ± 0.01 at $45.4°C$ and $[\Delta\nu_{10}/\Delta\nu_{ND}] = 0.29\pm0.001$ decreasing over the same temperature range to 0.255 ± 0.003. Other ratios are invariant.

Dramatic changes in the ratios of quadrupole couplings do occur if there is substitution of the counter ion. Phases prepared with decylammonium tetrafluoroborate have ratios, $[\Delta\nu_9/\Delta\nu_{ND}] = 0.994\pm0.008$; $[\Delta\nu_1/\Delta\nu_{ND}] = 1.97\pm0.05$ and $[\Delta\nu_9 \ \Delta\nu_{10}] = 4.63\pm0.05$. In decylammonium fluoride phases $[\Delta\nu_9 \ \Delta\nu_{10}] = 4.46\pm0.01$. The systematic dependence of these ratios on the nature of the counter ion has already been published (17).

The fact that the ratios of the deuterium quadrupole couplings in the chain are independent of the water content is indicative that the degree of order profile of the chain does not change in a relative sense. If the first segment C-D falls in S_{axis} value the last segment C-D has a corresponding change in S_{axis}. This implies that since the degree of order profile

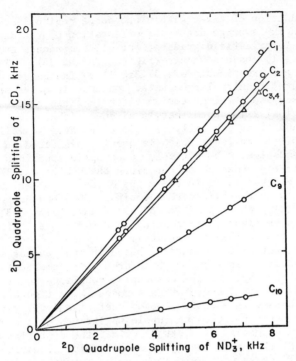

Figure 4. The quadrupole coupling constants in —CD₂—chain segments plotted against those of the —ND₃⁺ head group for mesophases of decylammonium chloride in which only the water content is varied. The indices C_i indicate the carbon number i in the chain numbered from the —ND₃⁺ head group.

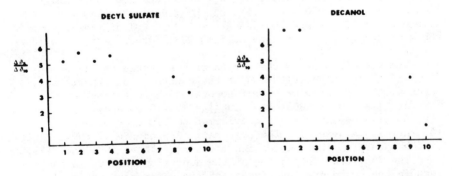

Figure 5. The ratios of deuterium quadrupole coupling constants in specifically deuterated decanol and decylsulfate ions which form part of the same lyomesophase in the oriented state. [$\Delta\nu x/\Delta\nu_{10}$] represents the ratio of the coupling in the xth carbon to the 10th carbon. The carbon position is indicated on the abcissae.

ratio is affected by changes in the interface, such as a change in counter ion, then the alteration of water concentration does not alter the basic structure of the interface in the systems studied, but merely changes the amount of local motion allowed in the structure.

Degree of Order Profile as a Function of Head Group

Many lyotropic mesophases which are oriented by magnetic fields, have two or more components in the hydrophobic compartment. A typical example is the sodium decyl sulphate/decylalcohol/ H_2O or D_2O/sodium sulphate system originally described by Flautt and Lawson (3). In these phases both decanol with an -OH polar head group and decylsulphate with $-SO_4^-$ head groups are co-aligned in the hydrophobic compartment. By suitable specific deuteration of both the decanol and decylsulphate ions it is possible to assign the deuterium doublets from the Type II mesophases and obtain the ratios of the deuterium quadrupole couplings in phases where only water content is varied. Although complete assignment of the 10 carbon chain in each case has not been achieved because of the lengthy synthetic work needed, enough of both chains were deuterated to get ratios [$\Delta\nu x/\Delta\nu_{10}$] for each chain, where x = 1,2,3,4,8 and 9 for the decyl sulphate ion and x = 1,2 and 9 for the decanol chain. In figure 5 these ratios are plotted against the numerator carbon number in the chain. It is quite clear from this result that the degree of order profile, and thus detailed chain segment motions are dependent on the head group. In the decylsulphate chain there is evidence of alternation of order, while in decanol chains they appear to have a more restricted motion near the -OH head group relative to the decylsulphate chain. This could not be accounted for by assuming that the $-SO_4^-$ group itself constituted an additional segment. That detailed chain motions of the hydrocarbon region should be dependent on the chemical nature of the head group, which in turn must influence local interface hydration and counter ion binding is not at all surprising when demonstrated experimentally, but both bulk theories of the interface and work in molecular biology exclusively ignores this concept. The degree of order profile of the chains is likely to become a means of accomplishing interface chemistry at the molecular level using an NMR spectrometer and homogeneously oriented lyomesophases.

Degree of Order Profile as a Function of Phase Type

A Type II phase described earlier of decylammonium chloride/ water/ammonium chloride sustains additions of cholesterol to form orienting phases at higher water content (figure 3). At a mole ratio per cent of cholesterol to detergent between 9 and 12, the Type II phase passes over into one in which a powder pattern deuterium doublet persists indefinitely in the spectrometer. By plotting the ratios of deuterium quadrupole couplings versus mole

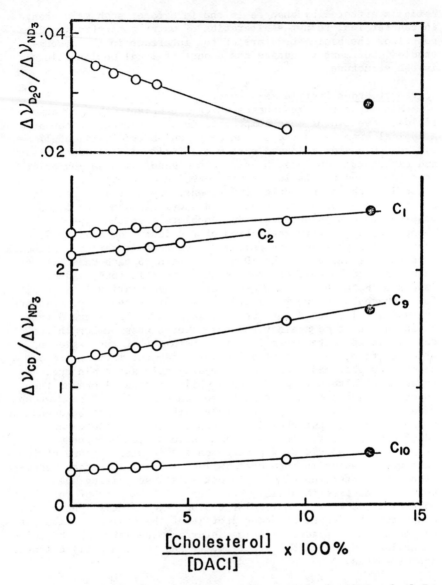

Figure 6. Ratios of deuterium quadrupole coupling constants $[\Delta\nu_{CDx}/\Delta\nu_{ND_3^+}]$, *where* x *denotes carbon number for decylammonium chloride mesophases with added cholesterol. The open circles correspond to Type II (probably hexagonal) mesophases and filled circles to a mesophase which renders a powder pattern (probably lamellar). Only in the case of the water doublet is no direct extrapolation possible, and in this case the geometry of the compartments changes drastically. The hydrocarbon chain motions are not perturbed by the phase change.*

ratio per cent cholesterol it is possible to compare the degree of order profile in Type II and the lamellar phases, which do not orient. It is clear from figure 6 that the hydrocarbon chain motions do not appear to depend on the phase type.

Abstract

Lyotropic mesophases oriented by a magnetic field can be used as tools to study interface science by NMR methods. Anisotropic parameters allow the study of internally divided aqueous, interface and hydrophobic regions of these phases. The proton dipole-dipole and deuterium quadrupole couplings measured in various oriented phases are useful in probing local degrees of order with respect to the magnetic field direction. The heterogeneity of local micro order of long axes in ions, molecules, head groups and hydrocarbon chain segments is determined and discussed. The order of ions in the electrical double layer is influenced by counter head group charges as well as the hydrophobicity of the ion and complex interface structure. The chemical flexibility of the mesophases is outlined. Water content changes affect the amount of motion in the interface but not the type of interface motions and structure. The surface motion is affected by counter ions and amphiphilic components. The degree of order profile of chains with different head groups in the same phase is quite distinct, but phase changes do not significantly affect this profile.

Acknowledgement

Work supported by the National Research Council of Canada, Conselho Nacional de Pesquisas do Brasil and Fundacão de Ampara à Pesquisa do Estado de São Paulo, Brazil.

Table I. Degrees of Orientation of Components of Lyotropic Liquid Crystals

Phase	Species	Axis	S_{axis}
DA	CH_3OH	C–O	$\pm 8 \times 10^{-3}$
DS	CH_3OH	C–O	$\pm 6 \times 10^{-3}$
DA	$CH_3CO_2^-$	C–C	-6.5×10^{-3}
DS	$CH_3NH_3^+$	C–N	-1.6×10^{-2}
DS	$CH_3SnCH_3^{++}$	C–Sn–C	-1.4×10^{-2}
DS	D_2O	O–D	$\pm 1 \times 10^{-3}$
DS	D_2O	O–D	$\mp 2 \times 10^{-3}$[†]
DA	$CH_3TlCH_3^+$	C–Tl–C	$\pm 8 \times 10^{-4}$
DS	$CH_3TlCH_3^+$	C–Tl–C	$\pm 7 \times 10^{-2}$
DS	O_2N–⬡–CO_2–	para	-2×10^{-2}
DA	O_2N–⬡–COOH	para	-8×10^{-2}
DS	O_2N–⬡–COOH	para	-4×10^{-2}
DA	H–$^+$N⬡–COOH	N–γ	$\pm 4 \times 10^{-3}$
DA	Cl–⬡–COOH	para	± 0.2
DA	–CD_2–	first segment axis[‡]	± 0.15
DA	–ND_3^+	first segment axis	± 0.17
DS	–CD_2–	first segment axis	± 0.14

All the phases are classified as type II except[†], which is a Type I. Degree of orientation for the first segment axis of the chain refers to the direction of a hypothetical all trans chain, which aligns perpendicular to the interface. This segment (cont.)

axis is approximately represented by the line joining the bisectors of the C_1-N and C_1-C_2 bonds or the C_1-C_2 and C_2-C_3 bonds. The S value of the $-ND_3^+$ head group is corrected for rotation of the moeity about the $-C-N$ bond. DA and DS represent phases based on the detergent cation decylammonium or the detergent anion decylsulphate respectively. All phases are about 90 mole % water and the balance detergent, alcohol and added electrolyte. In some cases as a consequence of the analysis the sign of S is known. The temperature of the measurements was 31±1°C and the S values correspond to mid-range compositions with a ±30% change with composition.

Literature Cited

1. Brown, G.H., Doane, J.W., and Neff, V.D., "A Review of the Structure and Physical Properties of Liquid Crystals," Pub. CRC. Cleveland (1971).
2. Luzzati, V., Mustacchi, H., Skoulios, A., and Husson, F., Acta Cryst., (1960), 13, 660, 668.
3. Lawson, K.D., and Flaut, T.J., J. Am. Chem. Soc., (1967), 89, 5489.
4. Saupe, A., Z. Naturforsch, (1965), 20a, 572.
5. Saupe, A., Angew. Chem. Int., Ed. (English), (1968), 7, 107.
6. Winsor, P.A., Chem. Rev., (1968), 68, 1.
7. Chen, D.M., "Ph.D. Dissertation", University of Waterloo, Waterloo, Ontario (1975).
8. Radley, K., Reeves, L.W., and Tracey, A.S., J. Phys. Chem., (1976), 80, 174.
9. Lee, Y., and Reeves, L.W., Can. J. Chem., (1975), 53, 161.
10. Lee, Y., and Reeves, L.W., Can. J. Chem., (1976), 54, 500.
11. Reeves, L.W., and Tracey, A.S., J. Am. Chem. Soc., (1974), 96, 1198.
12. Reeves, L.W., Suzuki, M., Tracey, A.S., and Vanin, J.A., Inorganic Chemistry, (1974), 13, 999.
13. Fujiwara, F.Y., and Reeves, L.W., J. Am. Chem. Soc., (In press).
14. Reeves, L.W., Suzuki, M., and Vanin, J.A., (unpublished work).
15. Fujiwara, F.Y., and Reeves, L.W., (unpublished work).
16. Cohen, M.H., and Reif, F., Solid State Phys., (1957), 5, 321.
17. Reeves, L.W., and Tracey, A.S., J. Am. Chem. Soc., (1975), 97, 5729.

7

Quadrupolar Echo Deuteron Magnetic Resonance Spectroscopy in Ordered Hydrocarbon Chains*

J. H. DAVIS, K. R. JEFFREY,** M. BLOOM, and M. I. VALIC

Department of Physics, University of British Columbia, Vancouver, B. C., Canada V6T 1W5

T. P. HIGGS

Department of Chemistry, University of British Columbia, Vancouver, B.C., Canada V6T 1W5

Phospholipid hydrocarbon chain fluidity and the liquid crystalline nature of the bilayer play an important role in the activity of biological membranes [1]. The molecular dynamics in these systems can be described by the order parameters S_i and the correlation times τ_i for the different positions on the hydrocarbon chains. The S_i provide a convenient measure of the mean square amplitudes of the motion while the τ_i give the time scales over which these motions occur. Since the molecular motions in liquid crystalline systems are anisotropic, in contrast to those in ordinary liquids, the static electric-quadrupole and magnetic-dipole interactions are not completely averaged out [2]. The residual quadrupolar or dipolar splittings in the nuclear magnetic resonance spectrum are easily measurable and can be directly related to the order parameters S_i. Relaxation time measurements can be used to determine the correlation times τ_i when the motions occur at frequencies less than or of the order of the Larmor frequency ω_0. NMR, then, is well suited to the study of the molecular dynamics of these systems.

The proton is, of course, the obvious nucleus to choose in an NMR study of these systems. However, since the dipolar interaction between protons along the chain produces a line broadening of roughly the same magnitude as the line splitting due to the nearest neighbour dipolar interactions, one normally observes a single broad absorption peak [3] and it is difficult to extract detailed information on the protons at different positions along the chain. The deuteron is uniquely suitable for studies of S_i since its quadrupole moment leads to a splitting which, in most cases, is much larger than the linewidths

*Research supported by the National Research Council of Canada and a special Killam-Canada Council Interdisciplinary Grant.
**Permanent address: Department of Physics, University of Guelph, Guelph, Ontario, Canada N1G 2W1.

due to dipolar interactions between deuterons or to field inhomogeneities. Deuterium is readily available and may be freely substituted for the hydrogen in these molecules without drastically altering the dynamics of the system. Since the degree of motion experienced by inequivalent deuterons is frequently quite different, the splittings, which depend on the amount of motional averaging, will differ and, in some cases, even in a system with many different deuteron sites, the individual lines may be resolved.

There are some experimental difficulties in doing deuterium magnetic resonance, however. Foremost among these are the problems in the chemical preparation of selectively deuterated samples. The preparation of perdeuterated samples is often significantly easier and, if the individual lines are well resolved, the use of these samples can provide very significant savings in time since all of the positions along the chain can be studied at once. Further, the absence of protons with their relatively large magnetic moments sharpens the lines considerably.

Another difficulty arises from the orientation dependence of the splittings. In many systems of biological interest, it is impossible to use oriented samples, so that information on the splittings must come from the study of unoriented samples. The resulting powder pattern spreads the signal intensity over a wide range. In principle, the study of such powder spectra does offer some advantages, however, providing that the broad spectra can be faithfully reproduced, since a single spectrum can give information on all orientations at once.

The instrumental difficulties encountered are due to the fact that in conventional electromagnets the maximum deuteron resonance frequency is about 15 MHz. This low frequency results in relatively poor signal-to-noise, necessitating the use of extensive signal averaging, in some cases requiring very long time to obtain a good spectrum. Often the deuterium powder spectra are very broad (up to $\sim$ 100 - 200 kHz) implying that much of the information is contained in the very early part of the free-induction-decay. Since the dead-time of most spectrometers operating in this frequency range is many tens of microseconds, much of the information about these broad components is lost and cannot be faithfully reproduced.

The usual application of pulsed Fourier transform spectroscopy involves the accumulation of the free-induction-decay (fid) following a single $\pi/2$ pulse applied at a frequency ω close to the sample's resonant frequency ω_0. The Fourier transform of this fid gives the frequency spectrum of the sample. This technique can, of course, be applied to a quadrupolar system, such as a deuterated sample. The quadrupolar spectrum of a given deuteron (I=1) is a doublet with lines at $\omega_0 \pm \Omega$. For $\omega = \omega_0$, the fid of a collection of isolated deuteron spins having quadrupolar interactions is given by

$$F_1(t) = \int_0^\infty g(\Omega)\cos\Omega t\, d\Omega. \tag{1}$$

The distribution function $g(\Omega)$ for the deuteron quadrupolar frequencies in an anisotropic liquid crystalline system is obtained by summing over splitting frequencies corresponding to all non-equivalent deuteron positions $i=1,2,\ldots,N$, and over the frequencies associated with the distributed of angles θ between the external magnetic field and the constraint vector or director determined by the symmetry properties of the oriented fluid.

$$\Omega_i(\theta) = \frac{3}{4} K_i S_i \left(\frac{3\cos^2\theta-1}{2} \right), \tag{2}$$

where $K_i = e^2 q_i Q/\hbar$ is the quadrupole coupling constant for the i'th deuteron and the order parameter S_i is given by the ensemble average

$$S_i = \tfrac{1}{2}(\overline{3\cos^2 \textcircled{H}_i -1}) \tag{3}$$

involving the angle $\textcircled{H}_i$ between the axis of symmetry of the electric field gradient [4], i.e. the C–D$^{(i)}$ bond direction***, and the director. For lamellar phases of the type studied here, the director is perpendicular to the planes of the lamellae.

Since $K_i \approx K \simeq 2\pi \times 1.7\times10^5$ s^{-1} for all the methylene deuterons [6] and $S_i \lesssim 0.25$ in the lamellar phases of soap solutions [7] and model phospholipid bilayer membranes [8], $F_1(t)$ decays appreciably in a time of order $(\overline{\Omega}_i)^{-1} \approx 40$ μs leading to some of the experimental problems described above.

Now, suppose that a second pulse which rotates the magnetization by an angle θ and having r-f phase ϕ with respect to the first pulse is applied at a time τ after the first pulse. A "quadrupole spin-echo" occurs at $t = 2\tau$ due to the refocussing of the nuclear magnetization [9]. This echo is maximum when $\theta = \pi/2$ and $\phi = \pi/2$. Indeed, the refocussing is complete under these conditions, aside from effects of relaxation and static magnetic field inhomogeneities, so that one can write the expression for the nuclear induction signal in the region $t \geq \tau$ as

$$F_2(t) = \left\{ \int_0^\infty g(\Omega)\cos[\Omega(t-2\tau)]\,d\Omega \right\} R(t) \tag{4}$$

*** The appropriate formula for S_i taking into account that the electric field gradient tensor is not quite axially symmetric about the C–D bond is given elsewhere [5].

where $R(t)$ represents the loss of transverse magnetization due to relaxation. Since the relaxation effects are produced primarily by magnetic dipolar interactions, which are much weaker than quadrupolar interactions for deuterons, or by very low frequency components of the fluctuating part of the quadrupolar interactions, which are also usually much less effective than the dephasing effects of the distribution $g(\Omega)$ of static, local, quadrupolar frequencies for the anisotropic systems discussed here, it is usually possible to select a value of τ which is much greater than the recovery time of a reasonably good r-f amplifier, but much less than the transverse relaxation time T_2. Under these conditions, it is possible to obtain a faithful reproduction of the spectrum by calculating the Fourier Transform of $F_2(t)$ starting at $t = 2\tau$.

It should be noticed that if the transverse relaxation time of the i'th deuteron is given by T_{2i}, i.e. $R_i(t) = \exp(-t/T_{2i})$, the powder pattern lineshape is given by

$$g(\Omega) = A \sum_{i=1}^{N} \frac{n_i}{T_{2i}} \int_0^{\pi} \sin\theta \left[\left\{ \frac{1}{T_{2i}^2} + \left[\Omega + \Omega_i(0) \right. \right. \right.$$

$$\left. \left. \left. \left(\frac{3\cos^2\theta - 1}{2} \right) \right]^2 \right\}^{-1} + \left\{ \frac{1}{T_{2i}^2} + \left[\Omega - \Omega_i(0) \left(\frac{3\cos^2\theta - 1}{2} \right) \right]^2 \right\}^{-1} \right] d\theta \quad (5)$$

where n_i is the number of deuterons having index i and A is a normalization constant. This spectrum consists of a superposition of N broad absorption curves, each with two sharp edges separated by a frequency $\Delta\nu_i = \Omega_i(0)/2\pi = 3K_iS_i/(8\pi)$ corresponding to the $\pm\pi/2$ orientations.

Figure 1-a shows the quadrupolar echo observed in a solution of 70% (by weight) perdeuterated potassium palmitate $(CD_3(CD_2)_{14}COOK)^\dagger$ - 30% H_2O at 100°C. The first pulse rotates the equilibrium magnetization by $\pi/2$ and is followed by the free-induction-decay. A time τ later the second pulse, also a $\pi/2$ pulse which is phase shifted by $\pi/2$ radians, is applied. At 2τ the magnetization has refocussed and the entire signal is observed (notice that the very sharp spike seen at $t = 2\tau$ in the echo has been lost from the fid due to the spectrometer dead time). The echo shown in the figure, the result of 200 scans, when transformed gives the spectrum shown in Figure 1-b.

$\dagger$ Palmitic-d_{31} acid and β, γ-di(palmitoyld_{31}) - L-α-Lecithin were prepared according to published procedures [11,12].

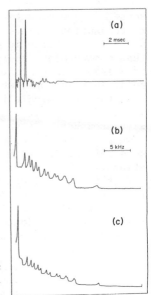

Figure 1. (a) The quadrupolar echo in the lamellar phase of KP-d_{31} at T = 100°C occurring at t = 2τ, where τ = 400 μsec, from 200 scans with a dwell time of 10 μsec. (b) The spectrum obtained from the Fourier transform of the quadrupolar echo. Spectral width is 50 kHz. (c) Computer simulation of the spectrum shown in b using Eq. (5).

Even though the spectrum is very broad and consists of a large number of overlapping powder patterns, 14 of the 15 possible peaks are clearly resolved and the baseline is flat. By carefully setting the phases of the two pulses and by starting the digitization of the signal at the exact center of the echo no phase adjustments need to be made on the transformed spectrum. Since the spectrum is symmetric about ω_0 the rf pulses are applied at this frequency causing the two halves of the spectrum to be superimposed resulting in a $\sqrt{2}$ improvement in signal-to-noise ratio.

Figure 1-c is a computer simulation of this spectrum from a superposition of 15 powder patterns with lineshapes given by Equation (5). This simulation gives a fairly accurate representation of the spectrum leading one to believe that, even when many of the lines overlap, an accurate characterization of the order parametersis possible.

From such a clearly defined spectrum, one can confidently obtain the splittings $\Delta\nu_i$ as a function of position on the chain, as shown in Figure 2, and from these directly determine the order parameters S_i. At lower temperatures the well established "plateau" characteristic of the order parameters

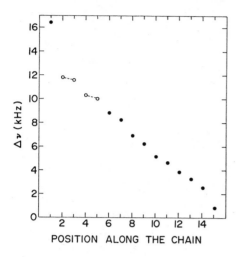

Figure 2. The peak positions relative to $\nu_o = 13.815$ MHz, at T = 100°C for the spectrum shown in Figure 1(b), plotted against position along the hydrocarbon chain

of these systems [4,7,8] is observed. However, the plateau is wiped out at high temperatures. The obvious pairing of the lines, as illustrated here, is a result of the "odd-even effect" [10]. A more detailed description of the temperature dependence of the order parameters is in progress and will be reported elsewhere.

An example of a situation where the lines are not clearly resolved is shown in Figure 3. This spectrum is for a 250 mg sample of chain-deuterated di-palitoyl phosphatidyl choline at 50°C. Since there are two chains in this molecule which are not necessarily undergoing the same motions, as many as 30 overlapping powder pattern spectra can be expected from this sample. In order to faithfully reproduce this spectrum it is important that no information be lost due to spectrometer dead-time. Figure 3-a shows the spectrum as obtained by transforming the quadrupolar echo after accumulating 4000 scans. Figure 3-b illustrates the effect of the loss of the initial part of the fid due to dead time. Here 4000 scans of the fid were accumulated with a dead time of 50 μsec at 13.815 MHz with a spectral width of 50 kHz. The spectrum obtained from the transform, as given in Fig. 3-b, is severly distorted. The first-order phase correction which should be applied to the spectrum is simply (dead time/dwell time) x 180° = 900°, and no zeroth order phase correction is required. If the first five points in the echo used for Figure 3-a are dropped, to correspond to a dead time of 50 μsec, and the transform is carried out, as well as the same first order phase correction (900°) as applied to the fid's spectrum, the spectrum appears as in Figure 3-c. As expected, Figures 3-b,c are nearly

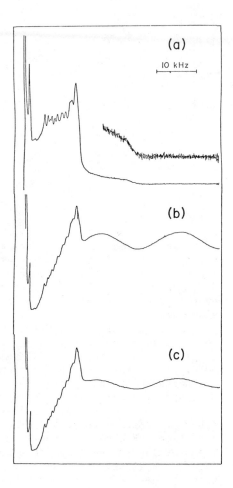

Figure 3. (a) The spectrum for DPL-d_{62} at T = 50° obtained from the quadrupolar echo after accumulation of 4,000 scans (spectral width of 50 kHz), showing several resolved lines as well as the characteristic plateau. The scale has been expanded by a factor of four in the base line region to illustrate the shoulder, the signal-to-noise ratio, and the flat base line. (b) The spectrum obtained under the same conditions as in (a) but using the fid with a dead time of 50 μsec, properly phase corrected. (c) The same spectrum, obtained by discarding the first five points in the quadrupolar echo (dwell time is 10 μsec) and making the same phase correction as in (b).

identical. The distortion apparent in these spectra is a
reflection of the information lost during the spectrometer
dead time, and further adjustments of the phase knobs can
only give misleading results.

   Because of the improvements in signal-to-noise and spectrum
fidelity a large variety of experiments which were previously
very time-consuming become quite manageable. Temperature
dependent studies of the quadrupolar splittings and relaxation
times in these systems are presently underway and will be
reported in the Symposium. An apparent relationship between
the temperature dependence of the hydrocarbon chain order
parameters and the water (D_2O) order parameters will be
discussed. Also, the variation of the relaxation times along
the chain indicates the influence of collective motions.
The application of this technique to natural membranes is
expected to be particularly rewarding.

ACKNOWLEDGEMENTS

   We wish to thank Dr. E.E. Burnell and Dr. A. MacKay for
helpful discussions.

LITERATURE CITED

[1] Chapman, D., Quart. Rev. Biophys., (1975), 8, 185.
[2] Lawson, K.D., and Flautt, T.J., J. Am. Chem. Soc., (1967),
      89, 5489; Charvolin, J., and Rigny, P., J. Chem. Phys.,
      (1973), 58, 3999.
[3] Lawson, K.D., and Flautt, T.J., Mol. Cryst., (1966), 1,
      241 and J. Phys. Chem., (1968), 72, 2066; Wennerstrom, H.,
      Chem. Phys. Letters, (1973), 18, 41.
[4] Charvolin, J., Manneville, P., and Deloche, B., Chem.
      Phys. Letters, (1973), 23, 349.
[5] Wennerstrom, H., Persson, N., and Lindman, B., ACS
      Symposium Series, (1975), 9, 253.
[6] Burnett, L.J., and Muller, B.M., J. Chem. Phys., (1971),
      55, 5829.
[7] Mely, B., Charvolin, J., and Keller, P., Chem. Phys. Lipids,
      (1975), 15, 161.
[8] Seelig, A., and Seelig, J., Biochem., (1974), 13, 4839.
[9] Solomon, I., Phys. Rev., (1958), 110, 61.
[10] Marcelja, S., J. Chem. Phys., (1974), 60, 3599; Pink, D.A.,
      J. Chem. Phys., (1975), 63, 2533.
[11] Dink-Nguyen, N., and Stenkagen, E.A., Chem. Abstracts,
      (1967), 67, 63814.
[12] Cubero-Robles, E., and Van Den Berg, D., Biophys. Biochem.
      Acta., (1969), 187, 520.

8

Self-Diffusion in Lyotropic Liquid Crystals Measured by NMR

GORDON J. T. TIDDY and ETHNA EVERISS

Unilever Research Port Sunlight Laboratory, Port Sunlight, Wirral, Merseyside, United Kingdom L62 4XN

ABSTRACT

The pulsed field gradient NMR technique has been used to measure the self diffusion coefficients of all the components in an aligned lamellar phase sample of lithium perfluoro-octanoate. Water diffusion coefficients were $\sim 8 \times 10^{-10}$ m^2s^{-1} and $\sim 6 \times 10^{-10}$ m^2s^{-1} for diffusion parallel and perpendicular to the lipid bilayers at 320K. In the same temperature region the comparable lithium diffusion coefficients were 3×10^{-11} m^2s^{-1} and 1.4×10^{-11} m^2s^{-1}. Perfluoro-octanoate ion diffusion within the bilayer was estimated to be 6×10^{-12} m^2s^{-1}, assuming zero diffusion perpendicular to the bilayer.

1. INTRODUCTION

Surfactants with perfluoroalkyl chains are similar to hydro-carbon surfactants in that they form micelles (1) and lyotropic liquid crystals (2,3). The liquid crystals are good materials for NMR studies since there is no possibility of overlap between the water and alkyl chain signals. The lamellar liquid crystal-line (l.c.) phase is often used as a model of biological membranes. In previous studies, a combination of neutron scattering and NMR measurements have been used to measure water translational diffu-sion across fluorinated lipid bilayers (4). The diffusion was less than one order of magnitude lower than that observed for normal water, and the anisotropy of diffusion was small. In this paper measurements of all the constituents (water, counter-ion and surfactant) of a lamellar phase are reported.

At 298K the system Lithium Perfluoro-octanoate (LiPFO) + water forms a uni-axial viscous l.c. phase containing 69% - 76% LiPFO (by wt) and this material melts in the range 303-13K to form a lamellar phase (3).

The viscous phase, originally thought to have the reversed hexagonal structure (3), can be aligned with uni-axis perpendicular to the surface of microscope cover-slips by heat cycling through the phase transition. In this way, samples with domains aligned in the same direction can be prepared. The alignment is unaltered by raising the temperature above the phase transition to the lamellar phase. Thus large, single alignment, lamellar phase samples can be prepared which give strong NMR signals for the measurement of self diffusion coefficients (D).

In order to use the pulsed field gradient NMR technique for measurements of D it is necessary to observe an echo with a $\pi/2-t-\pi$ pulse interval of $\sim$ 10ms. The residual dipole or quadrupole couplings that occur in liquid crystals (5) make this impossible for powder samples, with the exception of the water signal of H_2O where the residual dipole coupling is usually averaged by proton exchange. However, with counter-ions having a quadrupole moment and non-integer spin quantum numbers an echo is often observed due to the non shifted +1/2/-1/2 transition. In addition, in single alignment samples, echos modulated by the appropriate splitting frequency can be observed (e.g. for D_2O, Li^+, Na^+). So, for water and counter-ions, the pulsed field gradient technique can be used to measure diffusion parallel ($D_{||}$) and perpendicular ($D_{\perp}$) to the bilayer surfaces in the lamellar phase. For the surfactant the situation is more complex. If a single alignment sample is placed in the magnetic field at the magic angle then the non-averaged dipolar couplings vanish, and the resulting signal can be used for D measurements (6). Unless a quadrature field gradient coil is available, the measurements are restricted to one specific angle (7). These techniques have been used for the measurements of self-diffusion coefficients reported here. A single alignment lamellar phase sample of the LiPFO/water system has been prepared, and, for the first time measurements of self diffusion of all the components are reported. In addition, this is the first report of the anisotropy of counter-ion diffusion in a lamellar phase.

2. EXPERIMENTAL

Sample homogenization and materials were the same as those described previously (2,3). The single alignment sample was made from a mixture of 72% LiPFO (0.099 mol. fraction) and 28% water ($H_2O:D_2O$ 9:1 by wt., see below). The mixture ($\sim$1/2g) was placed in a 7mm o.d. NMR tube and nine glass rectangles cut from microscope cover-slips were added to form a series of parallel layers. The tube was cycled through the phase transition ca. 100 times over a period of 6 months. The degree of alignment was monitored using the free induction decays after a $\pi/2$ pulses of the ^7Li and ^2H resonances. The initial powder pattern was gradually replaced by the modulated signal of the single alignment sample,

until > 95% of the sample gave the modulated signal. No further
change was observed, and the 'non-aligned' material was thought
to be that adjacent to the wall of the NMR tube.

NMR measurements were made using a Bruker-Physik variable
frequency pulsed NMR spectrometer (B Kr 322s) operating at
60.0 Mhz (^{1}H, ^{19}F, $\pi/2$ pulse = 2 μs), 35.0 Mhz (^{7}Li, $\pi/2$ =
6 μs and 14.0 Mhz (^{2}H, $\pi/2$ = 10 μs). The Bruker field gradient
unit (BKr - 300 - Z18) and variable temperature unit were used as
reported previously (4), with the exception that the amplitude
of the field gradient pulses was monitored using the Bruker pulsed
gated integrator and a digital volt meter. The instrument was
calibrated with distilled water for H_2O measurements and 2.0M
lithium chloride and bromide solutions (for ^{7}Li measurements).
Because of the lack of a suitable standard for ^{19}F measurements
both the above standards were used, and this gives rise to the
larger error for the fluorocarbon self diffusion coefficient.
Also, the maximum T_2 value observed for the ^{19}F resonance was
$\sim$10 ms. This is much less than T_1 and is probably due to the
presence of a small distribution of alignment directions ($\sim$0.5°)
in the ordered material.

3.   RESULTS AND DISCUSSION

The pulsed field gradient method for measuring self-diffusion
coefficient consists of:

(i) The measurement of the echo height (h_0) following a
$\pi/2$-t-π r.f. pulse sequence.

(ii) The application of equal field gradient (f.g.) pulses
between the $\pi/2$ and π pulses and between the π-pulse and the
resulting echo, followed by measurement of the echo height (h_t).
The self-diffusion coefficient is given by Equation (1) (10).

$$\ln \frac{h_t}{h_o} = -D \quad \gamma^2\delta^2 \ (\Delta - \frac{1}{3} \delta)G_t^2 \quad = -DL \qquad (1)$$

where γ = gyromagnetic ratio
      δ = duration of the f.g. pulses
      Δ = time interval between f.g. pulses
      G_t = amplitude of f.g. pulses

D can be obtained from a graph of $\ln(h_t/h_o)$ against L, with
variation of δ^2, Δ or G_t. In practice the absolute determination
of G_t is tedious. Because of this, G_t was determined in the
present study by using standards with known D values, and measurements were made relative to these.

In disperse systems, the phenomenon of "restricted diffusion" often occurs (11). For small homogeneous regions self-diffusion can be fast, but the presence of impermeable barriers between different regions can result in slow diffusion over large distances. This causes a non-linear dependence of $\ln(h_t/h_o)$ on δ^2, G_t^2 and Δ. In the present system measurements were made over the ranges $h_t/h_o = 0.02 - 1.0$, $\Delta = 10\text{-}40\text{ms}$ for water, $h_t/h_o = 0.05 - 1.0$, $\Delta = 10\text{-}90\text{ms}$ for lithium ions, and $h_t/h_o = 0.1\text{-}1.0$, $\Delta = 10\text{ms}$ for perfluoro-octanoate ions. In every case the curves $\ln h_t/h_o$ against L were linear.

Water Diffusion

Measured diffusion coefficients are listed in Table 1. The values for water are considerably higher than those of Li^+ ions or perfluoro-octanoate ions, and are of the same order of magnitude as values reported for lamellar phase samples containing cesium and ammonium counter-ions (4). The anisotropy of water diffusion is lower than that reported previously, when values of $D_{/}/D_{} = 2$ were obtained. However, the most surprising feature of these results is the very high diffusion coefficient for water translation perpendicular to the lipid bilayers. Two explanations can be proposed for this:

(i)   The water travels through defects in the bilayer surface,
or

(ii)   the bilayer is highly porous to water.

While neither explanation can be proven, a limitation on the type of defect can be suggested. Since the anisotropy of water diffusion is lower than that of lithium ion diffusion then a large proportion (>50%) of the water transport must occur in regions small enough to prevent diffusion of hydrated lithium ions, i.e. less than $\sim$0.5nm. Assuming a D value equal to that of pure water for diffusion within defects, the surface area of the defects is calculated to be $\sim$20% of the total bilayer area. Thus the distance between adjacent defects must be $\sim$3nm. In a powder lamellar phase sample the size of individual crystallites is > 2 μm. (If the crystallites were smaller than this then deuteron quadrupole splittings would be averaged to zero.) Thus the defects referred to above cannot be those between individual crystallites.

It is possible that the fast diffusion is due to the occurrence of 'pores' within the bilayer surface through which water (and lithium ions) can move. By comparison with similar lamellar phase samples the water and fluorocarbon layers would be expected to have thicknesses of $\sim$1.5 nm and 1.2 nm respectively. If water penetrates one or two CF_2 groups along the chain from the head groups, then the thickness of the bilayer zone is reduced

Table I

Self-Diffusion Coefficients[a] of Water, Lithium Ions and
Perfluoro-Octanoate Ions in LiPFO Lamellar Phase

	Temp (K)	Δ (ms)	τ (ms)	$D_{\parallel}^{b}$ x10^{-10}m^{2}s^{-1}	$D_{\perp}^{b}$ x10^{-10}m^{2}s^{-1}	$D_{\parallel}/D_{\perp}$
Water	314.3	30.5	20.0	–	5.5 }	1.30
	314.8	25.3	20.0	7.1	– }	
	315.0	12.4	10.0	8.1	6.0	1.35
	320.5	12.4	10.0	8.5	6.0	1.42
	326.5	12.4	10.0	9.9	6.7	1.48
				x10^{-11}m^{2}s^{-1}	x10^{-11}m^{2}s^{-1}	
Li$^+$	316	11.2	10.8	2.6	1.34	1.94
	322	11.2	10.8	2.6	1.43	1.82
	328	11.2	10.8	3.0	1.51	1.99
				D(54.74°) x10^{-12}m^{2}s^{-1}	$D_{\parallel}$ (calc)[c] x10^{-13}m^{2}s^{-1}	
Perfluoro-	314.5	10.4	10.0	5.2	7.8	
Octanoate	320	10.4	10.0	6.0	9.0	
	326	10.4	10.0	6.8	10.2	

a) Values accurate to $\pm$ 10% (water, Li$^+$) and $\pm$ 25% (PFO).

b) Calculated from reference values of water (D = 2.3 x 10^{-9} m^{2}s^{-1} at 296K) (8) and Li$^+$ in 2.0M LiCl. (D = 8.4 x 10^{-10} m^{2}s^{-1} at 296K) (9).

Subscripts refer to diffusion parallel ($D_{\parallel}$) or perpendicular ($D_{\perp}$) to lipid bilayer.

c) Calculated assuming $D_{\perp}$ = 0 after ref. 7.

to ~1.0nm. Because of the absence of hydrogen bonding to neighboring molecules, water diffusion within the bilayer might be expected to be more rapid than in pure water resulting in the observed rapid diffusion.

Lithium Ion and Perfluoro-octanoate Ion Diffusion

The slow lithium ion diffusion along the water layers compared to that of aqueous lithium chloride is probably due to the high degree of counter-ion binding that is thought to occur in this system (3). The hydrated lithium ions spend most of the time associated with surfactant head groups, perhaps occupying the spaces between head groups. Their translational diffusion along the water layers is then determined by that of the surfactant ions. Diffusion perpendicular to the layers is due either to the presence of pores or cracks on the bilayers as described above. The low anisotropy of lithium ion diffusion may be due to the strong association between head groups and counter-ions giving rise to unusually low values of $D_{\parallel}$ together with a larger than expected value of $D_{\perp}$ because of the low value of the bi-layer thickness.

The values of $D_{\parallel}$ listed in the table for PFO diffusion should be regarded as upper limits since they were calculated assuming that $D_{\perp_{+}}$ was zero. In view of the low anisotropy of the water and Li^{+} diffusion this may not be valid. The values listed are two orders of magnitude less than values reported for potassium oleate (7) and potassium laurate (12) in lamellar phase under similar conditions. Since the molecular weight dependence of diffusion within the bilayer is expected to be rather small (7) the difference between the two sets of values may be due to differences in counter-ion binding between lithium and potassium ions. The stronger counter-ion binding of lithium ions could reduce the translational motion of the lipid. The D values for PFO ions are similar to those reported for the zwitterionic lipid lecithin from probe experiments (13). Further speculation on the origin of the differences requires information on the dependence of D on surface area and counter-ion.

REFERENCES

1. K. Shinoda, M. Hato and T. Hayashi, *J. Phys. Chem.* 76, 909 (1972); N. Muller and H. Simsohn, *Ibid*, 75, 942 (1971); I. K. Lin, *Ibid*, 76, 2019 (1972); P. Mukerjee and K. J. Mysels, ACS Symposium "Colloidal Dispersions and Micellar Behavior", 239 (1975).

2. G. J. T. Tiddy, *J.C.S. Faraday 1*, 68, 608 (1972); G.J.T. Tiddy and B. A. Wheeler, *J. Colloid Interface Sci.* 47, 59 (1974).

3.   E. Everiss, G.J.T. Tiddy, and B.A. Wheeler, J.C.S. Faraday Trans. 1, 72, 1747 (1976).

4.   G.J.T. Tiddy, J.B. Hayter, A.M. Hecht and J.W. White, Ber. Bunsenges. Phys. Chem., 78, 961 (1974); J.B. Hayter, A.M. Hecht, J.W. White and G.J.T.Tiddy, Faraday Disc. Chem. Soc., 57, 130 (1974).

5.   A. Johansson and B. Lindman, Ch. 8 in "Liquid Crystals and Plastic Crystals", Vol. 2, G.W. Gray and P. A. Winsor, eds., Ellis Horwood Ltd., Chichester, England, 1974.

6.   E.T. Samulski, B.A. Smith and C.G. Wade, Chem. Phys. Lett., 1973, 20, 167.

7.   S.B.W. Roeder, E.E. Burnell, An-Li Kuo and C.G. Wade, J. Chem. Phys. 64, 1848 (1976).

8.   J.S. Murday and R.M. Cotts, J. Chem. Phys. 53, 4724 (1970).

9.   E.A. Bakulin and G.E. Zavodnaya, Zh.Fiz.Khim. 36, 2261 (1962).

10.   E.O. Stejskal and J.E. Tanner, J. Chem. Phys. 42, 288 (1965).

11.   J.E. Tanner and E.O. Stejskal, J. Chem. Phys. 49, 1768 (1968).

12.   R.T. Roberts, Nature 242, 348 (1972).

13.   P. Devaux and H.M. McConnell, J. Amer. Chem. Soc. 94, 4475 (1972); E. Sackmann and H. Träuble, Ibid., 94, 4482, 4492, 4499 (1972).

Structure of Water and Hydrogen Bonding on Clays Studied by ^{7}Li and ^{1}H NMR

J. CONARD

Centre de Recherche sur les Solides à Organisation Cristalline Imparfaite, 45045 Orleans Cedex, France

Numerous papers (1...8) have been devoted to the study of the structure of water in the interlamellar space of clays which appear as negatively charged oxygen surfaces.

At low water content, a specific structure, characterized by a typical ^{1}H and ^{2}D nuclear magnetic resonance and infrared spectra has been reported. Hecht (2) has tried to obtain more information about the organization of water around the Na compensating cation in Montmorillonite and synthetic fluor phlogopite. Essentially the three lines observed in the ^{1}H spectra of water are originated from a single system, from the thermodynamic view point. Because of the impossibility of observing the structure of the ^{23}Na N.M.R. spectrum, a complete description of water arrangement was impossible. The ^{1}H spectra observed for θ = 1 and 2 were justified by a mixing of H_3O^+ and H_2O systems. We were able to observe the ^{7}Li signal of Li hectorite. The smaller quadrupolar constant of Li allowed us to measure the electrical field gradient (E.F.G.) at this nucleus. The lattice charge of hectorite is essentially produced by octaedral isomorphic substitution and it has the advantage of a low iron content.

EXPERIMENTAL

From ^{1}H N.M.R. spectra (Fig 1) it follows :
- A steady bond seems to exist between Li^+ and 3 H_2O since this hydrate is stable under vacuum at room temperature. It corresponds to half a monolayer, statistically speaking.
- A Pake's doublet, 2,7 G wide, seems to result from rotations around two almost orthogonal axes.
- A central line, with an angular dependence for oriented films (Kadi Hanifi) (9), has a relative intensity of about 7% in powders. As explained in (8) it shows the role of a magic angle between H_0 and the main axis of rotation Δ, of the whole ion, parallel to the c axis of the lattice.
- The temperature dependence of the three lines, corrected for Curie's law, appears as a combination of a thermally activated

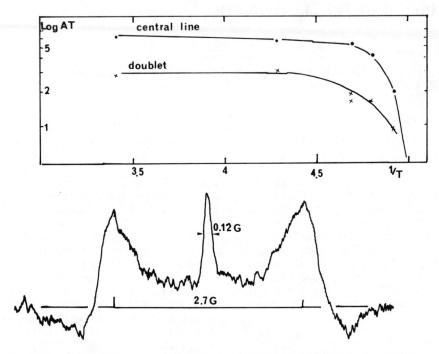

Figure 1.   N M R ^{1}H spectrum of Li hectorite dehydrated by SO_4H_2; $d = 1.8$, $fo = 15.1$ MHz.   a) derivative of the dispersion T $= 293°K$; b) thermal variation of the amplitude A of the line (the shape of the central line is invariant until it disappears). The activation energy measured on this curve Log $(A \times T) = f(1/T)$ is growing to values up to 40 kcal/mole when the Δ rotation stops at $190°K$. This change is characteristic of a cooperative phenomena.

(>6,5 kcal/mol) and of a temperature independent process as observed by Hecht in fluor phlogopite. The thermally activated contribution vanishes at - 80°C.

From ^7Li N.M.R. spectra (Fig 2) we obtain :
- A well defined axial gradient, acting on ^7Li, results from the electrical dipoles of the water molecules in (Li^+, 3 H_2O).
- A loss of symmetry or a distribution of E.F.G. strength occurs between 0 and - 80°C.
- The very small values of the ^1H dipolar local field is due to the time averaging resulting from the rapid rotation of water molecules above - 80°C.

DISCUSSION

From theoretical considerations presented by Clementi (10-11) it is known that free$_o$(Li^+, 3 H_2O) ion is essentially flat with Li - O = 1,95 ± 0,05 Å at room temperature (Zacchariasen's rule (12) would give also 1,95 Å). The H-H vector is not perpendicular to the plane of the ion in order to minimize the electrostatic energy between H atoms of adjacent water molecules. Fig 3 shows that it is easy to intercalate the (Li^+, 3 H_2O) using the 2 Å inter layer spacing calculated by Farmer (7) for a Mg^{2+} trihydrate. We point out that Mg^{2+} has almost the same radius (0,65/0,68 Å).

To form three H bonds with the oxygens of the hexagonal cavity of the clay (8), we have to put along a straight line, one OH of the water and the direction O_α(water)-O_1 (clay).The coincidence is good if each molecule of water rotates around an I axis choiced as one of the two free sp_3 orbitals of oxygen (instead of the symmetry axis of H_2O), put along the Li^+ - 0 direction. In addition, there is a favoured situation where a H bond is formed with either two neighbouring oxygens of the hexagon. This corresponds to a rotation of 60° around this I axis. The energy barrier for this movement is 0,9 Å wide. If we consider the two protons of the same water molecule, we find for the bonding with one face only, four possible orientations.

Combining the two faces, if possible, six orientations of water molecules around the I axis could be obtained. In this case one or two H bonds are formed. From a N.M.R. view point this corresponds to a free rotation.

Now its is possible to rotate by $\pi/3$, the complete hydrated ion, around the Δ axis, to obtain other equivalent positions. The energy barrier for the protons is now 0,7 Å wide only. Such a barrier can be easily permeated by tunneling. However the rotation of the hydrated ion needs a thermal activation corresponding to the motion of the three water oxygens above the oxygens of the hexagonal cavity. That seems to be the reason why the amplitude of the ^1H line has a thermal variation combining a temperature independant domain (at high temperature) and an activation energy > 6,5 kcal/mol between - 60° et - 80°C. Thermal activation and tun-

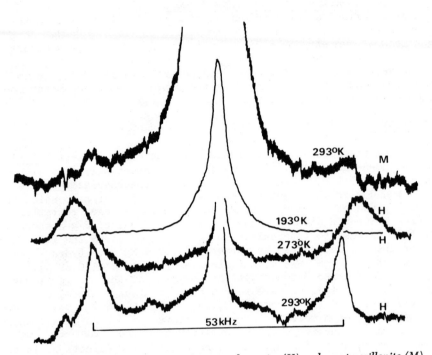

Figure 2. N M R ^7Li *spectra of the same hectorite (H) and montmorillonite (M). Fourier transform of the free precession at 34.9 MHz. A pure axial gradient, without protons' dipolar widening, is shown at room temperature. At 0°C some asymmetry is shown while the mean gradient grows up because of the thermal contraction of the hydrate. At −80°C, the Δ rotation is stopped; no gradient can be measured, but the dipolar field of protons is no more averaged and gives width to the central line. In montmorillonite the width of the central line results from a high iron content.*

neling of protons are simultaneously necessary and the slower step
is rate determining. At about - 60°C, where the two rates are equivalent, an order of magnitude for the average tunneling frequency
can be deduced : $\nu = k \, T/2h = 3.12 \times 10^{12}$ c/sec. This value would
be compared with the correlation time indicated by Hecht with Na
for $\theta = 1$: $\tau_c = 310^{-13} \exp (2470/T)$ sec,
        the value (5.510^{-13} sec) obtained for water
in vermiculite by Olejnik (13) with neutrons and the reference
value for a free water molecule 4.310^{-13} sec.

In summary, the trihydrated cation is fixed above the six
oxygens of the hexagonal cavity by three H-bonds in which the
three protons can be distributed among twelve equivalent positions
with respect to one face. Classicaly speaking, the positions occupied by the H atoms are coupled by electrostatic repulsion. However, because of the widths of the barriers between the 12 wells,
as compared with the wavelength of thermal protons (~ 0.5 Å), the
three protons, bonding the hydrate, could be completely delocalized by tunneling throughout the twelve sites. Nevertheless, this
delocalization is a cooperative phenomena since the three protons
cannot occupy the neighbouring wells at the same time.

The arguments in favor of this model are as follows :

1°) The width, intensity and anisotropy of the central line
are correctly explained as a result of the rotation of the trihydrate around its c axis (8).

2°) Thermal dependence supports the idea of a tunneling effect and it provides reasonable approximation for the tunneling
frequency.

3°) On the other hand if we assume that the central line is
completely originated from acidic free protons, it is necessary,
to account for the intensity of the ^{1}H line, to keep ionized, statistically speaking, about one half of the hydrated ions. This
would give electrical asymmetries in the ion, which are not compatible with the E.F.G. measured by ^{7}Li. The proposed model, which
keeps symmetry thanks to delocalization of protons, appears as a
better solution.

4°) The bonding of Li^+ to three negatively charged oxygens
through three polarized water molecules, gives a net charge of
- 0,33 e on each O atom in agreement with values proposed by Bernal for hydrated ions (14) and with the - 0,74 e obtained on ^{1}H
in hydrated lithium by Clementi (11).

5°) We would discuss now a little more, the geometrical requirements.
a) Relatively to the $Li-O_{(H_2O)}$ axis, laying in the (ab) plane, we

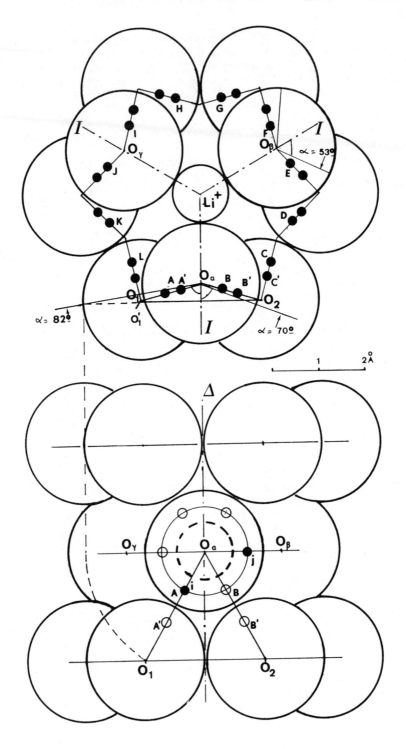

find for the angle α of the cone wich supports the OH bonds :
$\alpha = 53°$ if we use as rotation axis, the electrical symmetry axis
of the water molecule and $\alpha = 70°30'$ if we use one of the two sp_3
free orbitals of the water molecule. They would be compared with
$\alpha = 82°$ of the $O_{(H_2O)} - O_{(layer)}$ direction. Calculations from (11)
shows that the change in energy between $\alpha = 53°$ ($E = - 83,346$ eV)
and $\alpha = 82°$ ($E = -83,328$ eV) is only 18 m eV ie. 0,41 kcal/mol
which is small compared to the energy of H-bonding.
b) The strains produced by the three H-bonding would give a ditri-
gonal deformation of the six oxygens of the hexagonal cavity (16);
keeping the three bonded oxygens out of the hexagon, of about
0,2 Å, would give for the $O_{(H_2O)} - O_{(layer)}$, the 70°30' direction.
c) A shortening of the Li-O bond of about 0,15 Å would give about
the same result and Clementi describes for the length Li-O a soft
Gaussian function of half width 0,5 Å.
So a combination of these three effects permits an exact adjustment of the geometry of the bonding.

6°) In the first model proposed, classical estimation of the
E.F.G. from the static value of electric dipole for water has given a value looking somewhat exagerated (8). Using now the LiO
length 1,95 Å, i.e. 2,24 Å from the center of the dipole of water,
and a dipole moment 2,08 D.U. interpolated from (11), we obtain
$e^2q\ Q(1 - \gamma\infty)/2h = 89,5$k Hz (Exp.55k Hz) for the separation of the
triplet of ^7Li in powder hectorite with the hypotheses of 3 dipoles being in the a, b plane. But, may be, as noted in (12) the di-

*Figure 3. Bonding between (Li⁺, 3H₂O) and six hexagonal oxygen atoms
in hectorite. Top view: three molecules of water, centered at O_α, O_β, O_γ,
rotate around three I axes. Front view: the whole hydrated ion rotates
around a Δ axis. In this view we have represented an H bonding between
O_α of H₂O and O_1 of clay. The proton i is shared by tunneling in the OH
bond between ‐A and A'. But it can go from A to B through a barrier
0.7 Å wide. As the proton j can also go in (AA') and (BB') there exists
four possibilities of H bonding over one face (for one water molecule).
The top view indicates that the whole ion could rotate around Δ, of n
$\pi/3$, the protons of O_α coming for instance in CC' and DD'. The potential
barrier from B' to C' is now 0.7 Å wide, but oxygen O_α must overpass
O_2 with some thermal activation. The three protons are finally delocalized
throughout 24 wells (A, A' . . . L, L') which correspond to 12 possible
H bonds. If the two layers of clay are exactly superposed, six orientations are possible for the water molecule around an I axis, with one or
two H bonds with one or two layers. The long-time rearrangement pointed
out in (2)and(8) could be attributed to the relative ordering of the two
sheets. If true, the activation energy necessary to obtain a Δ rotation
corresponds to a transitory opening of the two opposite hexagons of
oxygens.*

pole direction pointing toward $^+$Li represents only a time average
situation. A rapid exchange between the two possible axis of rotation along the two sp$_3$ free orbitals of oxygen, tilts the dipole
by 36° with respect to the LiO direction. This is equivalent to
$\alpha = 60°$. This exchange could occur till the frequency of tunneling
of protons : the molecule overturns.

A shortening of the Li-O distance has a drastic effect on
the E.F.G. since it follows a $1/r^4$ law. For instance, we obtain
for a shortening of 0,15 Å as proposed in 5c, a ^{7}Li lines separation of 115k Hz. This corresponds to a tilt angle of the axis of
two water dipole, relative to Li-O vector, of about 73°.

As a function of the water content we have some preliminary
results :

For completely dehydrated hectorite (200°C, 10^{-6} torr)   The
^{7}Li N.M.R. spectra is reduced to a narrow single line. The T$_1$ va-
lue (217 m sec at room temperature instead of 28 m sec for the
(Li$^+$, 3 H$_2$O) system) proves the E.F.G. to be small. Presumably the
Li ion is exactly at the center of the hexagon of oxygens.

For higher water contents   The E.F.G. on Li decreases of a-
bout 20% and the T$_1$ grows up to 43 m sec in agreement with a quadrupolar relaxation mechanism. The ^{1}H spectrum goes to a triplet
structure, narrowing with water content and looks like to result
from a triplet of protons in rotation. Among different possible
organizations for this second hydration shell, which could be flat
also between the layers, H bonding could be formed with the molecules in the first shell, using the last unoccupied sp$_3$ orbital.
Thanks to tunneling this seems to be possible and explains the mixing of H$_3$O$^+$ and H$_2$O groups noted by Hecht. A more detailed study
of this spectra is going on.

In Li montmorillonite, ^{1}H spectra are widened by the presence of iron, but we did observe the same type of spectrum for ^{7}Li
with a lower intensity. This could result of a partial tetraedral
origin of the lattice charge.

CONCLUSION

For a small hydrated cation as lithium, in the interlamellar
space, the neutralization of the weakly electrically charged oxygens,
seems to use the flat trihydrate. Three H bonds give three local
charges of 0,33 e which are smeared out upon the six oxygens of
an hexagonal cavity. The geometry favours a tunneling of protons
throughout twelve possible sites for H-bonding. This delocalization would have two main effects : 1) the reduction of the number
of independant coordinates, due to the high symmetry of the system, would produce a stabilization of the cluster. 2) the equal
density of protons troughout the ring facilitates the output
(or the entrance) of one proton, exactly as it occurs for one electron in aromatic ring. The defect (or excess) of charge is temporarily smeared out over all the sites. This situation conciliates
the known local acidity of these systems and the N.M.R. results
which show an axially symmetrical cluster of three water molecules
rotating around two orthogonal axis. In agreement with vertical

begining of the isotherm this elementary cluster seems to be a
stable first step for the growing of water film.

ABSTRACT

Hydrated, Li balanced, powdered clays have been studied u-
sing ^1H and ^7Li N.M.R, at low coverages. In hectorite the Li trihydrate shows a ^1H spectrum made of a Pake's doublet, 2.7G wide,
and a narrow central line of relative intensity 7%. These two components disappear at 190°K. Doublet splitting in the ^7Li N.M.R.
spectra is a measure, through quadrupolar coupling, of the electrical field gradient generated by three water molecules pointing
toward Li. The very narrow central line indicates the high speed
of rotation of water molecules.

Using theoretical model proposed by Clementi for hydrated Li
ions, it is possible to locate the trihydrated Li ion at the center of the hexagonal cavity, the hydration water molecules being
H-bonded to three oxygen atoms of the lattice. On one face of the
sheet, twelve positions are possible for these three H.bonds, with
potential wells separated by barriers of 0,9 or 0,7A width. Proton
tunneling is therefore expected all over the hexagonal oxygen ring.

From the N.M.R. view point, the movement appears as the product of two orthogonal rotations : the first one around the C axis
and the secund one around one of the free orbitals of water oxygen
atom. Between 293°K and 210°K a temperature independent region is
observed for the variation of the intensity of the ^1H spectrum.
This supports the hypothesis for tunneling and it explains simultaneously the high speed of the molecular rotation, the existence
of a central line and the acidic properties of clays.

LITERATURE CITED

(1) Woesner D.E. J.Magn.Res. (1974) 16, 483
(2) Hecht A.M., Geissler E. J.Col.Int.Sci. (1973) 44, 1
(3) Hougardy J., Stone W., Fripiat J.J. J.Chem.Phys. (1976) 69, 9
(4) Calvet R. Ann.Agron. (1973) 24, 77 et 133
(5) Mamy J., Gaultier J.P. (1973) XXV, 43-51
(6) Prost R., Thesis Paris (1975) Ann.Agron. 26, 463
(7) Farmer V.C., Russel J.D. Trans.Far.Soc. (1971) 67, 2737
(8) Conard J. Proc.Int.Clay Conf. 1975 (Applied Pub) p.221
(9) Kadi Hanifi M.Thesis to be published
(10) Kistenmacher H., Popkie H., Clementi E. J.Chem.Phys. (1974)
     61, 799
(11) Clementi E., Popkie H. J.Chem.Phys. (1972) 57, 1077
(12) Kittel C. Int.Solid State Phys. (J.Wiley 4° Ed) p 130
(13) Olejnik S., Stirling G.C., White J.W. Disc.Far.Soc. (1970) 1,
     188
(14) Bernal J. J.Chim.Phys. (1953) 50 C1
(15) Resing H.A. in Adv.Mol.Relax.Proc. (1967) 1, 109
(16) Shirozu H., Bailey S.W. Ann.Mine. (1966) 51, 1124

Thanks, are due to C.Poinsignon, H.Vandamme, M.Kadi-Hanifi
R.Giese and J.J.Fripiat for stimulating discussions.

10

Orientation and Mobility of Hydrated Metal Ions in Layer Lattice Silicates

T. J. PINNAVAIA

Department of Chemistry, Michigan State University, East Lansing, Mich. 48824

The swelling layer lattice silicates known as smectites possess mica-like structures in which two-dimensional silicate anions are separated by layers of hydrated cations. Unlike the micas, however, the interlayer cations can be readily replaced by simple ion-exchange methods with almost any desired cation, including transition metals, organometallic, and carbonium ions. In addition, the interlayer space occupied by the exchange cations can be swelled from zero to several hundred Å, depending on the nature of the interlayer cations, the charge density on the silicate sheets, and the partial pressure of water in equilibrium with the solid phase. Similar swelling over a more limited range can be achieved by replacing the water molecules with a variety of polar solvents such as alcohols and ketones.

Our interest in these naturally occurring minerals has centered on their possible use as supports for homogeneous metal ion catalysts (1,2). To achieve this objective it was essential to determine the swelling conditions necessary to achieve the solution-like properties in the intracrystal environment. Earlier crystallographic studies have indicated that solvation of the exchange cations by two molecular layers of water resulted in a solid-like interlayer structure with the aquo complexes adopting well defined orientations in the intracrystal space (3). On the other hand, proton magnetic resonance studies (4-6) suggest that when the interlayers are swelled to several hundred Å units, the adsorbed water possesses appreciable solution-like properties. Under these conditions, however, the minerals are gel-like and are not especially well suited as solid supports for a homogeneous catalyst.

We have undertaken electron spin resonance investigations of the orientation and mobility of interlayer cations under different degrees of swelling to determine the minimum interlayer thickness necessary to achieve rapid tumbling of the exchange cations and diffusion of small substrate molecules into the intracrystal environment. Copper(II) or manganese(II) exchange forms of hectorite, montmorillonite and vermiculite have received

greatest attention. Considerable information on the nature of
diamagnetic exchange forms has also been obtained from the
effects which these cations have on the nature of the esr signals
of paramagnetic iron in orthorhombic environments in the silicate
sheet structure. The present discussion reviews only   the
nature of the Cu^{2+} and Mn^{2+} exchange forms.

The idealized unit cell formulas and cation exchange capacities of the minerals employed in these investigations are given
in Table I. Based on the experimental cation exchange capacities
and the theoretical surface areas (~ 800 m^2/g) the average distance between divalent exchange cations in the air-dried minerals
varies from ~ 6 Å (vermiculite) to 14 Å (hectorite).

Hydrated Cu^{2+} Minerals. When a copper(II) exchange form of
hectorite or montmorillonite is allowed to equilibrate in air
under ambient conditions the 001 spacing is 12.4 Å, and the water
to copper ratio is approximately 8:1. Since the silicate lattice
c dimension is 9.6 Å, the thickness of the interlayer region
(2.8Å) indicates that the copper(II) ions are hydrated by a monolayer of water. The esr spectra of the Cu(II) ions under these
conditions are illustrated in Figure 1. Randomly oriented powder
samples at room temperature and at 77°K consist of clearly defined g⊥ and g∥ components as expected for Cu(II) with tetragonal symmetry. When the spectrum of an oriented sample is
recorded with the magnetic field direction parallel to the silicate layers only the g⊥ component is observed. On the other
hand, when the sample is oriented with the magnetic field perpendicular to the silicate sheets, only the g∥ component is observed. These results indicate that the intercalated $Cu(H_2O)_4^{2+}$ ion
is oriented in the interlamellar region with the four-fold symmetry axis of the ion positioned perpendicular to the silicate
sheets. If water in outer spheres of coordination is removed by
heating the mineral to 110° the esr spectral features of the complex remain unchanged. Thus the amount of outer sphere water
under conditions where a monolayer is being formed does not alter
the orientation of the ion. Although the average copper-copper
distances among the montmorilonites and hectorite differ, oriented samples of each layer silicate show   the same esr spectral
changes when the film is positioned parallel and perpendicular to
the applied magnetic field. In each case the magnitude of g∥
is greater than g⊥ (see Table II).

Vermiculite has a higher charge density than hectorite and
montmorillonite and is more difficult to dehydrate. However, a
monolayer of interlayer water for the Cu^{2+} exchange form can be
achieved by dehydration of the mineral over P_4O_{10} at room temperature. As in the case of hectorite and montmorillonite an
oriented sample of the mineral shows that the tetragonal Cu^{2+} ion
adopts a restricted orientation on the silicate sheet with the
symmetry axis perpendicular to the layers.

Vermiculite is well suited for obtaining two molecular layers of water in the interlamellar region because the high surface

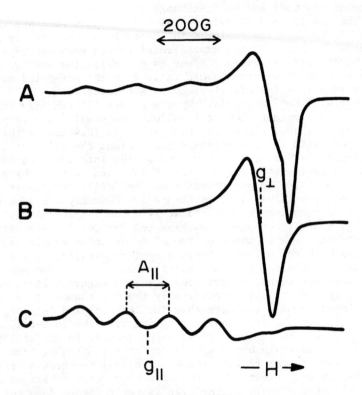

Figure 1. ESR spectra (X-band) for Cu(II) hectorite at room temperature: (A) powder sample, (B) oriented sample with H_o parallel to silicate sheets, (C) oriented sample with H_o perpendicular to silicate sheets

charge permits a maximum of only two layers of water when the
mineral is fully hydrated. The presence of two layers of water
in the interlamellar regions of an air dried sample of Cu^{2+} vermiculite is verified by observing x-ray reflections of several
rational orders that correspond to an 001 spacing of 14.2 Å. The
esr spectrum of an oriented sample at room temperature consists
of $g_{||}$ and $g_{|}$ components when the silicate layers are positioned
both parallel and perpendicular to the applied magnetic field.
The absence of an appreciable change in the relative intensities
of $g_{||}$ and $g_{|}$ indicates that the symmetry axis of the tetragonal
hexaaquo complex, $Cu(H_2O)_6{}^{2+}$, is inclined with respect to the
silicate sheets at an angle near 45°. Also the Cu^{2+}-water bonds
along the symmetry axis are longer than those in the X-Y plane of
the ion as $g_{||}$ is greater than $g_{|}$ ($g_{|}$ = 2.10, $\Delta H_{|}$ = 45G; $g_{||}$ =
2.40 A/c = 0.0115 cm^{-1}). Anisotropy in the g factor of $Cu(H_2O)_6{}^{2+}$
is rarely observed at room temperature. Isotropic thermal
motions are normally sufficiently rapid above 50°K to give a
single esr line (17). Rapid intramolecular exchange can occur
between three equivalent Jahn-Teller distorted states which correspond to axial elongation along the three possible sets of
water-copper-water axes (18). In absence of rapid tumbling this
motion will not lead to averaging of $g_{||}$ and $g_{|}$ when the ion is
adsorbed on a planar surface. It may be concluded therefore that
the interlayer ions adopt restricted orientations on the surface
even when the ion is hydrated by two molecular layers of water.
This result is consistent with the results of earlier x-ray diffraction studies of layer lattice silicates:
under conditions of two layer hydration the interlayer water is
highly structured.

The g values for interlayer $Cu(H_2O)_6{}^{2+}$ are larger than
those found for the planar aquo complex, and the calculated
average values of g are in good agreement with the observed
values of g_{av} for Cu(II) in aqueous solution (19) and for
$Cu(H_2O)_6{}^{2+}$ at the exchange sites of resins (20-21).

As the relative humidity is increased from approximately
40%   to 100%, the Cu^{2+} exchange forms of hectorite
and montmorillonite undergo a continuous transition from a monolayer phase (d_{001} = 12.4 Å) to a multimolecular interlayer water
phase (d_{001} = 20 Å) without forming a well defined intermediate
hydration state. However, the magnesium exchange forms of these
minerals exhibit an initial hydration state which is well ordered
with d_{001} = 15.0 Å. Complete replacement of magnesium by copper
causes the interlayer to collapse from three molecular layers of
water to a single water layer. However, when copper is doped
into magnesium hectorite at the 5% level, the 001 spacing remains
at 15.0 Å under air dried conditions.

The esr spectrum of Cu^{2+} doped into the 15.0 Å phase of
magnesium hectorite exhibits $g_{||}$(2.335) and $g_{|}$(2.065) values,
respectively, for the magnetic field oriented perpendicular and
parallel to the silicate sheets. It may be concluded that the

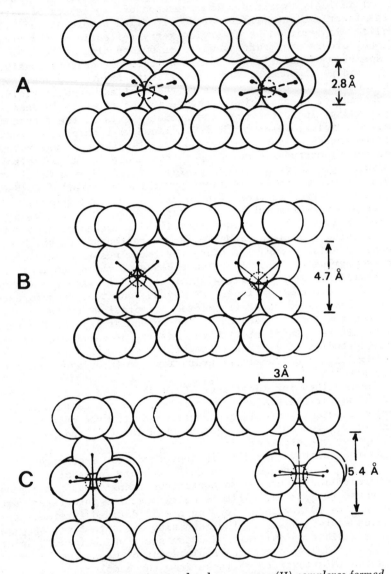

Figure 2.   Orientations of intercalated aquo copper(II) complexes formed by hydration with one, two, and three molecular layers of water. Open circles represent oxygen atoms of the silicate structure and ligand water molecules.

symmetry axis of the tetragonal $Cu(H_2O)_6^{2+}$ ion is perpendicular to the silicate sheets. The ion is not in a solution-like environment insofar as dynamic Jahn-Teller distortions or rapid tumbling do not average the anisotropy or broaden the signal. Well resolved splitting of $g_\perp$ ($B/C=-0.0022$ cm^{-1}) as well as $g_{||}$ ($A/C=-0.0156$ cm^{-1}) is observed.   Splitting of g is not well resolved when the copper ions exclusively occupy an inner layer, presumably because of Cu(II)-Cu(II) dipolar broadening.

The above spectral studies of Cu^{2+} ions hydrated by 1, 2, and 3 molecular layers of water indicate that the ions adopt restricted orientations in the interlayer environment. The restricted orientations are illustrated in Figure 2. The ions are best described as being in a solid-like environment rather than a solution-like environment. Under these conditions one should not expect a catalytically active ion to be accessible for reaction with substrate species diffusing from solution into the interlayer region.

The anisotropic esr signal of Cu^{2+} doped into magnesium hectorite is lost upon solvation of the mineral in liquid water or ethanol. Jahn-Teller distortion or random tumbling of the ion in the expanded interlayers ($d_{001} \sim 20$-22 Å) averages the anisotropy and relaxation due to modulation of the g tensor and spin rotation interactions result in a broad signal analogous to that observed for the ion in water or methanol solution. Hectorite fully exchanged with Cu^{2+} and fully swelled by water to a d_{001} value of 22 Å exhibits a single signal with a width near 120 gauss. These results suggest that when the interlayer region is swelled to a thickness of 10-12 Å the interlayer ions begin to adopt solution-like properties. In order to more fully estimate the tumbling motion of divalent ions in the restricted water layers, an esr linewidth study of Mn^{2+} exchange forms was undertaken.

Hydrated Mn^{2+} Minerals. It has been previously reported that the linewidths of the hyperfine lines of hydrated Mn^{2+} are broader on the exchange sites of montmorillonite than in bulk solution (22). The increase in line width was attributed to relaxation effects of the more restricted surface adsorbed ions. Also, Mn^{2+} montmorillonite has been reported to exhibit broader hyperfine lines when larger molecules (e.g. pyridine) replace water in ligand positions (23,24). Again, the result was interpreted in terms of reduced mobility of the Mn^{2+} ion. However, in addition to mobility effects other factors such as dipolar relaxation can also contribute to the observed esr linewidths.

The esr spectrum of Mn^{2+} in solution normally consists of 6 hyperfine lines due to coupling of the S=5/2 electron spin with the I=5/2 nuclear spin. Each hyperfine component consists of 3 superimposed Lorentzian lines due to the five $\Delta m_s=1$ transitions which are not resolved at X-band frequencies. The nondegeneracy of the $\Delta m_s=1$ transitions leads to inhomogeneous line broadening. The widths are the sum of two contributions: 1) solvent

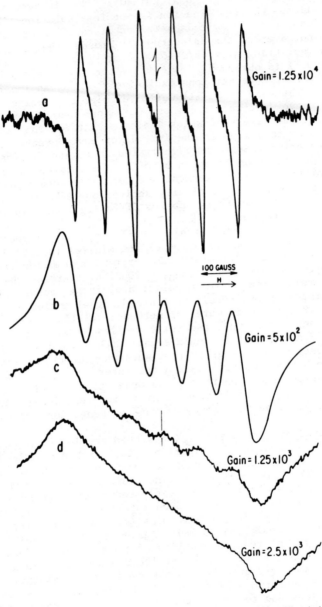

100 GAUSS

H →

Gain = 1.25 x 10⁴

Gain = 5 x 10²

Gain = 1.25 x 10³

Gain = 2.5 x 10³

American Mineralogist

Figure 3.   ESR spectra at room temperature for (A) MnCl₂ in methanol (5.0 × 10⁻⁵M) and for powder samples of hectorite (B) fully hydrated, (C) air-dried, and (D) dehydrated at 200°C for 24 hr. Vertical lines indicate the resonance position of standard pitch (g = 2.0028) (9).

collisional relaxation processes and 2) dipolar interactions between neighboring Mn^{2+} ions. The dipolar effect is concentration dependent and proportional to r^{-3} where r is the average manganese-manganese distance. In dilute solution the lines are narrow and determined exclusively by collisional relaxation processes. Increasing the concentration causes the six hyperfine components to broaden markedly until at a concentration of 2.3 $\underline{M}$ or greater ($r < 9.0$ Å), the hyperfine structure is lost and the spectrum appears as a single broad line.

A typical spectrum for Mn(II) in dilute solution is shown in Figure 3 along with spectra for Mn^{2+} hectorite under various conditions. It is seen that the fully hydrated mineral exhibits a "solution-like" spectrum, except that the hyperfine lines are broader. Reducing the amount of inner-layer water from several to two molecular layers by allowing the mineral to dry at approximately 50% relative humidity causes the line to broaden markedly. Thermal dehydration at 200° leads to still further line broadening and almost complete loss of hyperfine structure. Similar changes in line width are found for Mn^{2+}-saturated montmorillonite.

Since the average interlayer exchange ion distance for Mn^{2+} montmorillonite and hectorite is in the range 10 - 14 Å, the widths of the Mn^{2+} signals should be determined by dipolar relaxation effects. This is verified by the comparison in Figure 4 of the average width of the $m_I=\pm5/2$ lines for $MnCl_2$ in methanol solution and fully hydrated minerals of differing Mn^{2+}-Mn^{2+} distances. In Mn^{2+} vermiculite, where the exchange ion distance is approximately 7 Å, only a single broad line with a width of 710G is observed. The broadening is similar to that found for $MnCl_2$ salt (830G) and consistent with dipole-dipole coupling between magnetic ions 3-8 Å apart. Dipolar interactions between the Mn^{2+} exchange ions and iron ions in the silicate network may also contribute somewhat to the broadening of the manganese resonances. The line width of the $m_I=\pm5/2$ lines differs by 15G when an oriented sample is positioned parallel and perpendicular to the applied magnetic field. The difference in line widths is attributed to an isotropic dipolar relaxation by structural iron along the crystallographic c dimension. The smallest difference in line widths is observed for Mn^{2+} hectorite which has the lowest Fe^{3+} content ($< 0.14\%$). In this case the increase in line widths with decreasing hydration state can only be interpreted in terms of reduced mobility of the interlayer. However, it is difficult to assess quantitatively the interlayer mobility because the linewidths are still determined by an interionic dipolar relaxation mechanism involving neighboring Mn^{2+} exchange ions. In order to eliminate dipolar relaxation, Mn^{2+} was doped into Mg^{2+} hectorite at the 5% level.

In absence of dipolar interactions, spin relaxation of $Mn(H_2O)_6^{2+}$ in solution results from molecular collisions between the solvated ion and solvent molecules which cause random distortion

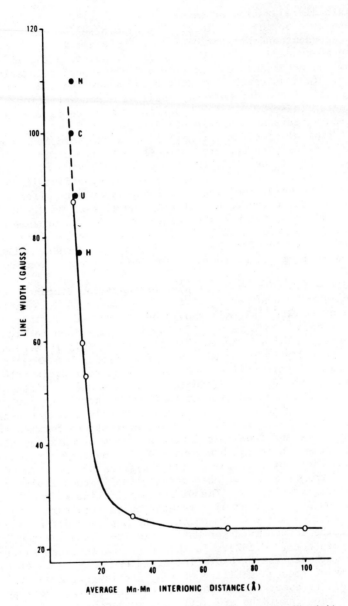

American Mineralogist

Figure 4. Average $m_I = \pm 5/2$ line widths for Mn^{2+} vs. interionic distance. Open points are for $MnCl_2$ in methanol, solid points are for nontronite (N), Upton (U) and Chambers (C) montmorillonites, and hectorite (H) under fully hydrated conditions (9).

of the complex (25,26). By observing the relative esr line widths in the two environments, one can estimate the correlation time τ, for the ion on the exchange surfaces relative to the ion in bulk solution. When $\omega_0\tau \ll 1$, which is generally true for $Mn(H_2O)_6^{2+}$ at room temperature and at X-band frequency ($\omega_0=0.58 \times 10^{11}$ radians/second), the width of the $-1/2 \longleftrightarrow +1/2$ transition is directly proportional to τ and the inner product of the zero field splitting tensor (27,28). If one assumes that the zero field splitting is the same in bulk solution and on the exchange surfaces of the mineral, then the relative correlation times should be directly proportional to the ratio of line widths. Figure 5 illustrates the esr spectrum of Mn^{2+} in magnesium hectorite in the fully hydrated, air-dried, and thermally dehydrated condition. The width of the fourth highest field component at room temperature, which is a reliable estimate of the width of the $+1/2 \longleftrightarrow -1/2$ transition (29), is 28.7 G for the fully hydrated mineral. In comparison, the width of $Mn(H_2O)_6^{2+}$ in dilute aqueous solution is 22 G. Therefore the value of τ, which can be taken physically to be the pre-collisional lifetime of the ion, is only *ca.* 30% longer in the interlayer than in bulk solution where it has been estimated to be 3.2 times 10^{-12} seconds. Thus, the fully hydrated interlayers approximately 12 Å thick are indeed very much solution-like. In contrast, τ for $Mn(H_2O)_6^{2+}$ has been estimated to be about 2.2 times larger in three-dimensional zeolites than in bulk solution.

Allowing the mineral to dry at room temperature should cause the mobility of the interlayer to decrease. This is verified by an increase in the line widths for the doped mineral as illustrated in Figure 5B. However, the lines are too broad and overlapping (average width 48 G) to obtain a simple quantitative estimate of the $-1/2 \longleftrightarrow +1/2$ transition.

Figure 5C shows the spectrum of the doped mineral under conditions where the interlayers are collapsed and the Mn^{2+} ions are coordinated to silicate oxygens in hexagonal positions within the silicate sheets. The spectrum consists of six main lines which represent the allowed $\Delta m_I=0$ transitions and five pairs of weaker doublets which are due to forbidden transitions with $\Delta m_I=\pm 1$. This spectrum is characteristic of Mn^{2+} in certain crystalline matrices and in frozen glasses. As expected under anhydrous conditions, there is no solution character to the interlayer Mn^{2+} ions. More recently, it has been shown that the allowed $m_I=0$ transition are split by about 14 and 10 G, respectively, when the silicate sheets are perpendicular and parallel to the applied magnetic fields (8). Although through space dipolar coupling of Mn^{2+} with structural hydrogen or fluorine atoms might be expected to split the spectral lines into doublets, this explanation is precluded because the splitting persists at the magic angle of 55°. However, the splitting is the same at X-band and Q-band frequencies. It is likely that the splitting arises because of a mixture of isotropic and anisotropic coupling. It

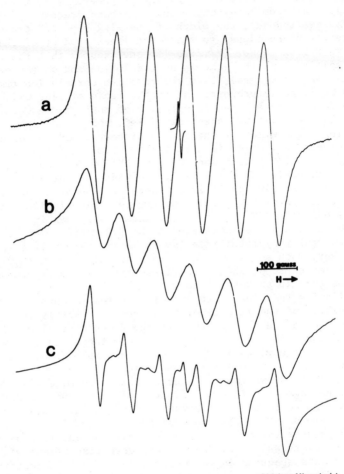

Figure 5. ESR spectra at room temperature of powder samples of Mn²⁺-doped Mg²⁺-hectorite: (A) fully hydrated, (B) air-dried, and (C) thermally dehydrated at 200°C (9).

has been suggested that the coupling reflects some degree of covalence between Mn^{2+} and fluorine in the fluorine containing cavities of the silicate lattice. A contribution due to Mn^{2+}-OH interactions in hydroxyl containing cavities cannot be ruled out. Nonetheless, the hyperfine coupling unequivocally establishes the positions of Mn^{2+} within hexagonal cavities in the dehydrated mineral.

Conclusions

Based on these studies two important conclusions can be drawn. First, when the interlayer thickness of the hydrated mineral is determined by the dimensions of the inner-coordination sphere of the hydrated exchange cations, the cations adopt restricted orientations in the interlayer region. Under these conditions the interlayer is solid-like in structure, and one should not expect a catalytically active exchange cation to be accessible for reaction with substrate molecules diffusing into the interlayer region from solution. Second, when the thickness of the interlayer region exceeds the dimensions of the inner-coordination sphere of the exchange cations, the exchange ions begin to tumble rapidly in the interlayer and take on appreciable solution-like properties. The accessibility of the exchange cation under conditions similar to these has recently been demonstrated by supporting homogeneous rhodium phosphine complexes in the intercrystal environment and by observing retention of their catalytic activity (1,2).

Acknowledgement

The original work reviewed in this paper was carried out in collaboration with Drs. D.M. Clementz and M.B. McBride and Professor M.M. Mortland of the Department of Crop and Soil Science, Michigan State University. Financial support by the National Science Foundation is gratefully acknowledged.

TABLE I

Idealized Unit Cell Formulas and Cation Exchange Capacities of Layer Silicates

Mineral	Unit Cell Formula[a]	Meq/100g	Reference
Hectorite (Hector, Calif.)	$M^{n+}_{0.64/n}[Mg_{5.42}Li_{0.68}Al_{0.02}](Si_{8.00})O_{20}(OH,F)_4$	73	12
Montmorillonite (Upton, Wyom.)	$M^{n+}_{0.64/n}[Al_{3.06}Fe_{0.32}Mg_{0.66}](Si_{7.90}Al_{0.10})O_{20}(OH)_4$	92	13
Nontronite (Garfield, Wash.)	$M^{n+}_{0.84/n}[Al_{0.26}Fe_{3.70}Mg_{0.04}](Al_{1.00}Si_{7.00})O_{20}(OH)_4$	104	14
Montmorillonite (Chambers, Ariz.)	$M^{n+}_{0.96/n}[Al_{2.84}Fe_{0.35}Mg_{0.85}](Al_{0.22}Si_{7.78})O_{20}(OH)_4$	116	15
Vermiculite (Llano, Texas)	$M^{n+}_{2.00/n}[Al_{0.30}Fe_{0.02}Mg_{5.66}](Al_{2.28}Si_{5.72})O_{20}(OH)_4$	200	16

a Water of hydration is omitted; M^{n+} is the interlayer exchange cation; elements in brackets fill octahedral positions in the silicate lattice, elements in parentheses fill tetragonal positions.

TABLE II

Esr Data for Intercalated $Cu(H_2O)_4^{2+}$

Mineral	Temp, °K	$g_\perp$	$\Delta H_\perp$, Gauss [a]	$g_\parallel$	$-10^4 A_{11}$, cm^{-1}	g_{av}
Hectorite	300	2.08	100	2.34	165	2.17
	77	2.08	100	2.33	175	2.16
Montmorillonite (Upton, Wyo.)	77	2.09	158	2.34	175	2.17
Montmorillonite (Chambers, Ariz)	77	2.09	140	2.23	175	2.16
Vermiculite	300	2.10	~80	~2.3		

[a] Maximum line width of the perpendicular component.

Literature Cited

1. Pinnavaia, T.J. and P.K. Welty (1975), J. Am. Chem. Soc., 97, 2712.
2. Pinnavaia, T.J., P.K. Welty, and J.F. Hoffman (1975), Proc. Intern.Clay Conf. (1975), Mexico City, 373.
3. Walker, G.F., (1956), Clays Clay Min., 4, 101
4. Graham, J., G.F. Walker, and G.W. West, (1964), J. Chem. Phys. 40, 540.
5. Boss, B.D. and E.O. Stejskal,(1968), J. Coll. Interface Sci. 26, 271
6. Fripiat, J.J., (1973), Ind. Chim. Belg., 38, 404.
7. Clementz, D.M., T.J. Pinnavaia, and M.M. Mortland, (1973), J. Phys. Chem., 77, 196.
8. McBride, M.B., T.J. Pinnavaia, and M.M. Mortland, (1975), J. Phys. Chem., 79, 2430.
9. McBride, M.B., T.J. Pinnavaia, and M.M. Mortland (1975), Am. Mineralogist, 60, 66.
10. McBride, M.B., T.J. Pinnavaia, and M.M. Mortland (1975), Clays Clay Minerals, 23, 103.
11. McBride, M.B., T.J. Pinnavaia, and M.M. Mortland (1975), Clays Clay Minerals, 23, 161.
12. Amer. Petr. Inst., Project 49
13. Ross, G.J., and M.M. Mortland (1966), Soil Sci. Soc. Am. Proc., 30, 337.
14. Amer. Petr. Inst., H-33A.
15. Schultz, L.G. (1969), Clays Clay Minerals, 17, 115.
16. Foster, M.D. (1961), Clays Clay Minerals, 10, 70.
17. McGarvey, B.R (1966), in "Transition Metal Chemistry", Vol. 3, R.L. Carlin, Ed., Marcel Dekker, New York.
18. Hudson, A (1966), Mol. Phys., 10, 575.
19. Fujiwara and H. Hayashi (1965), J. Phys. Chem., 43, 23.
20. Cohen, R. and C. Heitner-Werguen (1969), Inorg. Chim. Acta, 3, 647.
21. Umezawa, K. and T. Yamabe (1972), Bull. Chem. Soc. Jpn. 45, 56.
22. Furuhata, A. and K. Kuwata (1969), Nendo Kagaku, 9, 19.
23. Taracevich, Yu. I., and F.D. Ovcharenko (1972), Proc. Inter. Clay Conf., Madrid, 627.
24. Pafamov N.N., V.A. Sil'Chenko, Yu. I. Tarasevich, V.P. Telichkun, and Yu. A. Bratashevskii (1971), Urk. Khim. Zh., 37, 672.
25. Luckhurst, G.R. and G.F. Pedulli (1971), Mol. Phys., 22, 931.
26. Rubinstein, M., A. Baram, and Z. Luz (1971), Mol. Phys., 20, 67.
27. Burlamacchi, L. (1971), J. Chem. Phys., 55, 1205.
28. Burlamacchi, L., G. Martini, and E. Tiezzi (1970), J. Phys. Chem., 74, 3980.
29. Garrett, B.B., and L.O. Morgan (1966), J. Chem. Phys., 44, 890.

Characterization of the Small-Port Mordenite Adsorption Sites by Carbon-13 NMR

M. D. SEFCIK, JACOB SCHAEFER, and E. O. STEJSKAL

Monsanto Company, St. Louis, Mo. 63166

Complete characterization of adsorbate-adsorbent systems require the application of diverse analytical procedures. Adsorbents are usually characterized by what might be called steady state measurements (such as XRD, adsorption isotherms, chemical composition, etc.). These experiments determine macroscopic or static properties of the adsorbent. Adsorbates, on the other hand, are routinely analyzed in terms of their high frequency spectra. Infrared, Raman and ultraviolet spectroscopies measure vibrations and electronic transitions which occur in the frequency range of 10^{12} to 10^{14} Hertz. What is lacking from a complete picture of the adsorbate-adsorbent system is an analysis of the low-frequency dynamic properties of the system; that is, a description of any orientational influence the adsorbent may have on the adsorbate. Nuclear magnetic resonance has been applied to this problem with some success. Resing and coworkers [1] have used proton-NMR relaxation properties to study diffusion coefficients, jump times and the temperature of onset of molecular rotation in the zeolite system. Kaplan, Resing and Waugh reported the first carbon-13 NMR spectra of benzene adsorbed on charcoal and silica gel and demonstrated that the chemical shift aniso- tropy could be interpreted in terms of the molecular rotation and reorientation [2]. High resolution carbon-13 NMR has also been recently used to determine the extent of interactions between olefins and zeolytic cations [3,4].

We have found carbon-13 NMR to be particularly useful in obtaining information about the motions of molecules in the range of 10^2 to 10^6 Hertz. Analysis of the NMR lineshape produced by these slow moving molecules can provide information about the adsorbate rotational axis and the geometry of the adsorption site.

We will present here the results of a study of CO, CO_2, COS and CS_2 adsorbed on small-port Na^+- and cation-exchanged mordenites using ^{13}C NMR, which lead to a description of adsorption sites in this particular adsorbate-adsorbent system.

Lineshape Analysis

There are three sources of line broadening in the NMR experiment which limit the applicability of this technique in studying adsorbent-adsorbate interactions. The first is dipole-dipole interactions. If a molecule contains two nuclei with magnetic moments, then the extent of the magnetic interaction between those nuclei depends on their spatial separation, relative orientation and the size of the magnetic moments (5). These interactions may be between similar nuclei (homonuclear broadening) or unlike nuclei (heteronuclear broadening) and may be either inter- or intramolecular. In liquids these interactions are usually averaged to zero by the tumbling motions of the molecules, but this might not be the case for adsorbed molecules. Homonuclear dipolar broadening is a serious problem in proton NMR due to the high natural abundance and hence relatively short internuclear distance between protons. Multiple pulse experiments have been used to eliminate homonuclear interactions in proton NMR but will not be described here (a discussion of this technique can be found in ref. 6). By using rare-spin NMR (^{13}C, ^{15}N, ^{29}Si, etc.), homonuclear interactions are greatly reduced, while static heteronuclear dipolar broadening can be eliminated by double resonance techniques (7). Neither line-narrowing technique removes dipolar broadening associated with intermediate-frequency motions characterized by correlation frequencies on the order of 10^5 Hz.

A second source of unwanted line broadening arises from paramagnetic impurities in the sample. Paramagnetic broadening is similar to the dipole-dipole interactions discussed above except that it involves an electron dipole and can not be eliminated by instrumental procedures. Since the electron dipole is 657 times stronger than the ^1H nuclear dipole it is easy to see that very low concentrations of paramagnetic impurities can drastically affect the NMR lineshape. The effect of paramagnetic impurities in zeolites on molecular relaxation has been discussed in detail by Resing (8) and will not be presented here. In practice, paramagnetic impurities as great as 200 ppm in molecular sieve adsorbents are tolerable since many of the paramagnetic centers are apparently trapped away from the adsorption sites.

While both dipolar and paramagnetic interactions usually give rise to symmetric broadening of the nuclear magnetic resonance, molecular motions which are slower than about 10^4 Hertz allow a third major source of line broadening, a dispersion of resonance frequencies. The chemical shift or resonant Larmor frequency of a particular nucleus is proportional to the local magnetic field at that nucleus. The local magnetic field is, in the absence of dipolar interactions, dependent on the strength of the applied magnetic field, H_o, and the very small fields generated by electrons moving about the nucleus. In a molecule there is frequently an anisotropic distribution of electrons about the nucleus causing a directional dependence in the chemical shift. Theoretical resonance dispersions arising from chemical shift anisotropies for nuclei in randomly oriented molecules such as encountered here may be calculated for various nuclear site symmetries (9). A cubic nuclear site symmetry results in only one value for the chemical shift, a delta function shown in Figure 1 at σ_i. (The high frequency molecular motions encountered in liquids and gases effectively increase the nuclear site symmetry of the molecules to cubic resulting in the narrow lines characteristic of high resolution NMR). It is worthwhile to note at this point that any interaction between the adsorbate and the adsorbent which changes the electron density near the observed nucleus will result in a chemical shift, thus providing a mechanism for distinguishing chemi- from physisorption.

Molecules with axial symmetry have two principle values of the chemical shift, one perpendicular to the symmetry axis and one parallel to it. Figure 1 represents the chemical shift dispersion of a CO_2 molecule. The linear CO_2 molecule has three possible orientations with respect to the applied magnetic field, two of which have identical chemical shifts. The spectral dispersion which arises from a collection of randomly oriented CO_2 molecules, as in a frozen solid, is a broad line with the doubly degenerate chemical shifts (and hence greater intensity) on the downfield (10) side, as illustrated by the dashed curve in Figure 1.

If instead of a frozen solid the CO_2 molecules find themselves in an environment where they can execute anisotropic rotation, a very different chemical shift dispersion is observed. As shown by the drawing in Figure 1, rotation about one of the C_2 axes

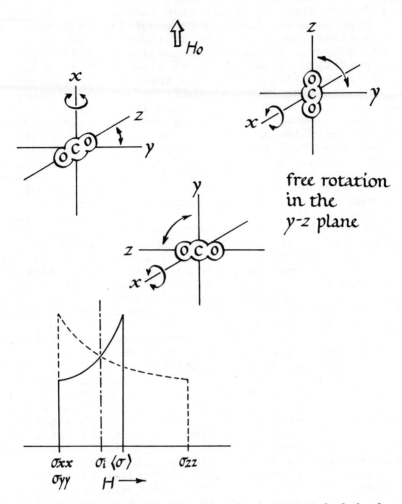

Figure 1. *Some theoretical spectra observed in the NMR of adsorbed carbon dioxide: a random array of nonrotating CO_2 molecules produces a chemical shift dispersion (dashed curve) with two principal values for the chemical shift tensors (σ_{zz} parallel to the molecular axis and σ_{xx}, σ_{yy} perpendicular to it). With rotation about the x-axis the CO_2 molecule experiences anisotropic rotation. This rotation leaves σ_{xx} unchanged but averages σ_{yy} and σ_{zz} to $\langle\sigma\rangle$ (solid curve). The areas beneath the two curves are not shown to scale. Isotropic rotation of the CO_2 molecule averages the chemical shift dispersion to its isotropic value, σ_i (delta function, broken line).*

perpendicular to the molecular axis leaves the chemical
shift for one orientation unchanged, but averages the
other two producing a narrower spectral dispersion
with the degenerate chemical shifts on the up field
side (solid curve).   With rotation about the other C_2
symmetry axis, the CO_2 molecule assumes isotropic or
"free" rotation and the chemical shift dispersion
collapses to the delta function, as mentioned earlier.
The fourth characteristic lineshape which can be
encountered in the NMR of solids is one in which all
three orientations of a molecule with respect to the
applied magnetic field have unique chemical shifts.
This arises from molecules which have lower than axial
symmetry (nonsymmetric in the molecular coordinate
system) or from axially symmetric molecular whose
anisotropic rotation is not sufficient to average the
principle chemical shift values (σ_{yy} and σ_{zz} in Figure
1).   These spectral dispersions are characterized by
their "tent" shape having the greatest intensity
between the two chemical shift extremes.
     In the following discussion of the experimental
results we have found it useful to refer to the
various chemical shift dispersion spectra acronymical-
ly.   There are four general lineshapes encountered in
the NMR of solids; Symmetric, Axially symmetric with
the degeneracy on the Left, Axially symmetric with the
degeneracy on the Right, and the "tent" shaped Non-
Symmetric chemical shift dispersion.   Thus, S, AL, AR
and NS describe the spectral shape.   In addition, the
spectral dispersions may be associated with molecules
which are nonrotating (NR), anisotropically rotating
(AR) or isotropically rotating (IR).   A combination
of both descriptors forms the complete acronym for the
shape and origin of the NMR spectra.   For example, the
spectra discussed in Figure 1 may be referred to as
S(IR), AL(NR) and AR(AR) spectra respectively.
     Since determination of the molecular motions
depends on the careful analysis of the broad spectral
dispersions arising from molecular chemical shift
anisotropy,   every effort should be made to eliminate
unwanted sources of line broadening and spectral overlap from molecules which have more than one resonance.
As mentioned previously, rare-spin NMR has the advantage of minimizing homonuclear dipolar interactions,
and the use of isotopically enriched samples may
simplify the spectra to that of only one nuclear
resonance.   As in the work presented here, the
carbon-13 NMR of carbon-containing inorganic gases
actually provide the most straightforward and accessible experiment since only single resonance

experiments need to be performed. Investigations of
adsorbed organic vapors and liquids require[13]C nuclear
magnetic double resonance techniques to eliminate
heteronuclear dipolar interactions.

Experimental Procedure

The sodium mordenite molecular sieves were prepared from high purity reagents by Leonard B. Sand
at Worcester Polytechnic Institute and contained
10-170 ppm iron. The x-ray diffraction patterns
indicated the samples to be 90-95% mordenite with the
remainder as analcime. The mordenites were of the
small-port variety; they adsorbed 15% by weight of CO_2
at 1 atm. and 0.1% by weight of benzene at 72 mm and
room temperature. Ion-exchanged mordenites were
prepared by the conventional procedure from 1 molar
salt solutions. Approximately 1/2 gram samples of the
molecular sieve were placed in 10 mm O.D. NMR tubes
and dried in vacuo under a programmed temperature rise
to 300°C. (The ammonium exchanged sieve was shown to
lose ammonia only above 415°C by DTA.) After cooling,
the sieves were allowed to adsorb 90% isotopically
enriched ^{13}C gases to the desired level. The loading
level was defined as a weight percent of the capacity
of the sieve at room temperature and 1 atm. (300 mmHg
for CS_2). The tubes were sealed with a Teflon plug
and Viton o-ring without exposure to the atmosphere
and were used reproducibly over several months.

The Fourier transform ^{13}C-NMR spectra were obtained on a Bruker spectrometer, equipped with a broadband
receiver and quadrature detector (11), and operating
at 22.6 MHz, with field stabilization provided by an
external time-share ^{19}F lock (11). The experiments
described here can be performed on any commercial
instrument, which is equipped with an external field
lock, and which is free from baseline artifacts (12).

Results and Discussion

The ^{13}C-NMR spectra of isotopically enriched [13]CO_2
adsorbed in varying amounts on the Na⁺mordenite are
shown in Figure 2. Three different spectral lineshapes appear in this series indicating that the
adsorbed CO_2 exists in at least three distinct states
(or a distribution of states), presumably differing in
their rotational freedom. Analysis of the resonance
lineshape for each of these states should provide
information about the local site geometry which
influences CO_2 adsorption.

At the highest loading level (upper spectrum in Figure 2) a narrow, relatively weak symmetric line appears superimposed upon the broad chemical shift dispersion. We believe this narrow line is due to CO_2 molecules which are freely rotating and translating in the large channel areas of the mordenite. (The large channel is comprised of 12-member rings perpendicular to the crystallographic c-axis.) This conclusion is substantiated by the observation of narrow line resonances for CO_2 adsorbed in the zeolites Y ([13]) and L in which the minimum internal pore dimensions (13 and 7.1 Å, respectively) are greater than the van der Waals length of the CO_2 molecule (5.1 A). The intensity of these narrow symmetric lines (which reflect the affinity of CO_2 for these sites) increases as the pore dimensions decrease to match the dimensions of the CO_2. These results imply that intercalated material or randomly located cations ([14]) frequently reduce the large channel size to less than the length of the CO_2 molecule.

The second spectral lineshape which is present in Figure 2 is best seen at low loading levels. At the 40% level only a broad chemical shift dispersion of type AL(NR)(see text) is observed, indicating that the adsorbed CO_2 molecules are not allowed to rotate in their absorption sites. Meier ([15]), in his definitive crystal structure analysis of the Na^+-mordenite, identified a series of smaller channels opening in the b direction which are circumscribed by 8-member rings having a free aperture of 2.9 x 5.7 Å. The small channels do not, however, interconnect with the next main channel but rather branch through distorted 8-member rings towards similar areas in the adjoining large channel. These distorted 8-member rings, which have a free aperture of only 2.8 Å, isolate the main channels, leaving them lined with two rows of side-pockets. Meier found that four sodium cations per unit cell resided in the center of these distorted 8-member rings (the location of the other four could not be determined). When the sodium cations are exchanged for larger cations such as Cs^+ or NH_4^+ these remain in the side-pockets, being unable to assume a position in the distorted 8-member ring ([16]).

To determine whether the adsorption sites responsible for the nonrotating CO_2 molecules were the side-pockets, we examined the effect of cation exchange on the ^{13}C NMR of adsorbed CO_2. When the sodium cations were exchanged, forming the Cs^+- and NH_4^+-mordenite there was no evidence of nonrotating CO_2 at any loading level. Thus, our results are consistent with the side-pockets being the first adsorption sites

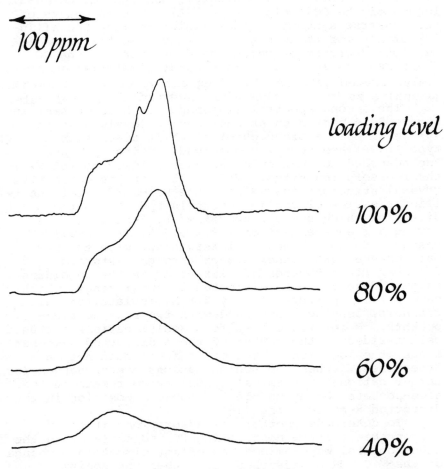

$^{13}CO_2$ adsorbed on
Na$^+$ "small port" mordenite

100 ppm

loading level

100%

80%

60%

40%

Figure 2.   Carbon-13 NMR spectra of CO_2 adsorbed on small-port Na$^+$-mordenite as a
function of loading level

to be filled in the Na$^{\pm}$mordenite and also providing the most severe rotational hinderance. This phenomena was first recognized by Barrer and Peterson (17) in their study of intercrystalline adsorption by synthetic mordenites. They reasoned that the first molecules adsorbed should preferentially occupy the side-pockets since these sites offered the highest coordination between the adsorbed molecule and the anionic oxygens. We have observed that CO, CO_2 and COS are readily adsorbed in the side-pockets of the Na$^{\pm}$mordenite but excluded from these sites in the Cs^+- and NH_4^+-mordenites. The resonance arising from molecules in the side-pocket sites is narrowed at higher loading level by collision induced rotation.

The broad resonance produced by CO_2 at high loading levels is of the type $\underline{AR}$(AR). Since CO_2 has axial symmetry, a resonance of this shape must be due to anisotropic rotation. The anisotropic rotation of molecules responsible for the $\underline{AR}$(AR) spectra is best characterized as a rotation in a single plane with a frequency greater than 10^4 Hertz. Anisotropic rotational behavior of adsorbed species on mordenites has been suggested in the literature. Gabuda and co-workers (18,19) have identified three types of zeolite water based on the proton -NMR linewidth: (1) rigidly bound, (2) anisotropically mobile, and (3) isotropically mobile. Water adsorbed on mordenite and other channel sieves such as chabazite, huelandite and laumontite was found to be anisotropically mobile. The motional behavior of carbon dioxide adsorbed on synthetic mordenite has been studied by Takaishi, et al. (20). These authors argued that the strong electric quadrapole moment of CO_2 interacts with the crystal field in the channel sieve, causing seriously hindered rotation. Based on statistical mechanics, Takaishi felt that at low coverage adsorbed CO_2 would be rigidly held below 50°C and become an anisotropic rotor above that temperature. Our results, at high coverage, indicate that CO_2 displays anisotropic rotation even at room temperature. Evidence will be offered below to suggest that steric rather than electronic effects are responsible for this behavior.

The adsorption sites in the mordenite which permit only anisotropic motion of adsorbed CO_2 are not located near the zeolitic cations. When synthetic Na$^+$-mordenite was converted to the Cs^+ and NH_4^+ forms, the $\underline{AR}$(AR) spectra remained unchanged, indicating that these larger cations did not inhibit the anisotropic rotation of the adsorbed gases. We conclude, therefore, that the adsorbed CO_2 which produces the $\underline{AR}$(AR) must be located in the occluded areas of the mordenite main channel.

Over the range of coverage studied here we have found evidence for nonrotating, anisotropically rotating and isotropically rotating adsorbed species at room temperature. The series of spectra in Figure 2 suggest the following sequence of events: at the lowest coverage CO_2 molecules are preferentially adsorbed in the side-pockets which line the main channel and are held with little rotational freedom. When all of the side-pocket sites are filled, adsorption occurs in the occluded or smallest sections of the main channel. Initially, the molecules which occupy these sites are relatively immobile but, with increasing concentration, intermolecular collisions induce rotation in a single plane of the small channel where the coordination number with respect to anionic oxygens is somewhat less than in the side-pockets. Finally, as these sites are filled, adsorption begins in some large pore areas of the mordenite, presumably the unblocked regions of the main channel. The affinity of the CO_2 for these large channel sites is low, consistent with the low coordination number and relatively unrestricted rotational behavior.

The ^{13}C-NMR spectra of CS_2 and COS adsorbed on the Na^+-mordenite are shown in Figure 3, again as a function of loading level. Due to the larger critical diameter of CS_2 (3.6 Å) one would not expect adsorption to occur in the side-pockets, and indeed, there was no AL(NR) type chemical shift dispersion indicative of this site, even at the lowest detectable loading levels. This conclusion is further supported by experiments with multicomponent systems. When the Na^+-mordenite was loaded to the 40% level with $^{13}CO_2$, and then allowed to adsorb $^{12}CS_2$ to its capacity, the ^{13}C-NMR spectrum (which detects only the CO_2) indicated that none of the CO_2 had been excluded from the side-pockets. In a control experiment, the addition of $^{12}CO_2$ to a sample partially filled with $^{13}CO_2$ resulted in complete scrambling of the label; the NMR spectrum was similar to that obtained with fully loaded $^{13}CO_2$, albeit less intense.

At all loading levels the NMR spectra of CS_2 adsorbed on the mordenite exhibit the AR(AR) spectral lineshape indicative of an anisotropically rotating axially symmetric molecule. As in the case of CO_2, the adsorption site which permits this rotation is believed to be the occluded areas of the main channel. It is interesting to note that the NMR spectra of CS_2 exhibit much more area (i.e., concentration) in the

narrow symmetric line than was observed for CO_2 or, as
can be seen in Figure 3, COS. The increased intensity
of the narrow line component on going from CO_2 to COS
to CS_2 suggests the larger molecules preferentially
adsorb in these large volume sites. The increased
affinity of COS and CS_2 in these areas may reflect
either a higher coordination with the zeolitic oxygens
or the greater polarizability of the adsorbates (21).

The NMR spectra for adsorbed carbonyl sulfide,
shown in Figure 3, have the greatest intensity to the
left of the spectrum at all loading levels. At low
coverage the chemical shift dispersion is of the type
AL(NR), similar to that observed for CO_2 at low
coverage. To establish the adsorption site responsible
for this spectrum we again turn to results of experiments on multicomponent systems. When the Na^+-
mordenite was filled to the 40% level with $^{13}CO_2$
(sufficient to fill the side-pockets), and then exposed
to an atmosphere of ^{12}COS, the resulting spectrum
indicated that the labeled carbon dioxide had been
excluded from the side-pockets and forced into the
less favorable adsorption sites in the large channel
area. Apparently, the small end of the carbonyl sulfide is able to fit into the side-pockets where it is
preferentially adsorbed in competition with CO_2.

At higher loading levels the NMR of COS does not
shift to the AR(AR) type spectra noted for CO_2 and CS_2.
Since the carbonyl sulfide is intermediate in size
between CO_2 and CS_2 we would have expected adsorption
in the occluded areas of the large channel which leads
to anisotropic rotation. To understand this apparent
anomaly it is necessary to refer to the illustrations
of the chemical shift dispersion in Figure 1. For the
linear symmetric molecule CO_2 (point group $D_{\infty h}$) a
rotation of 90° is sufficient to average the chemical
shift tensors and produce the AR(AR) type spectrum.
Carbonyl sulfide, which is an asymmetric linear
molecule (point group $C_{\infty v}$) requires a full 180° rotation to produce the same result. Based on their
van der Waals radii, CO_2, COS and CS_2 have molecular
lengths of 5.1, 6.0 and 6.6 Å and critical diameters
of 2.8, 3.6 and 3.6 Å, respectively. The main channel
aperature of the Na^+-mordenite, which is defined by
12-member rings perpendicular to the c-axis, measures
6.7 x 7.0 Å (15). The small-port mordenite, however,
does not adsorb molecules with critical diameters
larger than 4 Å. Our results indicate that CS_2 can
apparently execute at least 90° rotations in the main
channel of the small-port mordenite, but that COS can
not rotate a full 180°. This suggests that the

$^{13}CS_2$ and ^{13}COS adsorbed on Na⁺ "small port" mordenite

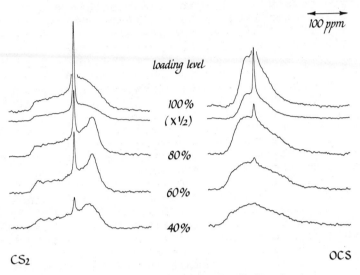

Figure 3. Carbon-13 NMR spectra of CS_2 (left) and COS (right) adsorbed on small-port Na^+-mordenite as a function of loading level

^{13}CO adsorbed on "small port" mordenite

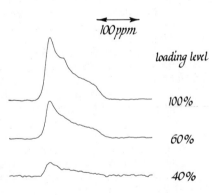

Figure 4. Carbon-13 NMR spectra of CO adsorbed on small-port Na^+-mordenite as a function of loading level

effective channel dimensions are reduced to approximately 4×5.5 Å by occlusions.
This channel size for the small-port mordenite is also supported by studies with carbon monoxide. The ^{13}C-NMR spectra of CO adsorbed on Na^{+}-mordenite are shown in Figure 4. At all loading levels the spectral dispersion remains of the type $\underline{AL}$(NR). Carbon monoxide absorption in the side-pockets was again verified by examining the spectra of CO when it was adsorbed on the ion-exchanged mordenites. With Cs^{+}- and NH_4^{+} mordenite the ^{13}C-spectra was relatively narrow, symmetric, and weak with an intensity only slightly greater than observed for gaseous CO at one atmosphere. The absence of any indication of main channel adsorption on the ion-exchanged sieves was unexpected. Since CO_2 is readily adsorbed into these areas we conclude that the molecular dimensions of carbon monoxide are sufficiently smaller than the small-port channel size so that no significant adsorption occurs. This is consistent with the reduced capacity of the Na^{+}-mordenite for CO compared to CO_2.
The results presented here demonstrate the information which can be obtained concerning the dynamic state of adsorbed species by rare-spin NMR. By judicious choice of probing gases, this technique not only allows one to view each adsorbed state essentially independent of the others but is also useful in defining the geometrical constraints of the various adsorption sites. Other NMR parameters, such as relaxation rates or temperature dependence of the spectral lineshape, may be useful for determining diffusion coefficients, exchange rates, jumping mechanisms and the energetics of each adsorbate-adsorbent interaction.

Acknowledgment

The authors wish to thank Leonard B. Sand for his assistance in preparing the synthetic mordenites used in this study and for his valuable discussion and criticism.

Literature Cited

1. H. A. Resing and J. S. Murday in "Molecular Sieves," Advan. Chem. Ser. 121, American Chemical Society, p. 414, Washington, D. C., 1973.
2. S. Kaplan, H. A. Resing and J. S. Waugh, J. Chem. Phys., (1973), 59, 5681.

3. H. Pfeifer, W. Schrimer and H. Winkler,"Molecular
    Sieves,"Advan. Chem. Ser. 121, American Chemical
    Society, p.430, Washington D. C., 1973.
4. D. Michel, W. Meiler and H. Pfeifer, J. Mol. Catal.,
    (1975), 1, 85.
5. A. Abragam, "The Principles of Nuclear Magnetism,"
    Chapter 8, Oxford University Press, London, 1961.
6. W-K. Rhim, D. D. Elleman and R. W. Vaughan,
    J. Chem. Phys., (1973), 59, 3740.
7. F. Block, Phys. Rev., (1958), 111, 841.
8. H. A. Resing, Advan. Molecular Relaxation Processes,
    (1972), 3, 199.
9. N. Bloembergen and T. J. Rowland, Acta Metall.,
    (1953), 1, 731.
10. All of the NMR spectra presented here are displayed
    with the field strength increasing from left to
    right.
11. E. O. Stejskal and Jacob Schaefer, J. Mag. Res.,
    (1974), 14, 160.
12. E. O. Stejskal and Jacob Schaefer, J. Mag. Res.,
    (1974), 15, 173.
13. E. O. Stejskal, Jacob Schaefer, J. M. S. Henis
    and M. K. Tripodi, J. Chem. Phys., (1974), 61,
    2351.
14. Y. Nishimura and H. Takahashi, Kolloid-Z. u.
    Z. Polymere, (1971), 245, 415.
15. W. M. Meier, Z. Krist, (1961), 115, 439.
16. L. C. V. Rees and A. Rao, Trans. Faraday Soc.,
    (1966), 62, 2103.
17. R. M. Barrer and D. L. Peterson, Proc. Roy. Soc.,
    (1964), 280A, 466.
18. I. A. Belitskii and S. P. Gabuda, Geol. Geofiz.,
    (1968), 6, 3.
19. E. E. Senderov, G. V. Yukhnevich and S. P. Gabuda,
    Radiospektrosk. Tverd. Tela, (1967), 149.
20. T. Takaishi, A. Yusa, Y. Ogino and S. Ozawa, Proc.
    Int. Conf. Solid Surf., 2nd (1974), 279.
21. D. W. Breck, "Zeolite Molecular Sieves," p. 664,
    Wiley, New York, 1974.

Use of Nitroxide Spin Probes in ESR Studies of Adsorbed Molecules on Solvated Layer Silicates

M. B. McBRIDE

Department of Agronomy, Cornell University, Ithaca, N. Y. 14853

Electron spin resonance (ESR) studies utilizing nitroxide spin probes doped into membranes (1,2) have produced information about the orientation and mobility of the probes in these systems, thereby revealing fundamental properties of membranes. Application of this spectroscopic technique to adsorption studies of a spin probe on layer silicates has elucidated certain principles of the organic-aluminosilicate interaction (3). Orientation data could be obtained in these studies because of the anisotropic hyperfine splitting constant (A) of the nitroxide probe and the ability of layer silicates to form well-oriented films. The layer silicates are composed of negatively charged aluminosilicate plates which maintain exchangeable cations between plates to balance the charge. Certain of these naturally occurring minerals are capable of swelling in water and other solvents by the adsorption of solvent between the plates.

Although information has been reported regarding the anisotropic motion and orientation of the protonated nitroxide probe (4-amino-2,2,6,6-tetramethylpiperidine N-oxide) adsorbed on ethanol- and water-solvated silicate surfaces (3), a systematic study of the molecule-surface interaction in the presence of numerous solvents is required for a more complete understanding of the processes of adsorption. Thus, in this study, solvents of widely varying chemical properties have been used to solvate a smectite doped with the protonated spin probe. In addition, similar experiments were carried out for a vermiculite in order to compare the surface properties of different layer silicates. Measurement of the orientation and mobility of the probe in these systems may permit the mechanisms of adsorption of organic molecules on solvated clay surfaces to be learned.

Materials and Methods

A California hectorite (<2μ particle size) with chemical formula and exchange capacity previously reported (4) was saturated with Na$^+$ ions using excess NaCl salt solution. The

salt was then washed from the clay suspension by repeated centrifuging and discarding the supernatant until a negative $AgNO_3$ test
for chloride was obtained. An aqueous solution of the spin probe
(4-amino-2,2,6,6-tetramethylpiperidine N-oxide), referred to in
this study as TEMPO, was titrated with HCl solution past the
equivalence point (determined with a pH meter) to protonate the
amine group of the probe. A known quantity of this cationic
form of the probe ($TEMPO^+$) was added to the Na^+-saturated
hectorite to produce a clay doped near the 10% level of exchange
capacity. The clay was then washed free of chloride and excess
probe molecules with distilled water. An aqueous suspension of
the doped hectorite was dried on a flat surface to produce an
oriented, self-supporting film for ESR studies. A $TEMPO^+$-doped
Na^+-vermiculite was similarly prepared for ESR studies. The
vermiculite, from Llano, Texas, had a chemical formula and
exchange capacity as reported previously (4).

The clay films were oriented in quartz tubes and ESR spectra
of the clays were recorded on a Varian V-4502 spectrometer
(X-band). The air-dry clays were equilibrated with various solvents by adding the liquids to the clay films in the quartz
tubes and sealing the tubes. The ESR spectra were then recorded
with excess solvent present.

A Norelco X-ray diffractometer was used to determine the
d(001) spacings of the hectorite films while the films were
wetted in the solvents.

Discussion of Results

The TEMPO$^+$-Hectorite System. The $TEMPO^+$ molecule has a
structure (Fig. 1) that permits varying alignments with layer
silicate surfaces depending upon the state of solvation of the
surfaces (3). The N-O bond axis is probably nearly colinear
with the C-N bond axis, since it is known that N-O and C=O are
colinear in a very similar probe molecule, 2,2,6,6-tetramethyl-
4-piperidone N-oxide (5), or TEMPONE. The axis system of the
molecule for purposes of ESR is based on the nitroxide group
(Fig. 1), and the largest hyperfine splitting value (A_{zz} is
observed when the magnetic field (H) is parallel ($||$) to the
z-axis of the molecule. Smaller splitting values (A_{xx}, A_{yy}) are
recorded by ESR when H is perpendicular ($\perp$) to the z-axis.
Splitting of the unpaired electron signal of the N-O π-orbital
into three resonances is caused by the nuclear spin of
$^{14}N (m_I = 1, 0, -1)$.

In most low-viscosity liquids at room temperature,
nitroxide spin probes rotate rapidly enough to average the
anisotropy of the hyperfine splitting (A) values. A three-line
symmetric ESR spectrum is thereby obtained for TEMPO in solution
which has a motionally-averaged hyperfine splitting value of
$A_o (A_o = 1/3 A_{xx} + 1/3 A_{yy} + 1/3 A_{zz})$. In addition, the anisotropic
g-tensor is averaged and the ESR spectrum in liquid is centered

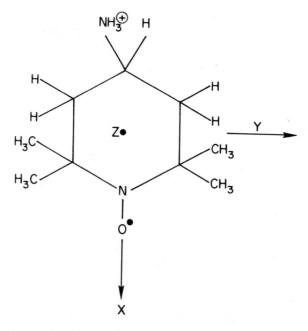

Figure 1.   *Diagram of protonated 4-amino-2,2,6,6-tetra-methylpiperidine N-oxide (TEMPO⁺) showing the molecular fixed-axis system*

at $g_0(g_0 = \frac{1}{3} g_{xx} + \frac{1}{3} g_{yy} + \frac{1}{3} g_{zz})$. This type of spectrum, demostrated for TEMPO in amyl alcohol (Fig. 2), was observed for TEMPO dissolved in all the solvents used in this study except for glycerol. Glycerol has a high enough viscosity at room temperature to prevent motional averaging of A and g (6). Calculation of rotational correlation times (τ_c) by a standard method (7), indicates that TEMPO has a correlation time in the range of 10^{-10} to 10^{-11} sec. when dissolved in the solvents (except glycerol) used in this study (Table 1). The values of A_0 for nitroxide probes dissolved in various solvents decrease as the solvent polarity decreases (6). This relationship is shown for the solvents in Fig. 3, demonstrating that A_0 increases with increasing dielectric constant. Data of other workers (8) for TEMPONE dissolved in several solvents are plotted (Fig. 3) to substantiate the relationship. From these data, it is apparent that the low-viscosity solvents of this study can be separated into three groups (a) low polarity:benzene (BZ) and carbon tetrachloride (CT) (b) medium polarity:iso-amyl (AM), n-butyl (BU), iso-propyl (PR), ethyl (ET), and methyl (ME) alcohols (c) high polarity:water (H_2O).

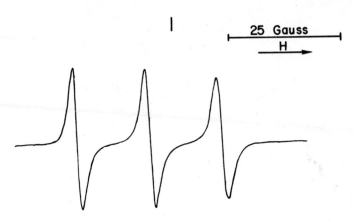

Figure 2. ESR spectrum of TEMPO in iso–amyl alcohol at 20°C ($\tau_c = 1 \times 10^{-10}$ sec, $g_o = 2.0053$, $A_o = 16.0$ gauss). The vertical line shown in this and later figures indicates the field position of the free electron (g = 2.0023).

When TEMPO$^+$-doped Na$^+$-hectorite films are solvated in the liquids listed above, the ESR spectra of the probe vary greatly (Figs. 4,5,6). These spectra can be described by the values of A||, A⊥, and ΔA (9), where A|| and A⊥ are the field separations in gauss between the low-field and middle resonances of the ESR spectrum for || and ⊥ orientations of the ab plane of the layer silicate relative to H. The difference between these two splitting parameters, denoted as ΔA = A⊥-A||, represents the degree of orientation of the probe molecule on the silicate. A value of ΔA=0 would indicate random orientations of the probe, while the maximum value of ΔA = A_{zz}-A_{xx}≈24 gauss (9) would describe strong alignment 'of the probe relative to the layer silicate. The correlation times (τ_c) of TEMPO$^+$ in these systems is calculated from the equations (7):

$$\tau_c = 0.65 \ W_o (R_+ - 2)$$

$$R_+ = (\frac{h_o}{h_{+1}})^{\frac{1}{2}} + (\frac{h_o}{h_{-1}})^{\frac{1}{2}}$$

where W_o is the line-width (gauss) of the central line of the spectrum, and the symbols h_{+1}, h_o, and h_{-1} represent the peak heights of the low, middle, and high field lines of the spectrum, respectively. This equation is intended for calculation of random-motion τ_c values in the fast-motion region (τ_c<10^{-8} sec.), and is therefore not totally accurate for the non-random motion observed in this study. However, calculation of τ_c values for the || and ⊥ orientations of the silicate relative to H allows general comparisons with τ_c values in solution. The anisotropy of motion on surfaces often results in different τ_c values with

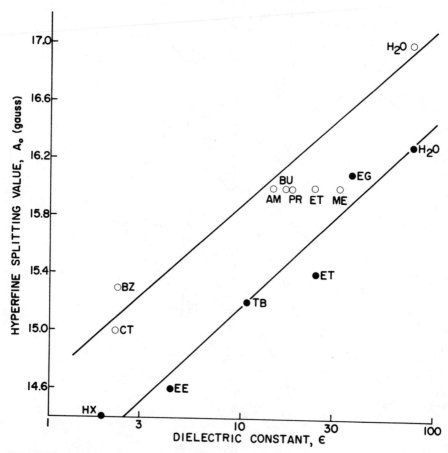

Figure 3.   Relationship of hyperfine splitting values (A$_o$) *of the spin probe in solution* (20°C) *to the dielectric constant* (ϵ) *of the solvent. Open circles* (○) *represent present work while solid circles* (●) *represent data of Snipes et al. (8) for TEMPONE. Symbols for solvents not described in the discussion are: EG (ethylene glycol), TB (t-butanol), EE (ethyl ether), HX (n-hexane).*

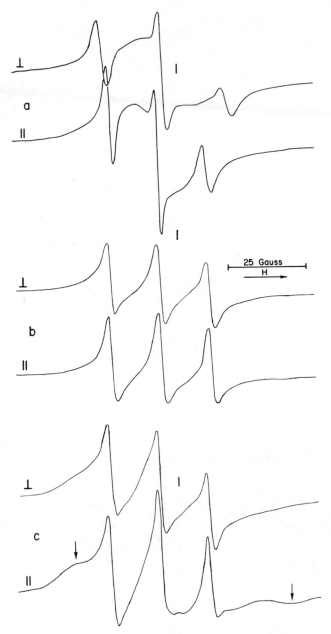

Figure 4. ESR spectra of TEMPO⁺-doped Na⁺-hectorite films
solvated in excess (a) H₂O, (b) methanol, (c) ethanol. Films are
oriented perpendicular (⊥) and parallel (||) to the magnetic field,
H, for each solvent.

different orientations of the sample relative to H. The values of τ_c for $||$ and $\perp$ orientations, as well as the values of $A||$, and $A_\perp$ are presented in Table I for TEMPO$^+$ adsorbed on solvated Na$^+$-hectorite.

The TEMPO$^+$ spectrum in excess H_2O exhibits considerable anisotropy as well as a 20-50 times increased τ_c value (Fig. 4a, Table I). There is little evidence of immobilized TEMPO$^+$ in the interlamellar regions of the hectorite, although spectra previously obtained on the same system below 100% relative humidity have shown the broad, "rigid-glass" spectrum ($\underline{3}$), probably caused by probe molecules trapped in partly collapsed interlayers. The basal spacing of the wet hectorite is near 18 Å (Table II), apparently large enough to permit considerable interlamellar tumbling of the probe. However, the anisotropy of the spectrum is evidence that the probe is tending to align with the surface as diagrammed in Fig. 7a. This orientation is considered to be a consequence of the attraction between the hydrophobic portion of the molecule and the layer silicate surface ($\underline{3}$). The surface oxygens of smectites have only limited attraction for polar molecules such as H_2O ($\underline{10}$).

The three-line, nearly isotropic ESR spectra of TEMPO$^+$ in ET- and ME-solvated Na$^+$-hectorite (Fig. 4b, 4c) are evidence of considerable motional freedom of the probe in these systems, despite the basal spacings of 17 Å (Table II). The values of ΔA for these spectra are near zero, while τ_c is estimated at about 20 times that for TEMPO$^+$ in solution (Table I). This mobility suggests a reduced probe-surface interaction as a result of the less polar nature of the solvents in comparison with H_2O. The ET and ME molecules may effectively compete with the probe for surface adsorption sites or may interact more strongly with the probe, preventing its alignment with the silicate surface. Although the spectra in ME show little evidence of immobilized TEMPO$^+$, the ET spectra demonstrate a broad signal partially obscured by the sharper three-line spectrum of solution-like TEMPO$^+$ (Fig. 4b, 4c). Since the largest splitting value of this signal occurs when the hectorite films are oriented $||$ to H (see arrows, Fig. 4c), the immobilized probe must be oriented with the N-O bond axis $\perp$ to the $\underline{ab}$ plane of the layer silicate. This alignment is diagrammed in $\overline{\text{Fig}}$. 7b, suggesting that the probe molecule "bridges" across the $\sim$7.5 Å wide distance between layer silicate plates. Such an orientation would allow the charged end of the molecule to approach the surface charge sites while permitting the hydrophobic methyl groups of the probe to interact with ET molecules in the interlayers. The broad spectrum of Fig. 4c represents probes with reduced molecular motion ($\tau_c >> 10^{-8}$ sec.), a probable result of motional hindrance and surface interactions in the interlayer which has a width similar to the diameter of the probe ($\sim$7Å).

The higher molecular weight alcohols with lower dielectric constants than ET and ME, when used to solvate TEMPO$^+$-doped

Na$^+$-hectorite, demonstrate spectra with higher proportions of immobilized probe relative to the more freely rotating probe (Fig. 5) in comparison to the spectra of the probe in ET- and ME-solvated hectorite. The spectra of the PR-solvated system have a relatively narrow, nearly isotropic, three-line signal (Fig. 5a) produced by near random tumbling of adsorbed TEMPO$^+$ at a rate about six times slower than in solution (Table I). The weak, broad spectrum of immobilized TEMPO$^+$ (Fig. 5a) is orientation-dependent, with the largest value of A when the hectorite film is $\perp$ to H (see arrows). This dependence of A is indicative of immobilized TEMPO$^+$ aligned with the N-O axis $\parallel$ to the ab plane of the silicate, as shown in Fig. 7c. The basal spacing of Na$^+$-hectorite solvated in PR (Table II) is only large enough to permit the TEMPO$^+$ molecules in interlamellar regions to lie "flat", thereby explaining the immobilized probes. The more mobile TEMPO$^+$ must be considered to occupy external surfaces of the solvated hectorite or interlamellar regions expanded to spacings greater than the dimensions of the probe molecule.

The spectra of BU-solvated hectorite (Fig. 5b) are again composed of signals from two distinct probe environments--the more mobile probe undergoing near-random tumbling at a rate reduced nearly six times relative to solution, and the immobilized probe (denoted by arrows) aligned in interlamellar regions. The basal spacing of the clay (Table II) prevents the probe from rotating in the interlamellar sites, and the relative intensity of the signal due to immobile probe appears greater for BU than PR (Fig. 5b). In comparison, the spectra of AM-solvated hectorite (Fig. 5c) demonstrate essentially no mobile TEMPO$^+$, while the strongly oriented, immobilized TEMPO$^+$ is dominant. Again, the interlamellar spacing of the AM-solvated hectorite prevents free rotation of TEMPO$^+$ and produces strong alignment (Fig. 7c).

The ESR spectra of TEMPO$^+$ in Na$^+$-hectorite solvated with the least polar solvents, BZ and CT, are produced by well-oriented, immobilized probes (Fig. 6a, 6b, Table I). The basal spacings of the layer silicate (Table II) indicate the limited ability of these solvents to expand the interlamellar regions, thereby allowing essentially all of the probe molecules to be trapped and strongly aligned (Fig. 7c).

It should be pointed out that the addition of the probe to the Na$^+$-hectorite at the $\sim$10% level of exchange somewhat alters the properties of the system. Table II demonstrates that the probe tends to increase the basal spacing of the hectorite in solvents which normally produce <15Å spacings for Na$^+$-hectorite. Conversely, the addition of probe decreases the basal spacings in solvents which produce >15 Å spacings for pure Na$^+$-hectorite. As previously shown (3), TEMPO$^+$ holds the hectorite plates apart, preventing collapse of interlamellar regions. However, the hydrophobic nature of the probe tends to reduce the swelling of

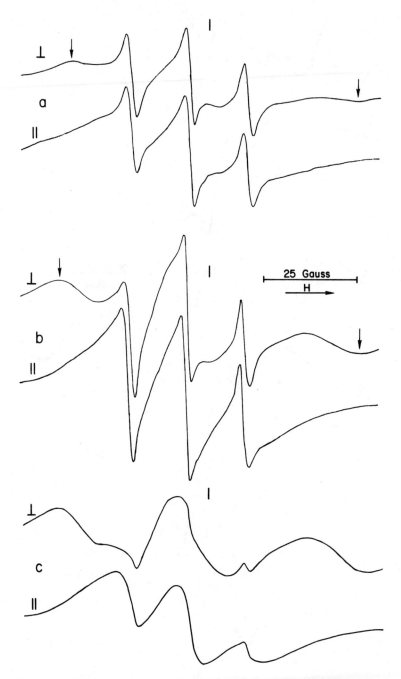

*Figure 5.   ESR spectra of TEMPO⁺-doped Na⁺-hectorite films solvated in
excess (a) isopropyl alcohol, (b) n-butanol, (c) iso-amyl alcohol*

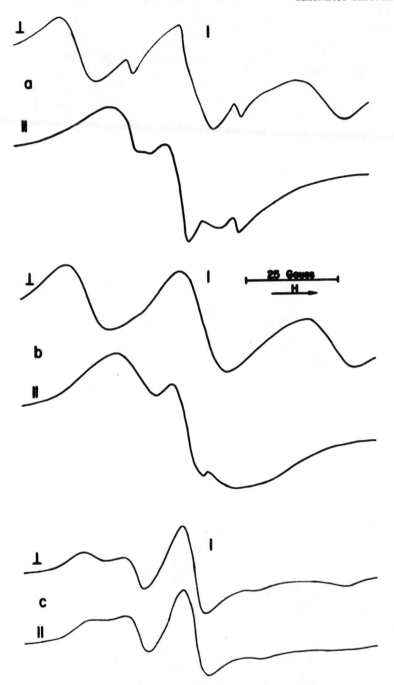

Figure 6.   ESR spectra of TEMPO⁺-doped Na⁺-hectorite films solvated in excess (a) benzene, (b) carbon tetrachloride, (c) glycerol

the hectorite in water and other polar solvents (3). This fact, coupled with the possible segregation of Na^+ and $TEMPO^+$ ions in the interlayers similar to alkylammonium-metal ion segregation (11,12) could produce $TEMPO^+$-enriched, non-expanded interlayers that might account for some probe immobilization.

In summary, the spectra produced in the various solvents show some consistent trends. The lower the polarity of the solvent, the less the interlayers of hectorite are able to expand, and the greater the proportion of $TEMPO^+$ that is immobilized. Thus, hectorite fully solvated in H_2O or ME contains little or no immobile probe, while ET-, PR-, BU-, and AM-solvated hectorites contain increasing proportions of the immobile probe. The solvents of very low polarity (BZ,CT) allow $TEMPO^+$ molecules to be almost completely restricted in interlayers. The results support the concept that solvent adsorption and interlamellar expansion is largely a result of cation-dipole interactions (13), since polar molecules are capable of solvating the interlamellar Na^+ ions, while non-polar molecules have little attraction for metallic cations and probably adsorb on the silicate surface. It is likely that the Na^+ ions in the hectorites solvated with BZ or CT are non-solvated and localized in hexagonal holes of the silicate surface (14).

Although the apparent basal spacings for ET- and ME-solvated hectorites are essentially the same, as are the spacings of PR-, BU- and AM-solvated systems (Table II), the relative proportions of immobilized probe vary (Figs. 4,5). This suggests that collapse of silicate interlayers is more complete in solvents of lower dielectric constant if it is assumed that the more mobile probes are adsorbed on external surfaces or in partially expanded interlayers that permit molecular rotation. For example, the lack of mobile $TEMPO^+$ in the AM, BZ, and CT systems suggests that virtually all layer silicate platelets collapse together to a 14-15 Å spacing, leaving very few $TEMPO^+$ molecules on external surfaces.

In general, the degree of restriction of the more mobile fraction of adsorbed $TEMPO^+$ varies with the polarity of the solvent. The adsorbed probe in H_2O-solvated Na^+-hectorite has a rotational correlation time about 50 times that of the probe in aqueous solution (Table I). In ET- and ME-solvated systems the adsorbed probe is considerably more mobile, with a 20-fold increase in τ_c relative to free solution. In the PR- and BU-solvated systems, τ_c for the mobile probe is only increased about six times. The data suggest that interaction of $TEMPO^+$ with the silicate surface is controlled by the polarity of the solvent. Highly polar solvents permit strong probe-surface interactions because of the lack of attraction between polar molecules (eg., H_2O) and the silicate oxygens as well as the weak interaction of polar molecules with the hydrophobic groups of $TEMPO^+$. This strong surface attraction of $TEMPO^+$ in aqueous systems might be compared to the strong adsorption of alkyl-

ammonium ions on smectites to form non-expanding, hydrophobic
clay systems (15).

Although the glycerol-solvated hectorite produces a
"rigid-glass" type of spectrum as expected for solvents of very
high viscosity (Fig. 6c), there is some alignment of TEMPO[+] in
the $\sim$8 Å wide interlayer as inferred from the orientation-
dependent spectrum. The alignment of the N-O axis is imperfect,
but appears to be along the ab plane of the silicate. Therefore,
the probe molecule is tending to lie "flat" as it did in the
hydrated systems. The spectra (Fig. 6c) are virtually identical
to those previously reported for Mg^{2+}-hectorite equilibrated at
100% relative humidity and cooled to 0°C (3). The relatively
high dielectric constant of glycerol (ε=42.5) probably accounts
for the similarity in orientation of TEMPO[+] in the glycerol and
H_2O systems. However, the interlamellar mobility of TEMPO[+] is
much lower in the glycerol system because of the high viscosity
of glycerol.

   The TEMPO[+]-Vermiculite System. As a basis for comparison
with the hectorite, TEMPO[+]-doped Na^+-vermiculite was oriented
in films and equilibrated at 100% relative humidity. The spectra
of the probe (Fig. 8) are composed mainly of the isotropic,
solution-like signal with $A|\sim A||\sim$16.5 gauss. Some evidence of a
broad background signal (immobile probe) seems to be present.
Apparently, because the Na^+-vermiculite does not expand beyond a
basal spacing of $\sim$14.7 Å in water, essentially all of the surface
area exposed to the probe molecule can be considered to be
external. A part of the TEMPO[+] may actually penetrate the narrow
interlayers with some difficulty, producing the broad background
signal. However, the external surfaces of the vermiculite behave
much differently from the "external" surfaces of H_2O-expanded
hectorite, permitting very rapid, solution-like, random motion
of TEMPO[+]. The vermiculite surfaces appear to interact weakly
with the probe, despite the electrostatic attraction between
TEMPO[+] and the surface. It is likely that the relatively strong
hydration of oxygen atoms on the vermiculite surface (10)
prevents direct probe-surface contact and preferred alignment of
the probe.

   Very similar spectra to that obtained for the hydrated
Na^+-vermiculite are produced for ET-, BU-, and BZ-solvated
vermiculite with $A|\sim A||\sim$15.3, 15.0, and 14.5 gauss, respectively,
for the surface-adsorbed TEMPO[+]. The polarity of the solvent
influences the splitting value as expected from Fig. 3, but has
little influence on the strength of the probe-surface interaction.
It is concluded that the vermiculite surface has less attraction
for TEMPO[+] in polar solvents because of the affinity of polar
molecules for the surface. In non-polar solvents, the probe is
attracted to the solution phase and does not orient on the
vermiculite surface. Similar solution-like isotropic spectra of
TEMPO[+] have been obtained on finely-ground micas equilibrated at

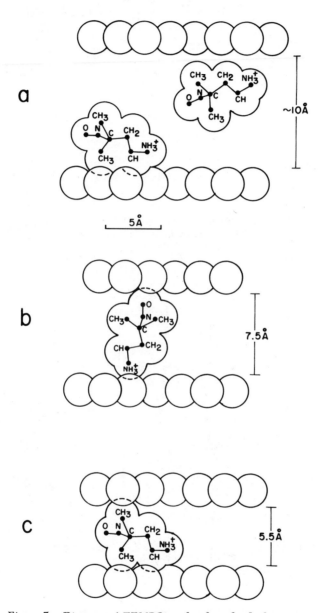

*Figure 7. Diagram of TEMPO⁺ molecules adsorbed in inter-
lamellar regions between the surface oxygen atoms of hectorite plates. The situations depicted are: (a) partially oriented
TEMPO⁺ in expanded interlayers of hydrated hectorite; (b)
"immobilized" TEMPO⁺ in interlayers of ethanol- and metha-
nol-solvated hectorite. (c) "immobilized" TEMPO⁺ in interlayers of hectorite solvated in liquids of low polarity.*

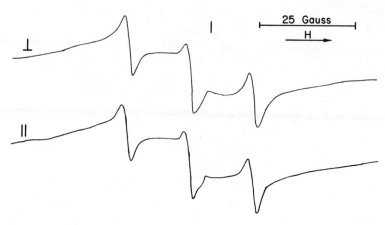

*Figure 8. ESR spectra of TEMPO⁺-doped Na⁺-vermiculite films equili-
brated over H_2O (100% relative humidity) and oriented ⊥ and ∥ to H*

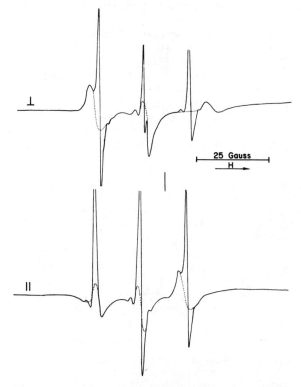

*Figure 9. ESR spectra of Na⁺-hectorite films equili-
brated with a dilute (∼ 10⁻⁴M) solution of TEMPONE
and oriented ⊥ and ∥ to H. The dotted line outlines the
spectrum of adsorbed TEMPONE.*

100% relative humidity. The position of negative charges of vermiculite and mica near the surface of the silicate, as well as the greater charge density and reduced swelling of these minerals compared to smectites, accounts for their similar behavior.

The TEMPONE-Hectorite System. Although studies of the cationic probe, TEMPO+, have been shown to provide useful information on the mechanisms of adsorption, use of a neutral probe molecule permits the attractive forces between the molecule and surface to be evaluated in the absence of electrostatic interactions. For this reason, 2,2,6,6,-tetramethyl-4-piperidone N-oxide (TEMPONE) was dissolved in various solvents and equilibrated with Na+-hectorite.

The ESR spectra of aqueous TEMPONE ($\sim 10^{-4}$M) added to oriented Na+-hectorite films (Fig. 9) consist of an isotropic three-line spectrum due to TEMPONE in free solution as well as a superimposed spectrum of adsorbed TEMPONE (dotted line) with an orientation dependence similar to the spectrum of adsorbed TEMPO+ on hydrated hectorite (Fig. 4a). Evidently, the lack of charge on the probe molecule does not influence the surface-molecule interaction when adsorption occurs, but does allow many of the probes to remain in solution. The TEMPONE dissolved in ET($\sim 10^{-4}$M) appears to be adsorbed by Na+-hectorite, since the spectrum demonstrates a nearly isotropic signal ($A_\perp$=14.8 gauss, $A_{||}$=15.3 gauss) with values of τ_c similar to those for ET-solvated TEMPO+-doped hectorite (0.3-0.5 x 10^{-9} sec.). The probe apparently occupies interlayers, having about a 20 times reduction in mobility compared to solution. In addition, a small fraction of the probe is immobilized with the N-O axis oriented $\perp$ to the ab plane of the silicate, as indicated by the weak broad signal (Fig. 10a) similar to the immobilized TEMPO+ present in ET-solvated hectorite. The adsorption of TEMPO+ and TEMPONE on the ET-solvated system have similar characteristics, again suggesting that the similar chemical nature of the two probes determines the mechanisms of surface interaction, while the charge of TEMPO+ simply insures that it is not free to enter the solution phase. The spectrum of TEMPONE dissolved in BZ ($\sim 10^{-4}$M) and added to dry Na+-hectorite (Fig. 10b) further demonstrates this principle. The spectrum consists of a broad, orientation-dependent signal due to TEMPONE aligned in inter-lamellar regions that are partially collapsed in BZ. This signal is very similar to that of TEMPO+ in BZ-solvated hectorite (Fig. 6a). However, an isotropic three-line spectrum of solution-like TEMPONE is superimposed on the broad signal, probably representing weakly adsorbed or solution TEMPONE in equilibrium with the immobile interlamellar probe.

Conclusions

The use of nitroxide spin probes for adsorption studies on

surfaces is a relatively new technique that promises to provide
fundamental information on molecular adsorption processes. This
study has shown that the surface-adsorbent interaction is modified by the solvent present and the nature of the surface. The
adsorbed molecule has two features which distinguish it from the
solution probe--a reduced rotational correlation time and
non-random tumbling at the liquid-surface interface. The reduced
correlation time has implications for the reactivity of adsorbed
molecules and the ability of these molecules to diffuse.
Non-random tumbling is a result of direct molecule-surface
contact. Such contact occurs on hydrated smectite surfaces
because of the weak attraction of these surfaces for water molecules. However, on the more highly charged vermiculite surfaces,
the probe molecules do not adsorb directly because of the
solvation of surface oxygens by water. Future adsorption
studies on various types of surfaces using spin probes with
different chemical characteristics are likely to produce further
understanding of the processes of physical adsorption.

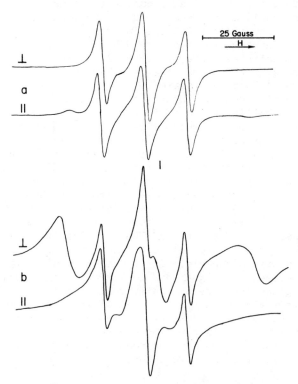

Figure 10. ESR spectra of Na⁺-hectorite films equili-
brated with dilute (~ 10⁻⁴M solutions of TEMPONE in
(a) ethanol, (b) benzene and oriented ⊥ and ‖ to H

TABLE I: Rotational Correlation Times $(\times 10^9$ sec.) and Hyperfine Splitting Values (gauss) for Adsorbed and Solution $TEMPO^+$.

Solvent*	Rotational Correlation Time[†]			Hyperfine Splitting Value[‡]		
	Solution	Adsorbed		Sol'n (A_0)	Adsorbed	
		$\tau_c(\perp)$	$\tau_c(\|\|)$		$A\perp$	$A\|\|$
H_2O	0.053	2.63	1.48	17.0	19.2	15.5
ME	0.012	0.25	0.24	16.0	15.7	15.8
ET	0.023	0.44	0.55	16.0	15.5	15.6
PR	0.061	0.38	0.55	16.0	15.3	15.5
BU	0.087	0.60	0.47	16.0	15.2	15.4
AM	0.103	immobile		16.0	33.6	15.9
BZ	0.036	immobile		15.3	31.2	13.9
CT	0.011	immobile		15.0	31.2	12.8

*Solvents are listed in order of decreasing dielectric constant.

[†]Values of τ_c were calculated by the method described by Sachs and Latorre([7]). τ_c of the mobile adsorbed probe was determined for both $\perp$ and $\|\|$ orientations of the silicate.

[‡]Splitting values were determined for the predominant form of adsorbed $TEMPO^+$.

TABLE II: Basal (d_{001}) Spacings of Na^+-hectorite in Solvents with Different Dielectric Constants (ϵ).

		Basal Spacing ($\mathring{A}$)	
Solvent	ϵ	Na^+-hectorite	$TEMPO^+$-doped hectorite
H_2O	78.5	>>25	~18*
ME	32.6	17.0	17.0
ET	24.3	17.0	16.9
PR	18.3	14.2	14.7
BU	17.1	13.6*	14.6
AM	14.7	14.5*	14.8
BZ	2.3	~14*	14.2
CT	2.2	~13.0*	13.8

*Broad diffraction peaks indicating randomly interstratified spacings of hectorite interlayers.

Literature Cited
1. Keith, A. D. and Snipes, W., Science (1974), 183, 666.
2. Griffith, O. H., Libertini, L. J., and Birrel, G. B., J. Phys. Chem. (1971), 75, 3417.
3. McBride, Murray B., J. Phys. Chem. (1976), 80, 196.
4. McBride, M. B., Pinnavaia, T. J. and Mortland, M. M., American Mineralogist (1975), 60, 66.
5. Hwang, J. S., Mason, R. P., Hwang, L. and Freed, J. H., J. Phys. Chem. (1975), 79, 489.
6. Chignell, C. F., Aldrichimica Acta (1974), 7, 1.
7. Sachs, F. and Latorre, R., Biophys. J. (1974), 14, 316.
8. Snipes, W., Cupp, J., Cohn, G., and Keith, A., Biophys. J. (1974), 14, 20.
9. Jost, P., Libertini, L. J., Hebert, V. C. and Griffith, O. H., J. Mol. Biol. (1971), 59, 77.
10. Farmer, V. C. and Russell, J. D., Trans. Faraday Soc. (1971), No. 585, 67, Part 9, 2737.
11. McBride, M. B. and Mortland, M. M., Clay Minerals (1975), 10, 357.
12. Barrer, R. M. and Brummer, K., Trans. Faraday Soc. (1963), 59, 959.
13. Theng, B. K. G., "The Chemistry of Clay-Organic Reactions", pp. 32-135, Wiley and Sons, New York, 1974.
14. Berkheiser, V. and Mortland, M. M., Clays and Clay Minerals (1975), 23, 404.
15. McBride, M. B. and Mortland, M. M., Clays and Clay Minerals (1973), 21, 323.

ESR Studies of Radicals Adsorbed on Zeolite

J. SOHMA and M. SHIOTANI

Faculty of Engineering, Hokkaido University, Sapporo 060, Japan

Introduction

It has been established that ESR studies on radicals adsorbed on a zeolite provide important information on interactions between the adsorbed radicals and the adsorbent.(1-6) Interesting finding of an extra-coupling was reported on the ethyl radical adsorbed on the zeolite.(7) Furthermore, observed changes in anisotropy of hyperfine couplings and g factors may be helpful to elucidate motion of the radicals trapped on the zeolite. Thus it seems interesting to study the details of ESR spectra from simple radicals, such as methyl or amino, trapped on zeolite and to discuss natures of the trapping sites of the zeolite in relation to behaviors of the trapped radicals.

Experimental

The zeolites used in the experiments were Linde molecular sieves, 4-A supplied by the Union-Carbide Corp in U.S.A. and Nishio Industry in Japan. The zeolites were heat-treated under vacuum (ca. 10^{-4} Torr) for five hours usually at different temperatures from room temperature to 550°C after initial pre-heat-treatment in air. Gases were introduced through either a break-seal or a cock to the zeolite contained in a spectrosil ESR sample tube at -196°C and allowed to adsorb on the zeolite up to room temperature. The samples used were: normal methane, 99.7% purity, obtained from Gaschro Industry; deuterated methane, CD_4, 99 atomic %, from Stobler Isotope Chemicals, ethane and normal ammonia, 99.5% purity, from Takachiho Chemical Co. and isotopic ammonias, $^{15}NH_3$ and ND_4, 97% purity, from Merk Sharp Dohome. The ammount of gas adsorbed was controlled and checked by measuring the pressure drop in a known volume. Radicals were produced at -196°C by γ-irradiation from ^{60}Co source to the gas adsorbed zeolites. Total dose per sample was 0.2-0.5 Mrad. ESR spectra were recorded in the temperature range from -196°C to +50°C by JEOL JES-PE-1 spectrometer operating at X-band with 100 KHz

142

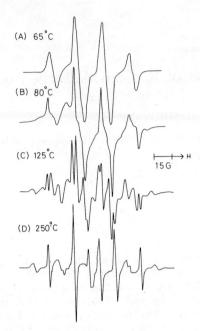

(A) 65°C

(B) 80°C

(C) 125°C

15G ⟶ H

(D) 250°C

Figure 1.

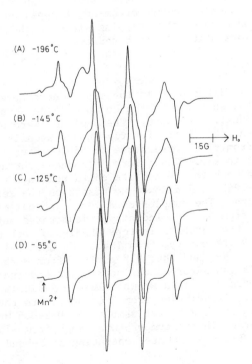

(A) −196°C

(B) −145°C

15G ⟶ H₀

(C) −125°C

(D) −55°C

Mn²⁺

Figure 2.

modulation. Temperature of a sample in the ESR cavity was controlled by an attached variable temperature unit.

Types of Methyl Radicals Adsorbed on the Heat-Treated Zeolites

Methane on Very Weakly Heat-Treated Zeolite (Heat-Treatment Below 50°C).

The ESR spectrum observed at -196°C after γ-irradiation
consists of a broad central band and a doublet with the separation
of 508G from a hydrogen atom. This broad spectrum is completely
different from any spectrum of the methyl radicals and identical
to that observed from a γ-irradiated zeolite adsorbing no methane.
Apparently no methyl radical is stabilized at -196°C in the
zeolite heat-treated below 50°C.

Methane Adsorbed on Mildly Heat-Treated Zeolite. (Heat-Treatment at ca. 80°C).

The ESR spectrum observed at -196°C in this case is reproduced as (B) in Fig. 1. It appears to be nearly a quartet with
several satellites. Apparently the spectrum is different from
that of the normal methyl, but the separation of the main quartet,
22.3G, is close to the normal one. It was found that the spectrum
changed its shape with rising temperatures and the spectrum
observed at -55°C was identical with normal one, as shown in Fig.
2. The temperature variation of the line shape was completely
reversible except for the small decrease in the total intensity.
The spectrum shown in Fig. 3 (A), which was observed at -196°C
after warming to -55°C, may be analysed as a quartet modified with
the anisotropic coupling originating from an additional proton.
The spectrum derived from this assumption is represented by the
stick diagram in Fig. 3 (B). In this spectrum parameters used are
following: 18.9G and 8.4G for $A_{\parallel}$ and $A_{\perp}$ of the extra proton
coupling and 22.3G for the methyl protons, respectively. The
correspondence of the main peaks of the observed spectrum with the
stick diagram seems satisfactory. These experimental results
indicate strongly that the radical species responsible for the
spectrum is the methyl radical having coupling with an additional
proton, namely $H_3C\cdot \cdots H^+$, in which the unpaired electron interacts with the extra proton mainly by the dipolar coupling. This
species is called an abnormal methyl radical (I).

Methane Adsorbed on Strongly Heat Treated Zeolite (Heat-Treatment at ca. 250°C).

The ESR spectrum observed from irradiated methane adsorbed on

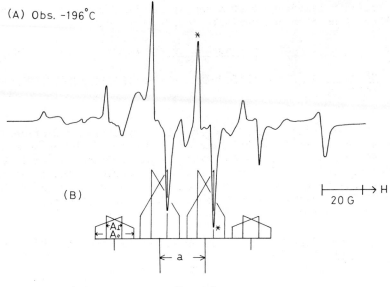

Figure 3.

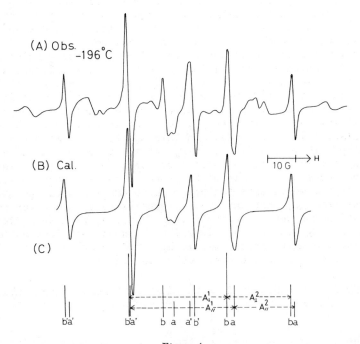

Figure 4.

strongly heat-treated zeolite is shown as (D) in Fig. 1.   The
pattern is different from either one of the normal methyl radical
and the abnormal methyl radical (I).   Although there are several
diffuse peaks in the spectrum, it is principally a double triplet.
This simulated pattern, which is derived from the assumption that
the methyl radical has an anisotropic g factor with two equivalent
protons and one non-equivalent one, is shown in as (C) in Fig. 4
together with the observed spectrum (A).   The used parameters are
$g_{\parallel}$ = 2.0023, $g_{\perp}$ = 2.0032, $A_{\parallel}^2$ = 21.8, $A_{\perp}^2$ = 23.2 for equivalent two
protons, $A_{\parallel}^1$ = 37.8G, $A_{\perp}^1$ = 35.6G for the other proton and ΔH_{msl}
(Lorentzian) = 1.3G.   The agreement between the observed double
triplet and the simulated one (B) is satisfactory, if one neglects
the diffuse peaks.   The methyl radical giving this spectrum is
called an abnormal methyl radical (II).   This radical is unstable
on warming and begin to decay above -160°C near boiling point of
CH_4 (-161.7°C).

The ESR spectra observed for the methyl radical trapped in
the zeolites heat-treated in the temperature range between 80°C
and 250°C are mixture of the two spectra of the abnormal methyl
radicals (I) and (II).   An example of this spectrum is shown as
(C) in Fig. 1.   No ESR spectrum was observed for γ-irradiated
methane, which had been trapped in the zeolites heat-treated at
temperature above 500°C.   The detailed identifications of these
spectra was published in the other paper. (9)

Ethyl Radical Adsorbed on Zeolite (7)

The zeolite used in this experiments was heat-treated for
three hours at 150°C.   ESR spectrum observed at -125°C from γ-
irradiated zeolites adsorbing ethane showed clear multiplets as
shown as A in Fig. 5.   The spectrum is definitely different from
that of the ethyl radical either in liquid phase, (8)or in
adsorbed state on silica gel. (10)   The main features of the
spectrum agree with that obtained by assuming that each line of
the spectrum (8) (Fig. 5-C) of ethyl radical is split into a
doublet, as shown in Fig. 5-B.   A spectrum simulated from the
above-mentioned assumption is shown as "D" in the same figure.   In
this simulation the couplings of 28.7G and 21.4G were taken for
the three β protons and the two α protons, respectively, in the
ethyl radical, and 8.0G for the separation of the extra doublet
and 7.0G for the line-width.   The above assumed values for the
couplings of both β and α protons in the ethyl radical are nearly
equal to 26.9G and 22.4G, respectively, which are the reported
values (8) for the ethyl radical.   From the close similarity
between the observed spectrum and the simulated one it is concluded that the observed spectrum is the spectrum of the abnormal
type of ethyl radical having the coupling with an additional
proton, namely $H_3C-CH_2 \cdot \cdots H^+$.   The spectrum was observed from
the ethane adsorbed on the zeolites, which had been heat-treated
at higher temperature, such as 550°C, although no spectrum from

the γ-irradiated methane was adsorbed on such a very strongly
heat-treated zeolite. The spectrum from the trapped ethyl radical
obtained under this condition was interpreted similarly as that
from the abnormal methyl radical (I) mentioned.

Types of the Adsorbed Amino Radicals

Ammonia Adsorbed on Mildly Heat-Treated Zeolite.

a) High coverage with ammonia (ca. 1.7×10^{-2} mol/gr.):
Nearly 1.7×10^{-2} mol. of ammonia was adsorbed per one gram of the
zeolite. In such a higher coverage each cavity in zeolite is
almost full with ammonia molecules. An example of the ESR spectrum observed at -196°C from the irradiated ammonia highly
adsorbed on the mildly heat-treated zeolite is shown as A in Fig.
6, which is a broad quintet. This is a typical pattern observed
for the radical possessing anisotropic hyperfine and g tensors
which is trapped in either amorphous or polycrystalline matrix.
This spectrum is quite close to that (11) observed from the irradiated aqueous solution of ammonia in glassy state. From the
close similarity between the observed spectra one may taken $NH_2 \cdot$
radical surrounded with water molecules as the radical responsible to the spectrum. It was found that this radical was unstable
and almost decayed out at -80°C. This unstable, water-surrounded
$NH_2 \cdot$ radical is called $NH_2 \cdot$ (I). The well-resolved spectrum,
shown as B in Fig. 6, was observed at -75°C after the decay of
the radical $NH_2 \cdot$ (I) although the spectral intensity decreased to
nearly one tenth of the initial one. This spectrum is almost
isotropic and apparently a triple-triple (Fig. 6-C), in which the
coupling constant a_{iso}^{H} of the two protons is 23.5G and that of
the nitrogen a_{iso}^{14N} is 11.7G. This identification was reconfirmed
by using the deuterated ammonia ND_3. The spectra observed at
-30°C is reproduced in Fig. 7. As shown in same figure the
observed pattern is explained by a superposition of the two
spectra from $ND_2 \cdot$ and $\cdot NDH$. This stable radical survived above
-80°C is called amino radical (II).
b) Low coverage of ammonia (ca. 8.5×10^{-4} mol/gr.): Approximately 8.5×10^{-4} mol of ammonia was adsorbed on one gram of the
zeolite. In such a low coverage only one ammonia molecule or
less exists, on average, in a cavity in the zeolite. The ESR
spectrum observed at -196°C under these conditioned looked
similar to A in Fig. 6. After decaying the unstable NH_2 radical
(I) by annealing the sample at room temperature, the line-shape
observed at -196°C was a little but clearly changed to that shown
as A in Fig. 8. The radical, which gives the spectrum and is
survived after annealing, is so stable that one could follow the
temperature variation of the spectrum up to 50°C, as shown in
Fig. 8. One can identify the radical as $NH_2 \cdot$ by comparing the

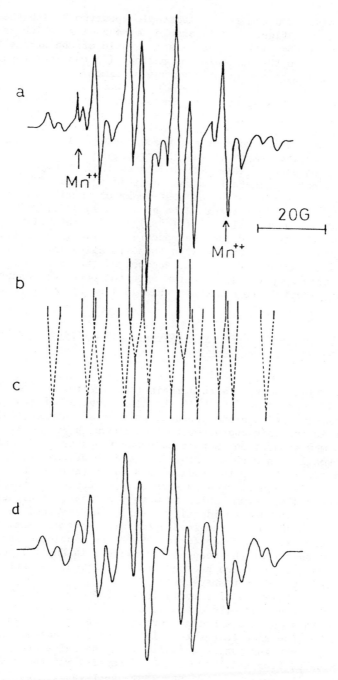

Figure 5.

well resolved and completely isotropic spectrum D with the stick
diagram in the figure. This stable amino radical which is identical to the $NH_2 \cdot$ (II). The relative ratio of the stable NH_2
radical (II) to the unstable NH_2 radical (I) was increased with
decreasing in the amount of coverage of ammonia.

The spectrum observed at -65°C is shown as A in Fig. 9. The
major part of this spectrum is almost same to the spectrum C in
Fig. 8 but there is apparently other component consisting of the
thirteen lines with a separation of 14.6G, which were not clearly
apparent at the low gain of the spectrometer. It was found that
the same 13-line component was observed even when the isotopic
$^{15}NH_3$ was adsorbed. This fact indicates that this spectral component does not originate from the adsorbed ammonia. The observed
relative intensities of outer most peaks of the 13-line component
were experimentally determined to be 1:3:8:15:?:?:?:?:?:20:10:4:1.
If one assume a radical Na_4^{3+} ($I(^{23}Na)$ = 3/2), the relative intensities of the expected 13-line spectrum is 1:4:10:20:31:40:44:40:
31:20:10:4:1. Agreement between these relative intensities is
rather satisfactory if one take account of the fact that the main
part of this spectrum is obscured with the other component. Thus,
it seems plausible to attribute this 13-line spectrum to the
radical Na_4^{3+}.

Ammonia Adsorbed on Very Strongly Heat-Treated Zeolite (Heat-Treatment at ca. 350°C).

a) High coverage with ammonia (ca. 1.7×10^{-2} mol/gr.): ESR
spectrum observed at -196°C in the high coverage of ammonia on
this heat-treated zeolite appeared as a broad quintet, which is
similar in the main character to that shown in Fig. 6-A. However,
the maximum separation in this case was found as 121G, which is a
little smaller than that, 126G, in the case of the mildly heat-
treated zeolite. And the radical was so unstable as to decay out
at lower temperature, such as -150°C. One could not observe a
well-resolved spectrum, which is merely obtained at higher temperatures. It was hard to identify clearly the responsible radical
on the basis of such unresolved spectrum. However, similarity
between the patterns observed at -196°C, as mentioned above, leads
one to presume that the responsible radical is $NH_2 \cdot$(I), characteristic behaviors of which are the smaller maximum separation and
rapid decay at lower temperatures.

b) Low coverage of ammonia (ca. 8.5×10^{-4} mol/gr.): The ESR
spectrum observed at -196°C from the irradiated zeolite covered
thinly with ammonia was again a broad quintet similar to those
observed at the same temperature in the various cases. Resolution
of the spectrum increased gradually with the raised temperatures
up to -100°C, from which the signal intensity was found to
decrease. The well-resolved spectrum observed at -120°C showed
additional four line structure on each hyperfine line of the amino

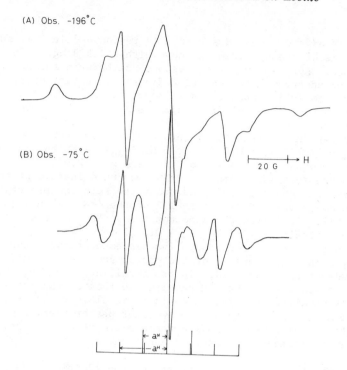

(A) Obs. -196°C

(B) Obs. -75°C

$\vdash$ 20 G $\longmapsto$ H

Figure 6.

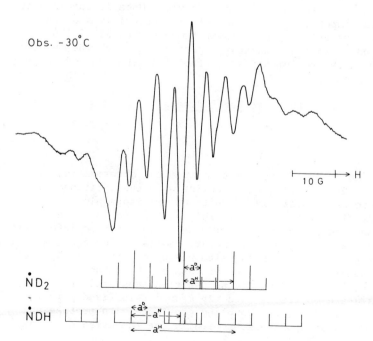

Obs. -30°C

$\vdash$ 10 G $\longrightarrow$ H

$\overset{\bullet}{N}D_2$

$\overset{\bullet}{N}DH$

Figure 7.

radical, $NH_2 \cdot$.

An example of the observed spectrum is reproduced in Fig. 10
and the analysis of the spectrum is shown with stick diagram in
the same figure. Apparently the correspondence between the peaks
of the observed spectrum with the stick diagram is satisfactory.
Thus, the spectrum is interpreted as being due to a $^{14}NH_2 \cdot$ radical
interacting with a neucleus of 3/2, which is most likely to be the
sodium ^{23}Na in the zeolite A, namely $NH_2 \cdot \cdots Na^+$. The ESR parameters, which were experimentally determined from the correspon-
dence, are $A_{//}$ = 37.7G, $A_{\perp}$ = 0±0.5G for ^{14}N, A_{iso}= 24.8G for the
two equivalent protons a_{iso}= 3.2G for ^{23}Na and $g_{//}$ = 2.0041, $g_{\perp}$ =
2.0048. When the isotropic ammonia, $^{15}NH_3$ was used instead of the
normal ammonia, the spectrum observed under the same experimental
conditions became simpler due to spin 1/2 of the ^{15}N nucleus, as
shown in Fig. 11. Analysing the spectrum from $^{15}NH_2 \cdot$ in the
similar way one obtain $A_{//}$ = 52.1G, $A_{\perp}$ = 0±0.5G for ^{15}N, a_{iso}=25.1G
for the protons, a_{iso}=3.2G for ^{23}Na. The agreement between the
corresponding coupling constants of the proton and the sodium in
these two kinds of the amino radicals are excellent. The ratio of
the coupling constant with the ^{15}N neucleus to that of the ^{14}N was
experimentally obtained as 1.44, which is quite close to the ratio
(1.43) of the magnetic moments of the two isotopes of thenitrogen.
These quantitative agreements confirm the assignment of the spectrum to the amino radicals interacting additionally with the ^{23}Na
nucleus.

It is worthy to note that appearance of the additional quartet by the heat-treatment was irreversible. Once it appeared at
the elevated temperatures and the main characteristics of the
spectra were preserved even at lower temperatures like -196°C.
This fact indicates that the trapping site, at which the super-
hyperfine with the sodium appears, stabilizes the trapped amino
radical.

Molecular Motions of the Adsorbed Radicals

Methane.

In the analysis of the spectrum from the abnormal methyl
radical (I), $CH_3 \cdot \cdots H^+$, the coupling with the additional proton
is assumed to be anisotropic in spite of the isotropic coupling
with the methyl protons and ^{13}C. This difference in the anisotropy on these couplings may provide us an interesting information
on the molecular motion of this methyl radical. Suppose a methyl
radical trapped on an adsorbing site having a proton which gives
additional coupling. The center of mass of the methyl radical is
assumed to be fixed at 77°K at the position separated γ from H^+.
(see Fig. 12) and to rotate randomly around this fixed point.
From these assumptions, the coupling constants with protons and
^{13}C are averaged to appear isotropically but the extra coupling

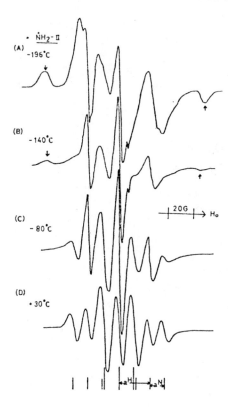

Figure 8.

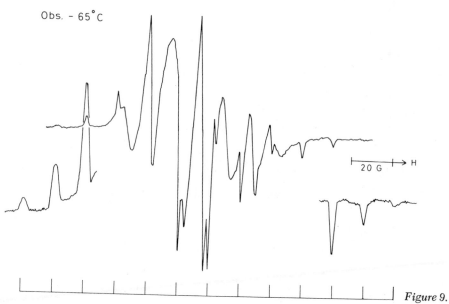

Obs. − 65°C

Figure 9.

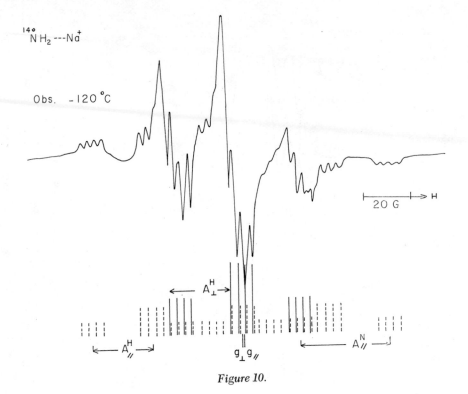

Figure 10.

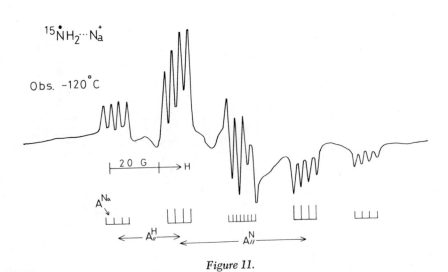

Figure 11.

with the additional proton shows anisotropic coupling because the angle between the static field Ħ and ř is not time dependent. Thus, the experimental results indicate that the methyl radical randomly rotates around the fixed point which is sufficiently close to produce an extra coupling with the proton in the zeolite matrix.

It was also found that the spectrum attributed to the abnormal methyl ($CH_3 \cdot \cdots H^+$) changed its line-shape from the complicated shape to the simple quartet upon warming, which is identical to the normal methyl spectrum, as shown in Fig. 2. This change in the line-shape was completely reversible. Based on the model mentioned above the disappearance of the additional coupling with the extra proton means an averaging of the anisotropic hf coupling with the extra proton. Because, comparing the stick diagram with the observed spectrum in Fig. 3, one can determine the perpendicular component, $A_\perp$, of the extra proton from the separation of the maximum slope of the main peak marked with asteroids and the temperature variation of this separation from Fig. 2. The rotational correlation time τ_c of this molecular motion can be determined from this averaging of this anisotropy by using the equation derived by Freed and his collaborators. (13)

$$\tau_c = 5.4 \times 10^{-10} (1-S)^{-1.36}$$

$$S \equiv A(T)/A_\perp(-196°C)$$

where $A(T)$ means $A_\perp$ component at a temperature T. An arrhenius plot of the τ_c was found to be linear in the temperature range between $-196°C$ and $-120°C$. The activation energy of this molecular motion is obtained as 1.5 Kcal/mol from this plot.

Ammonia.

It was mentioned that the $NH_2 \cdot (I)$ radical, which was produced in the high coverage, is more unstable and decays at lower temperature. And also the extreme separation in this case is smaller than the other. These facts suggest that the $NH_2 \cdot (I)$ radical is more mobile than the other trapped in the different conditions. Spectral change of the stable radical, $NH_2 \cdot (II)$, which is shown in Fig. 8, is used to estimate the temperature variation (Fig.13-A) of the extreme separation (marked with the arrows). Inserting the observed separations at various temperatures into the above equation the correlation time τ_c at each temperature was evaluated for the radical(II) as well as the radical(I). From the Arrhenius plots, shown in Fig. 13-B, the activation energies for the $NH_2 \cdot (I)$ and $NH_2 \cdot (II)$ are obtained as 0.7 Kcal/mol and 2.8 Kcal/mol, respectively. It is worthy to note that the activation energy is parallel to the stability of the trapped radicals; the smallest for the most unstable $NH_2 \cdot (I)$, medium value for the unstable $CH_3 \cdot \cdots H^+$, and the highest for the most stable $NH_2 \cdot (II)$.

Natures of the Various Trapping Sites

It is well known that a stocked zeolite adsorbs water in air
and the surface of the zeolite is covered with water molecules if
it is not sufficiently heat-treated. By heat-treatment the water
molecules are desorbed and degree of desorption may be dependent
on severity of the heat-treatment. No ESR spectrum observed from
the methane adsorbed on no treated zeolite means that such a
zeolite can not stabilize the methyl radical. This is probably
because surface of no treated zeolite is covered with water molecules which inhibit adsorptivity for methane. ESR spectrum, which
was observed from the abundantly adsorbed ammonia on the mildly
heat-treated zeolite, is almost same to that from the ammonia
radical trapped in glassy aqueous solution. This similarity
strongly suggested that ammonia molecules trapped in the mildly
heat-treated zeolite are surrounded with water molecules as in
the aqueous solution. This model may be supported by the fact
that some of the radicals produced from deutrated ammonia ND_3 were
found to be the type of $NDH\cdot$, which is formed by an exchange with
a hydrogen from the surrounding molecule, presumably water molecule.
In such trapping circumstances, the trapped amino radical ($NH_2\cdot(I)$)
is unstable and mobile and the potential barrier for detrapping is
rather small. The activation energy 0.7 Kcal/mol obtained from
the correlation time might be taken as an energy depth for libera-
tion from the trap. In same heat-treatment the stable amino
radical NH_2(II) was found as minor component in the high coverage
and major component in the low coverage. This fact seems to
suggest that the sites, at which the amino radicals are trapped
more strongly, are limited in number and covered selectively with
ammonia molecules in early stage of the adsorption. It is very
likely that this kind of site is that near silanol group, $-Si-OH$,
on the zeolite surface, at which ammonia molecule interacts rather
strongly to the surface. The activation energy obtained for the
molecular motion of this stable $NH_2\cdot$(II) radical might be considered as a measure of energy depth for trapping. If so, 2.5Kcal/
mol is larger than that for unstable trapping.

On the additional couplings with the extra proton, which were
found for both methy radical (II) and ethyl radical, the experiments were designed to determined what is the origin of the extra
proton. In the similar experiments with deutrated methane no
superhyperfine with the extra proton was observed. (9) This fact
demonstrates that this extra proton originates from the adsorbed
methane molecule. It was also concluded that the extra proton
coupling with ethyl radical comes from the adsorbed ethane. (7).
And the extra proton is not mobile but fixed on the zeolite surface, as shown in Fig. 12. These results can be used to build up
a picture for trapping as follows; Methyl molecule detach hydrogen atom by receiving the energy from the zeolite matrix which
primarily absorbs the energy of radiation, and this detached
hydrogen is trapped at the site which oxidizes the trapped hydrogen to the proton. This proton trapped and stabilized on the
oxidizing site on the zeolite matrix acts as a trapping site for

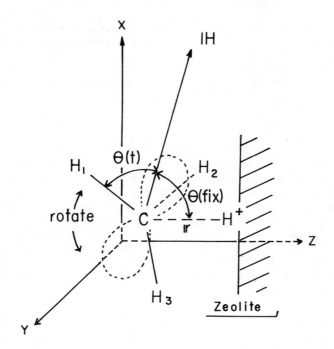

Figure 12.

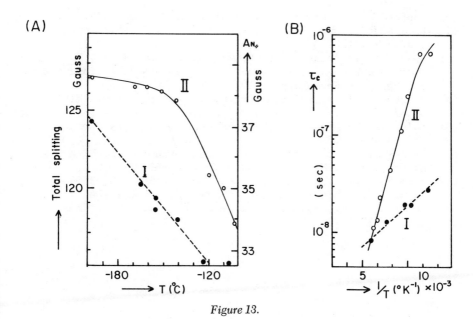

Figure 13.

a methyl radical, which experiences the super-hyperfine interaction with the trapped proton. The activation energy, 1.5 Kcal/mol, of the molecular motion of the radical, which is trapped at such site, might be regarded as an approximate energy depth for the trapping. By warming up the zeolite having such methyl radical (II), the methyl radical is now liberated from the trapping site and become the normal methyl radical. By cooling down the sample the spectrum changed reversibly and the radical seems to be trapped again at the same type of the trapping site.

Ionic sites, like Na^+, on the zeolite may become a strong trapping site for a polar molecule like an ammonia. However, such an ionic site is presumably hiding deeply in the adsorbed water layers and is only accessible when the adsorbed water molecules have been sufficiently desorbed by very strongly heat treatment, such as 350°C. And density of such a site is so small, probably one site in one cavity, that the sites are occupied with ammonia molecules even when the coverage of ammonia is less than monolayer. This kind of argument could explain the fact that $NH_2 \cdot \cdots Na^+$ radical was observed only in the low coverage at the very strongly heat-treatment.

Literature Cited

(1) Turkevich, J. and Fujita, Y., Science (1966) 152, 1619.
(2) Fujimoto, M., Gesser, H. D., Carbut, B. and Cohen, A., Science (1966) 154, 381.
(3) Katsu, T., Yanagita, M., and Fujita, Y., J. Phys. Chem. (1970) 43, 580.
(4) Lunford, J. H. et. al., J. Catalysis (1972) 24, 262.
(5) Komatsu, T., Lund, A., and Kinell, P. O., J. Phys. Chem. (1972) 76, 1721.
(6) Huang, Y. and Vansant, E. F., J. Phys. Chem. (1973) 77, 663.
(7) Kudo, S., Hasegawa, A., Komatsu, T., Shiotani, M., and Sohma, J., Chem. Letters, Japan (1973) 705.
(8) Fessenden, R. W. and Schuler, R. H., J. Chem. Phys. (1963) 39, 2147.
(9) Shiotani, M., Yuasa, F., and Sohma, J., J. Phys. Chem. (1975) 79, 2669.
(10) Kazanskii, V. B., Pariskii, G. B., Alexandrov, I. V. A., and Zhidomivov, G. M., Solid State Phys. (1969) 5, 649.
(11) Al-Naimy, B. S. et al., J. Phys. Chem. (1966) 70, 3654.
(12) Suzuno, Y., Shiotani, M., and Sohma, J., Unpublished data.
(13) Goldman, S. A., Bruno, G. V., and Freed, J. H., J. Phys. Chem. (1972) 76, 1858.

14

Recent Progress in Semiconductor Surface Studies by EPR

D. HANEMAN

School of Physics, The University of New South Wales, P.O. Box 1, Kennsington, N.S.W. 2033, Australia

The use of electron paramagnetic resonance techniques in studies of surfaces, interfaces and adsorbed species has been described (1-3). In this paper we review some recent and current studies of interest. Perhaps the most significant advance has been the elucidation of the gas-sensitive resonance from silicon, with the highlighting of several new properties and concepts, and the concomitant evidence regarding the nature of amorphous films. There is also considerable interest in the surface of GaAs, and EPR has been useful in helping to elucidate aspects of the surface structure. It has long been wondered whether the rich many-line EPR spectrum of molecular oxygen would show detectable effects in the adsorbed state and we describe recent studies on this. In the cases of Si and GaAs, no new lines were observed but some broadening effects on the O_2 lines seem to be present. Surface studies are particularly sensitive to contamination and an interesting and initially unsuspected source of this was discovered in the case of semiconductor samples vacuum crushed in a container with a stainless steel lid. A few microscopic fragments of Fe_3O_4 (magnetite) became mixed with the sample and displayed remarkable effects including a field induced phase (Verwey) transition. Direct evidence of the slowness of powders to attain equilibrium temperatures was obtained from these studies by taking advantage of the temperature sensitivity of the phase transition in the magnetite. A new category of information is obtainable from spin dependent conductivity effects using EPR techniques. At least two paramagnetic centres in the surface region of Si have been detected which are below the detection limits of conventional EPR techniques.

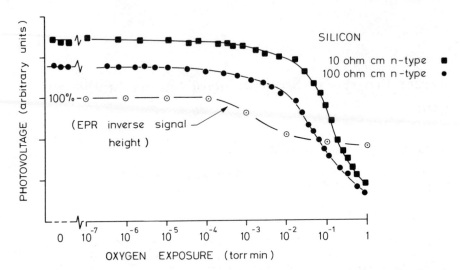

Figure 1. *Effects of exposure to oxygen upon maximum photovoltage that is developed when split in Si is scanned by fine light spot. Also shown is effect upon inverse of EPR signal height, obtained from vacuum-crushed Si.*

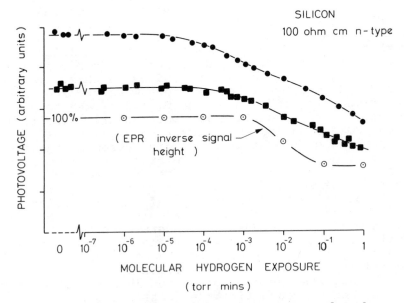

Figure 2. *Effects of exposure to hydrogen upon maximum photovoltage that is developed when split in Si is scanned by fine light spot. Also shown is effect upon inverse of EPR signal height, obtained from vacuum-crushed Si.*

Gas-Sensitive Paramagnetism in Si and Ge

It has long been known that Si gives an EPR signal
(g = 2.0055, width about 0.65 mT) when crushed, abraded,
chipped, cleaved, cracked or heavily ion bombarded (3).
When crushed or fractured in ultra high vacuum, the
signal is affected by molecular oxygen, hydrogen and
water vapour, but not noticeably (few percent changes)
until exposures of order 10^{-4} torr min are used. This
is in contrast to the behaviour of surface sensitive
properties such as photoemission which are affected at
exposures of order 10^{-8} torr min and yet are only
slightly affected by molecular hydrogen. The latter,
surprisingly, has a strong effect on the EPR signal.

These phenomena remained a puzzle for many years.
Many attempts were made to explain them but all were
open to serious objection in the form of incompatible
experimental data. Recently however the matter has
been cleared up and the explanation (4) accounts
for a wide variety of previously apparently incompatible results.

We present here a brief summary of some of the
key data. The relatively slow gas response of the EPR
signal was reminiscent of the gas response of surface
barriers at the surfaces of fine cracks studied
separately previously (5). The parameter measured was
the photovoltage which is the maximum voltage developed
at ohmic contacts on either side of a split which is
scanned by a fine light or electron beam, and is
related to surface barrier height. This in turn is
related to gas adsorption. Figure 1 shows changes as
a function of oxygen exposure of both the photovoltage
across a carefully prepared crack in Si, and also
the inverse of the EPR signal height from vacuum
crushed Si. Figure 2 shows a similar correspondence
when molecular hydrogen exposures are used. The
explanation of the slow response in the case of the
cracks is straightforward - the effective exposure of
surfaces in a narrow fissure is less than that of
external surfaces by a factor which is roughly the
ratio of the surfaces of the crack to the surface of
the jaw opening. This factor (6) can be in the range
10^4 to 10^6. Hence an exposure of 10^{-4} torr min may
correspond to only 10^{-8} torr min in the crack, i.e.
the surfaces at the crack are just as sensitive to gas
as free surfaces, but see an effectively much lower
amount of gas per unit area than external surfaces.

This explanation not only accounts qualitatively
for the magnitude of the exposure effects on crack
surface photovoltages but also accounts for the effects

of molecular hydrogen. The latter has a low sticking coefficient on free surfaces and bounces off, but molecules heading into the crack become temporarily trapped, as shown in Figure 3, with a correspondingly longer residence time in contact with the crack surfaces.

The behaviour of the EPR signal in these two respects (relatively slow for oxygen, slow but relatively large for hydrogen) is quite analogous to that of the crack surfaces barriers and strongly suggests that EPR centres are present in cracks. A single prepared crack has too low an EPR signal to be detected. However the hypothesis was tested by creating a large number of cracks in a specimen by multiply indenting it with a diamond point. This produces some crushed particles, which were removed by ultrasonic cleaning, and in addition a set of cracks. The EPR signal from such specimens was measured as a function of the load P applied to the point, and the results are shown in Figure 4. Note that the signal S is proportional to $P^{4/3}$. Now studies of transparent specimens have shown both theoretically and experimentally (7) that the crack radius r is proportional to $P^{2/3}$. Therefore

$$S \propto P^{4/3} = (P^{2/3})^2 \propto r^2 \propto A$$

i.e. the EPR signal is proportional to the crack area A. This is good confirmation of the hypothesis.

It was shown (4), by scanning electron microscope studies of fracture surfaces, that microcracks were more prevalent than previously realised, and in particular were not infrequently found to be present under steps, as shown schematically in Figure 5. By comparing the number of spins with the surface areas of the cracks in the indentation experiments, the spin density was shown (4) to be in the region of 1 spin per few crack surface atoms. From comparison of crushed powder signal intensities and surface areas, one requires cracks to be present in a surface region of area F such that the crack surface area is in the region of F/4. From inspection of scanning electron microscope pictures, this is readily possible.

The temperature behaviour of the EPR signal was close to T^{-1} which is usually characteristic of localised centres. However no such centres could be postulated which fitted the evidence. It was concluded that the centres were in fact localised states on the surfaces of the cracks. The reasons for localisation taking place were as follows.

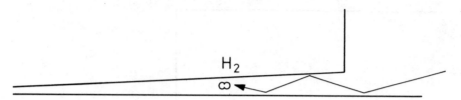

Figure 3. *Schematic showing entry and temporary entrapment of hydrogen molecules in split*

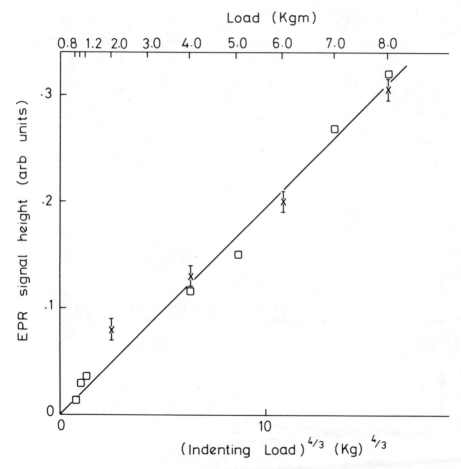

Figure 4. *Graph of EPR signal height, averaged per indentation, vs. 4/3 power of load of diamond point indenter*

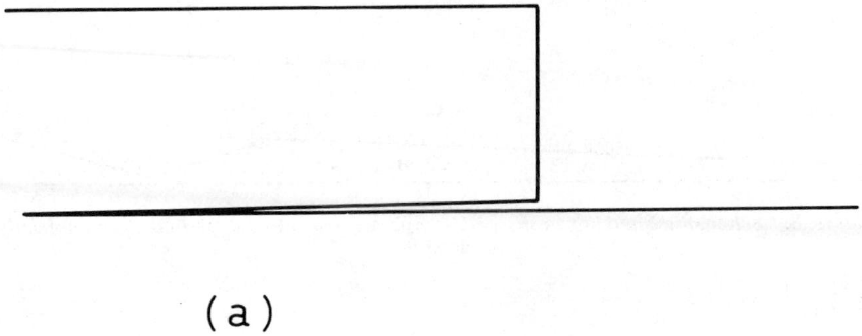

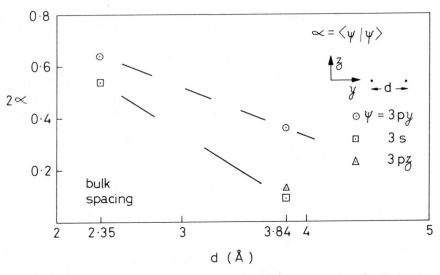

(a)

Figure 5.   Schematic of crack under step

$\propto = \langle \psi | \psi \rangle$

$\psi = 3py$ (⊙)

$3s$ (□)

$3pz$ (△)

bulk spacing

d (Å)

Figure 6.   Wave function overlap (10) as function of spacing of nuclei, for various atomic wavefunctions. At bulk spacing of Si, 0.235 nm, $2\alpha = 1.34$ for sp^3 functions.

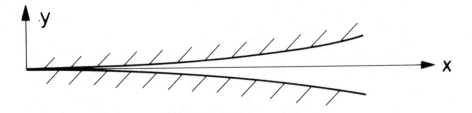

Figure 7.   Schematic of crack

When two surfaces are placed opposite each other in close proximity as in a crack, the wave functions of the surface electrons on one side can overlap with those on the other side. The charge density contours of outer electrons on Si surface atoms have been computed (8,9). They become very small (few percent of maximum value) at a distance of about 0.25 nm from the nucleus, so that overlap at a spacing of more than 0.5 nm becomes very small. A similar conclusion is reached from calculations (10) of wave function overlap at two spacings performed previously. As shown in Figure 6 the overlap is most for the case of in-line p orbitals, but even in this case it becomes slight after about 0.5 nm. Hence in order for appreciable effects to occur, the spacing between the surfaces must be less than about 0.5 nm. (These remarks refer to <ψ/ψ> whereas the important quantity is <ψ/H/ψ> where H is the interaction Hamiltonian. The latter is harder to calculate but its range will be similar to that of <ψ/ψ>).

This figure enables us to make a quantitative check. For special controlled cracks studied by X-ray transmission topography (6), the jaws of an approximately 0.5 mm long crack could be as little as 1.5 nm apart, due presumably to slight step mismatch (there was a measurable shear of about 2 nm at the jaws), or perhaps to microscopic (order Angstroms) debris. The shape of the crack sides for such cases is not readily calculable exactly, but lies between straight and parabolic. Let half the spacing of the sides be y, at a distance x from the crack tip, as in Figure 7. Then for straight sides $y = \alpha x$ where α is a constant. Taking the above case, $y = 0.8$ nm at $x = 0.5$ mm, we obtain $\alpha = 1.6 \times 10^{-6}$. Then at $y = 0.25$ nm, $x = 0.16$ mm i.e., 30% of the crack has a spacing of less than 0.5 nm. This case of straight sides is extreme. For a more likely parabolic case we put

$$y = ax^2$$

and, with the former boundary condition, one deduces $a = 3.2 \times 10^{-3} m^{-1}$. At $y = 0.25$ nm, $x \simeq 0.28$ mm, i.e. 60% of the crack has a spacing of less than 0.5 nm. Hence in general about 30-60% of the crack is in a condition of wavefunction overlap. If the microcracks are not much different from the controlled cracks, there is therefore ample scope for an EPR signal corresponding to spins on about 10% of the crack area, as discussed, provided there is something like 1 unpaired spin per several surface atoms in the overlap region. Most microcracks are much shorter than 0.5 nm, so that the overlap extent would be even greater. However they are produced under rougher conditions than those used in the above experiments where the material was very carefully separated. Hence in most naturally occurring microcracks, shear and other distortions are likely to be greater, so that the jaw openings might be relatively larger than by extrapolating from the above figures. Even so, a sizable proportion of a crack must have sides within about 0.5 nm, if it is only a small fraction of a mm long.

A. Properties of Overlap Regions. We now consider the properties of the overlap regions in detail. At the very base of the split, as indicated schematically in Figure 8(a), we have a transition from a healed region (5) to one with a finite gap. Now the two sides of the split are subject to three effects: (a) the separation increases towards the mouth; (b) the original registry between them on an atom-to-atom basis becomes lost due to shear, Figure 8(b), since even the carefully prepared controlled splits showed measureable shear in the non-healed region. In those cases values ranged from 2.4 to 8×10^{-6} radians, giving about 3 nm displacement at the mouth of a 0.5 mm split, and thus more than 0.1 nm over most of it. Hence atoms are no longer opposite their pre-cleavage neighbours; (c) contact regions exist at the edges of topographical irregularities such as steps on the faces of the split, and these are centers of pressure causing deformation of the material, Figure 8(c). The result of these three effects is that, even in the 0.5 nm region of separation, the set of displacement vectors to opposite surface neighbours for any surface atom varies from site to site. Any individual atom is thus subject to forces from atoms on the opposite surface, but these forces vary from site to site since the shear displacement and separation vary (increase towards the jaw mouth), and the stress displacement varies also, being centred at somewhat random points and lines. This is a situation of varying potential which is of the kind considered by Anderson (11) and others. It is hence a possible practical example of Anderson localised sites.

B. Localised States. The criterion for Anderson localisation has been discussed by various authors (11-16). The consensus is that localisation occurs if the half width of the distribution of

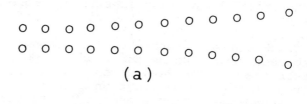

(a)

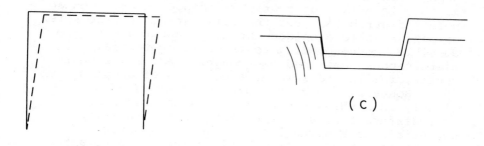

(b)

(c)

Figure 8. (a) Schematic of base of crack showing transition from healed to separated region. (b) Top view of crack, showing shear of one side with respect to other. (c) End view of crack showing pressure at contact between schematic protrusion on one side and corresponding gap on other.

potentials is about greater than the width of the band
resulting if all potentials were the same. The normal
state band has a width of about 0.3eV (8).The potential
disturbance due to the varying overlap can be
estimated for comparison with this figure. At the base
of the crack the overlap is strong, close to bulk, and
will be in the region of the single bond energy in
bulk Si, namely 2.37eV (17). When the separation of
the crack sides is more than 0.5-0.6 nm, the overlap
approaches zero. Hence the range of potential
disturbance is over 2eV, which easily exceeds 0.3eV
and thus fits the localisation criterion. Although
the disturbance varies somewhat monotonically as one
moves in a direction x along the split, since
separation and shear displacement are both functions of
x, this is broken up by the various pressures contact
points distributed over the split area. At the stress
origin, elastic displacement up to about 0.1 nm are
possible, corresponding to order 1eV potential
disturbance. This reduces roughly radially from the
center, falling to zero at a distance of about 0.2 mm.
(5). However, the stress centers are spaced more
closely than this, so that every site is subject
to a different resultant stress. Hence the site
potentials at a given x are no longer identical.

If the above concepts are correct, then to agree
with experiment they must lead to a temperature
dependence of the paramagnetism χ that is close to T^{-1}.
This was considered in detail (4). Without repeating
the mathematical details, the result (18) was obtained,
taking into account correlation corrections to one
electron theory, that

$$\chi = \tfrac{1}{4} \, g^2 \mu_B^2 \; \frac{N_S}{\Delta} \left(\frac{w_o}{kT} + 2\ln 2 \right)$$

where w_o is the self interaction energy of two electrons on the one site (correlation correction), Δ is
the width of the (assumed constant-density) band, g
is the g value, μ_B the Bohr magneton and N_S the number
of states (orthogonal, localised). The result holds
also when interactions between electrons on neighbouring sites are taken into account (19), provided
this term is not unreasonably large. Since w_o/kT
is much greater than $2\ln 2$, the formula gives the
paramagnetism as inversely proportional to T, as
required by the measurements.

A variety of other properties of the signal were
also explained such as its slight variability in width
between samples, which is now due to differences in

distributions of cracks between samples. Since
localised states still have a spread over a number of
sites, the explanation is consistent with the spread
nature of the wave function of the EPR centre as
deduced from the absence of discernible hyperfine
structure [expected from 5.7% abundant Si (29)] and
the effects of alloying Ge into the Si (20).

In the case of amorphous films the EPR signal (21),
both for Ge and Si, is usually identical with that
from crushed samples. This is now explained as due
to localised centres on the surfaces of the microscopic
aggregates making up the films. Here of course
localisation is endemic through the bulk whereas in
the crystalline materials the localisation on the
crack surfaces is due to the spatially varying
overlap of forces between opposite faces.

Gallium Arsenide

Some time ago it was found that vacuum crushed
samples of GaAs and AlSb, when exposed to oxygen at
liquid nitrogen temperatures, displayed an EPR signal
due to O_2^- ions adsorbed on the surfaces (22). In the
case of AlSb, hyperfine structure was detectable, which
enabled identification of the adsorption sites as Al,
and the conclusion that the dangling bond wave function
on the Al was over 90% p like. Although hfs was not
detectable from the GaAs, the signal was overall
very similar to that from AlSb, and similar conclusions
were made.

We have attempted similar experiments on single
crystal GaAs cleaved at uhv and exposed to oxygen at
low temperatures. A signal was at first not detectable but after a low pressure microwave frequency
discharge occurred in the oxygen, an EPR signal was
detected, which looked similar to that from GaAs
powder but a little narrower. Some preliminary
results, yet to be confirmed, suggest hyperfine
structure also was observable. The O_2^- signal occurs
readily on n type crushed GaAs but only weakly on p
type material, which indicates strongly that the Fermi
level at the surface is different for n and p type
material. Recent evidence suggests that it is pinned
with respect to the band edges for n type but not for
p type material (23). Our results are thus consistent
with this since the O_2^- energy level is apparently
close to the Fermi level on n type material, about
midway between the band edges at the surface. Hence
electrons can transfer from the bulk to the O_2, but
not in p type material where the Fermi level at the

surface must be well below the O_2^- level, i.e. near
the valence band edge. The latter is expected if
there is no pinning for p type material, and this was
concluded from emission data (23).

Recently various theoretical calculations have
been carried out for GaAs (110) surfaces. (These are
the easy cleavage surfaces, and presumably predominate
in crushed material). A self consistent pseudopotent-
ial calculation for an unreconstructed surface (24)
finds the dangling bond charge to be localised on the
As surface and the empty state on the Ga, which thus
can accept the O_2^- adsorbate. However the wavefunction
was more s than p like, contrary to the EPR data (22),
and other estimates based on reconstruction (25,26).
It seems from this that reconstruction probably occurs
on GaAs surfaces and it would be useful to carry out
a self consistent calculation for such a case.

Molecular Oxygen Spectra

The EPR spectrum of molecular oxygen contains over
100 lines due to transitions in rotational levels.
Some of these may be affected if the gas is in the
adsorbed state. An attempt was made to check this but
unequivocal effects could not be established. However
by working at pressures lower than those used previously, many new lines were discovered (27). In all, over
220 lines were observed at 10^{-3} torr in the magnetic
field range 0-1 Tesla. Some of these broadened
rapidly as the pressure increased, accounting for
their non detection in previous experiments at higher
pressures.

It has been reported (28) that for oxygen adsorbed
on amorphous carbon, two new lines with the same g
value but greater linewidths appeared at the positions
of the previous free oxygen lines. The authors
attributed these to chemisorbed and physisorbed
oxygen. However no such effects were found with
crushed silicon. A reassessment of the results on
carbon showed that the observed effects could be
explained by taking into account the broadening of
oxygen lines when the gas is in the fine pores of the
carbon, together with the effects of overmodulation.

Magnetic-Field-Induced Transitions in Small Magnetite
Fragments

In some EPR experiments on semiconductors in
ultra high vacuum, a stainless steel cap covered a
small glass bowl in which a glass-coated rod crushed

a specimen while in ultra high vacuum. The powder was
then tipped out in vacuo into a quartz appendage tube
protruding into a microwave cavity. On some occasions
a large background signal was observed ($\underline{29}$) with
peculiar properties. It was traced eventually to
contaminant microscopic particles coming from the steel
cap. The particles turned out, somewhat unexpectedly,
to be Fe_3O_4, magnetite.

An example of the unusual phenomena is shown in
Figure 9. Note that a portion of the broad resonance
totally disappears at a certain critical value of the
sweep magnetic field. This critical value varies over
a large range for a few degrees change in temperature.

The phenomena were explained ($\underline{29}$) as due to a
Verwey transition, i.e. the ordering of Fe^{2+} and Fe^{3+}
ions on octahedral sites below a critical temperature,
giving ferrimagnetism. This transition is apparently
field dependent. The magnetite (Fe_3O_4) presumably
forms during vacuum bakeout of the Fe_2O_3 normally
present on the stainless steel surfaces, and tiny
particles are knocked off by contact with the semiconductor powder. This is of interest since the
nature of the oxide on stainless steel surfaces in
baked vacuum systems has not previously been
determined.

The temperature sensitivity of the value of the
critical field made it possible to estimate when
equilibrium temperature was attained in powder samples
containing the magnetite fragments. Volumes of Si or
GaAs of about 0.1 cc took at least 20 or more minutes
to equilibrate when cooled by about 50 K degrees via
a sapphire rod.

Spin Dependent Conductivity in Silicon

In the presence of a magnetic field both current
carriers and paramagnetic centres are subject to
Zeeman energy splitting. The interaction between
current carriers and fixed centres then contains a
spin-spin term which depends on whether the incident
and fixed spins involved are parallel or antiparallel.
If microwave radiation is used to alter the
population (spin up and spin down) of the current
carriers then the parallel and antiparallel
contributions to the interaction are altered in
relative magnitude, leading to a slight change in the
conductivity ($\underline{30}$). By monitoring the conductivity as
a function of magnetic field, while irradiating with
microwaves and using optical pumping to get sufficient
current carriers, it was found possible to detect

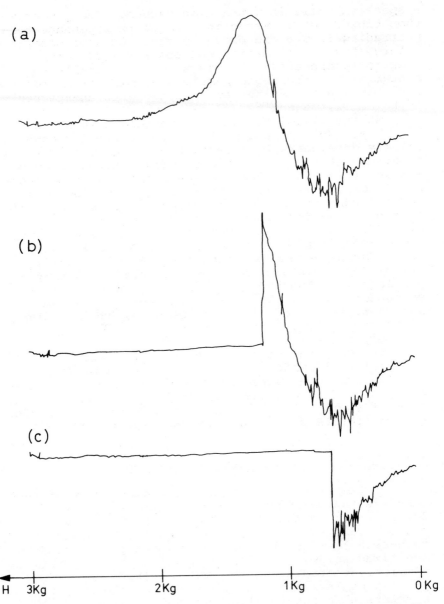

Figure 9. Temperature variation of broad signal (12 db). (a) 103°K, (b) 98°K. Note signal disappearance at critical field. (c) 95°K. Critical field has shifted. (d) For T = 90°K, no signal was observed, i.e. critical field shifted to near 0 Kg.

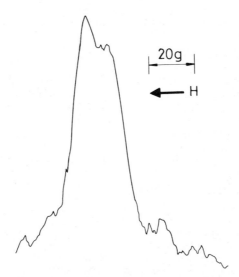

Figure 10. Lock-in amplifier detection of two peaks in conductivity of illuminated Si sample during magnetic field scan. (Microwave power modulated at 35 Khz, 500 nV amplifier sensitivity).

paramagnetic recombination centres in very low concentration at Si surfaces. We have used this method on samples of different surface treatment and found at least two different centres to be present, as shown by the split peaks in Figure 10. Neither of the centres is detectable by normal EPR techniques. A number of interesting phenomena have been found in the behaviour with strong illumination which have not yet been clarified, but the evidence shows that the centres are at or very near to the surface, possibly associated with topographical irregularities induced by the surface finishing treatment.

Acknowledgements

Excellent experimental work was carried out by B.P. Lemke and J. Menendes-Cortinas. This work was supported by the Australian Research Grants Committee and the U.S. Army Research and Development Group (Far East) under Grant No. DA-CRD-AFE-S92-544-71-6168.

Literature Cited

1. Haneman, D. "Characterization of Solid Surfaces", Chapter 14, p. 337. ed. P.F. Kane and G.B. Larrabee (Plenum Press, 1974).

2. Adrian, F.J., J. Colloid. Interface Sci. (1968),
    26, 317.
3. Haneman, D., Jap. J. Appl. Phys. Suppl. (1974),
    2, (Pt.2), 371.
4. Lemke, B.P. and Haneman, D., submitted to
    Phys. Rev.; Phys. Rev. Lett (1975), 35, 1379.
5. Haneman, D., Grant, J.T.P. and Khokhar, R.L.,
    Surface Sci. (1969), 13, 119.
6. Khokhar, R.U. and Haneman, D., J. Appl. Phys.
    (1973). 44, 1231
7. Lawn, B.R. and Fuller, E.R., J. Mat. Sci. (1975),
    10, 2016.
8. Schluter, M. Chehkowsky, Louie, S.G. and Cohen,M.L.
    Phys. Rev. (1975), B12, 4200.
9. Appelbaum, J.A. and Hamann, D.R., Phys. Rev. Lett.
    (1974), 32, 225; Phys. Rev. (1973), B8, 1777.
10. Haneman, D. and Heron, D.L., "The Structure and
    Chemistry of Solid Surfaces", Chapter 24, ed.
    G.A. Somorjai, (Wiley, New York, 1969).
11. Anderson, P.W., Phys. Rev. (1958), 109, 1492.
12. Mott, N.F., Phil. Mag. (1970), 22, 7; (1969), 19,
    835.
13. Cohen, M.H., J. noncryst. Solids, (1970), 4, 391
14. Ziman, J.,J. Phys. C. (1969), 2, 1230.
15. Ball, M.A., J. Phys. C. (1971), 4, 1747.
16. Cohen, M.H., Fritzsche, H. and Ovshinsky, S.R.,
    Phys. Rev. Lett. (1969), 22, 1065.
17. Sanderson, R.T., "Chemical Bonds and Bond Energy",
    (Academic Press, New York, 1971).
18. Kaplan, T.A., Mahanti, S.D. and Hartmann, W.M.,
    Phys. Rev. Lett. (1971), 27, 1796.
19. Miller, D.J., to be published.
20. Miller, D.J. and Haneman, D., Surface Sci. (1972)
    33, 477.
21. Brodsky, M.H., Title, R.S., Weiser, K. and
    Pettit, G.D., Phys. Rev. (1970), B1, 2632.
22. Miller, D.J. and Haneman, D., Phys. Rev. (1971),
    B3, 2918.
23. Gregory, P.W. and Spicer, W.E., Phys. Rev. (1971),
    B13, 725.
24. Chelikowsky, J. and Cohen, M.H., Phys. Rev. Lett.
    (1976).
25. Pandy, K.C. and Phillips, J.C., Phys. Rev. Lett.,
    (1975), 34, 1450.
26. Harrison, W.A., Surface Sci. (1976), 55, 1.
27. Lemke, B.P. and Haneman, D. to be published.
28. Seymour, R.C. and Wood, J.C., Surface Sci, (1971),
    27, 605.
29. Lemke, B.P. and Haneman, D., J. Magn. Res. (1976),
    22.
30. Lepine, D., Phys. Rev. (1972), B6, 436.

Paramagnetic Defects in the Surface Region of Processed Silicon

PHILIP J. CAPLAN

US Army Electronics Technology & Devices Laboratory (ECOM), Fort Monmouth, N.J. 07703

One of the most significant trends in the evolution of electronics has been the development of large-scale integrated circuits based on the metal-oxide-semiconductor (MOS) structure. During the growth of this technology many physical and chemical processes have been perfected which make it possible to fabricate and evaluate satisfactory devices. As a result much has been learned about the nature of the silicon or oxidized silicon surface (1,2). Sensitive electrical measurements such as capacitance-voltage measurements reveal the presence of very small concentrations of impurities at the Si-SiO$_2$ interface. Yet much of the information is qualitative and more specific spectroscopic techniques such as magnetic resonance have been explored to throw additional light on this important area. As long ago as 1960 (3), it was proposed that electron spin resonance (ESR), nuclear magnetic resonance (NMR), and even dynamic polarization experiments might add significantly to the knowledge of the silicon surface.

Process Steps Generating ESR Signals

Let us consider some basic preliminary steps in the processing of a MOS device. A slice of monocrystalline silicon with a given crystallographic orientation is cut off from a boule and given a high polish. It is etched with a mixture of HF, HNO3, and acetic acid, which removes the remaining damage region of the surface. It is known that very shortly after the sample is etched and exposed to air, an oxide layer forms of the order of 20 Å in thickness. Further processing requires, however, a thicker oxide layer of perhaps 2000 Å, which is grown by heating in oxygen in a furnace at 1000° to 1200°C. We will not go further with the device fabrication, but will now show that the aforementioned processes display associated ESR signals, as shown in Table I.

Before etching, the damaged surface region produces an ESR signal with g = 2.0055, and about 5 gauss wide, which is readily observable even at room temperature. After etching, the original damage signal disappears, but a photoinduced ESR signal having

Table I
Processes and ESR Signatures

Process	g-value	Other characteristics
Surface Damage	2.0055	Damage region is 1 micron deep
Standard Si Etch	2.007-2.008 2.003-2.004	Optically stimulated
Thermal Oxide	P_A 2.000 P_B 2.002-2.010 P_C 2.06-2.07	In SiO_2 or Si near surface In SiO_2 layer near interface Extends many microns into silicon

two components appears at an observation temperature of 77°.
After the oxidation procedure as many as three new resonances may
appear. They were denoted by Nishi, who first described them, as
P_A, P_B and P_C. They also were observed at 77° and were shown to
depend on the details of the thermal oxidation.

Surface-Damage Signal

After this brief survey, we shall now consider the individual resonances in detail. The surface-damage signal was the
first to be discovered (4,5). It occurred when a silicon surface
was sandblasted, polished, or crushed. By crushing into finer
powder an increasingly large signal is obtained. The signal was
independent of dopants in the silicon, occurring equally well in
n- or p-type. It disappeared, however, when about 1 micron of
the surface was removed by a chemical etch, thus indicating its
surface nature. A series of carefully designed experiments were
carried out to investigate the atomic nature of these ESR centers,
particularly by Haneman (6,7,8,9). In order to start out with a
chemically simple system, high-purity single-crystal specimens of
silicon were sealed off together with a glass slug in a high-
vacuum system with pressures in the range $10^{-8} - 10^{-9}$ torr, and
they were crushed in vacuum by shaking the glass slug within the
enclosure. The resulting powder, about 5 microns in size, displayed the typical strong surface-damage resonance. Subsequently,
various gases were introduced into the vacuum chamber, and significant increases in signal amplitude were noted for H_2 adsorbed
($\sim$50%) and for adsorbed O_2 ($\sim$100%), whereas water vapor caused a
decrease of $\sim$20%. From this sensitivity to the introduction of
gases, Haneman proposed that most of the spins contributing to the
surface-damage signal were localized at the surface, and that they
were in fact the dangling bonds to be expected at the surface of
the silicon crystallites.

How then can one explain the often observed fact that in order to remove the surface-damage signal you have to etch a distance of 1 micron into the silicon surface? This is due to the fact that within this damaged region there is a vast labyrinth of cracks and fissures all contributing to a much greater effective surface area. Thus etching this away reduces the signal to a negligible level.

Pursuing this hypothesis further, Haneman constructed an apparatus for cleaving silicon crystals in high vacuum and studying the weak ESR signal from the bare silicon surfaces thus generated. From this work he estimated that about 1 spin per 10 surface atoms was paramagnetic.

Other workers have, however, not been convinced that the signal observed in crushed samples is explained simply by a large effective surface, but attribute the ESR signal to defect structures. A recent interesting contribution (10) to this controversial question was made when evidence was found that the ESR signals obtained by Haneman from cleaved samples were spurious. It was shown that in fact cleaving created a fine dust at the cleavage surface and when this was wiped off the signal became considerably weaker. The ESR signal was thus attributed to amorphous regions in severely damaged powders rather than dangling surface bonds.

An interesting by-product of the ultrahigh vacuum experiments on crushed silicon has little to do with silicon itself, but is worth mentioning in the general context of this conference. Heat treatments in vacuum of the silicon powder produced a sharp resonance at g = 2.0028, and the properties of this signal as a function of ambient and temperature were studied (11,12). Hypotheses in terms of the structure of the silicon surface were proposed to account for this signal, but it eventually turned out that the identical signal was seen when other materials aside from Si were studied. The cause was traced to leakage of small amounts of carbon compounds from the oil in the vacuum systems which adsorbed on the silicon surface and at elevated temperatures formed radical compounds. Indeed the properties of this signal were found to be similar to those of the signal from heated carbonaceous materials (13).

Effects of Thermal Oxidation

We now take up the important area of thermal oxidation of the silicon surface. There are several types of defects that have been identified as influencing the behavior of MOS devices (1). First there are positively charged defects permanently located in the oxide near the oxide-silicon interface. These are thought to be associated with the nonstoichiometry at the interface region. A second type of defect also consists of positive charges in the oxide. However, these are due to mobile alkali metal ions, and their distribution in the oxide can be altered by the application of electric fields. Great care is taken to eliminate sodium

contamination from the processing, since such a variable charge
system causes the parameters of the final device to be unstable.
A third category is the so-called fast surface states which exist
at the interface. These are traps which can rapidly be charged
or discharged by electrons from the silicon. These are assumed
to be due either to structural defects at the interface, impurity
atoms in this region, or both. Finally, there is the special case
of positively charged defects in the oxide caused by ionizing
radiation.

It is thus important to study the ESR signals (see Figure 1)
that are generated by thermal oxidation (14,15), to interpret them
with respect to the structure of the oxidized silicon surface and,
if possible to relate them to the electrically observed defects.
Nishi (15), having observed three distinct resonances, wanted to
determine their location within the surface. He made successive
slow etches into the oxide layer and then into the silicon, and
in between etches recorded the ESR signal strengths. In this way
he determined that the P_A and P_B signals were in the oxide. He
succeeded in profiling the stronger P_B signal, and determined that
the peak of the P_B distribution is near the oxide-silicon interface. The P_C signal was also profiled, and it extended about 10
microns into the silicon bulk. In our repetitions of this work
(16), we found P_C actually extending much further into the bulk.
Various auxiliary experiments led us to conclude that P_C was actually due to iron impurities (Fe^0) that were distributed through
the silicon. The heat treatment either converted previously present iron ions to the Fe^0 state, or caused iron atoms from the
outside to diffuse into the material.

The P_B signal appeared to be the most significant with regard
to its relevance to the interface structure. It was noted (15)
that the spatial distribution of the P_B center is the same as that
measured for the fixed surface change, i.e. concentrated near the
oxide-silicon interface. Just as these electrically charged
species were attributed to defects in the stoichiometry near the
interface, so it seemed likely that P_B resulted from a defect in
the SiO_2 structure. Nishi concluded that it was due to a defect
where a silicon atom has only three electrons bonded and one
hanging loose. This conclusion was reinforced by comparison with
the resonance exhibited by silicon monoxide (17). This amorphous
material has many defects in its bulk structure, and its strong
resonance with g = 2.0055 has been ascribed to trivalent silicon
with one unpaired electron. Now the P_B signal in the oxide-silicon transition region is anisotropic with respect to the angle between the magnetic field and the silicon surface, ranging from
about g = 2.002 to g = 2.010, the average value being ∿2.005.
Thus the resonance at the oxidized silicon surface may be the same
as that in bulk silicon monoxide.

In our laboratory we were concerned with one flaw in this
identification. The silicon monoxide resonance is readily observable at room temperature, although of course a few times weaker

than at 77°. Therefore, it was puzzling that Nishi reported that none of his ESR signals were visible at room temperature. We therefore ran some of our samples containing P_B at room temperature, and indeed found that the P_B signals were visible and of the expected magnitude (11). Thus this objection against the identification of the P_B centers and SiO centers is removed.

It is remarkable that although the thermal oxide region is characterized as noncrystalline, the P_B signal is anisotropic with respect to the surface direction. This indicates that at least the portion of the oxide near the interface region has a macroscopic symmetry determined by the silicon substrate.

Photoinduced ESR Due to Etching

About three years ago a new and unusual kind of silicon surface resonance was announced by a group at Tohoku University (18). It had been assumed that etching silicon merely serves to remove the damage region and its attendant ESR signal. However, it was found that after etching, a new resonance appeared if the sample was simultaneously exposed to light (see Figure 2) with energy greater than the silicon band gap. The resonance appeared to have two components and displayed some anisotropy with respect to the angle between the surface and the external magnetic field. It was also noted that the photosignal response depended on the conductivity type, n or p, and the effect was not observed in samples with too high a conductivity.

It was suggested that these resonances were due to water being incorporated during the etching and rinsing process. In order to test the hypothesis, we decided in our laboratory to perform various chemical treatments on the surface (16). First we verified that treatment with HF alone did not produce the photosignal. This was true whether the HF was used on a fresh silicon surface or a thermally oxidized surface. Of course HF does not etch into the silicon itself. Therefore, we tried other types of chemical etch that do attack silicon. These included hot potassium hydroxide, an etch consisting of HF mixed with H_2O_2, and a procedure of treating the sample successively in HF and hydrazine. Although these etches encompass a variety of chemical types, none of them were effective in producing the photosignal. Since in all of them water was in contact with the surface, the supposition that water was responsible for the photosignal seemed doubtful. We additionally checked that acetic acid was not necessary to produce the photosignal, so we suggested that the presence of HNO_3 was the crucial factor, and that some kind of nitrogenous radical was responsible for the resonance.

Another question we addressed was the location of the ESR centers with respect to the surface. One would guess that they are superficially adsorbed atoms. One would like to strip away the outer atomic layers gradually and check the ESR every time. Certainly one cannot use for this purpose a weak version of the

standard silicon etch, since it is this etch that generates the
photosignal. However, one can safely try HF alone, since we know
that by itself it does not generate the signal. It was found
that etching in concentrated HF does not visibly reduce the photosignal, even though it presumably dissolves the outer 20-angstrom
layer of oxide. It was found, however, that repeated cycles of
HF treatment and air exposure, i.e., oxidation and dissolution of
successive 20-angstrom layers, do diminish the photosignal remarkably. This indicates that the photosignal center may extend
to about 100 Å into the silicon.

In a recent publication (19) the Tohoku University group have
explained this photoinduced signal as being quite different from
the normal ESR phenomenon. To understand the mechanism let us
refer to a previous study by Lepine (20). He etched a silicon
wafer, inserted it into an ESR cavity, and simultaneously irradiated it with light and with a microwave field. As the magnetic
field was varied he monitored the change in resistivity of the
sample (a few parts in 10^6) as the resonant field was traversed,
and a resonance in the resistivity curve was traced out. His
explanation was that the optical radiation, by generating elec-
tron-hole pairs, contributed to the conductivity of the sample.
The recombination rate of the electrons and holes, however, is
influenced by recombination centers at the surface of the silicon.
These centers are also paramagnetic, and the recombination rate
depends on the relative spin orientation between the recombination
centers and conduction electrons. Thus when these surface defects
are paramagnetically saturated, the concentration of electrons and
holes is altered and a change in the resistivity is observed.

Now we return to our photoinduced signal observed by ESR.

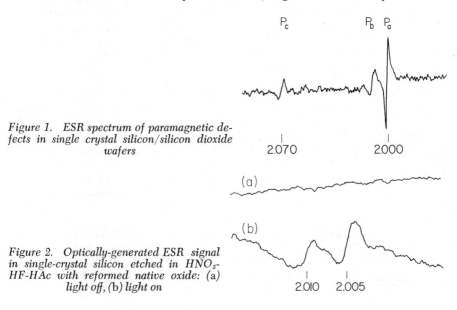

Figure 1. ESR spectrum of paramagnetic defects in single crystal silicon/silicon dioxide
wafers

Figure 2. Optically-generated ESR signal
in single-crystal silicon etched in HNO₃-
HF-HAc with reformed native oxide: (a)
light off, (b) light on

The proposed theory (19) is that we are not observing the usual
ESR effect, i.e. a change in the cavity Q due to paramagnetic
losses, but a change in the electric losses due to the changing
sample conductivity, which is in turn caused by saturation of
the surface paramagnetic centers. This interpretation is supported by certain of our observations (21). First, we have often
been perplexed by the fact that the photoinduced ESR depends on
just how the sample is positioned in the dewar vessel. According
to the theory that we are actually observing electric losses in
the cavity this is reasonable, since the normal sample position
is in a region of maximum magnetic field and low electric field,
but if the sample is displaced it enters a region of higher electric field which couples more effectively with the changing resistivity. Another pertinent experimental fact is our work with
silicon-on-sapphire samples. One of the newer developments in
integrated current technology is the fabrication of circuits on
very thin silicon single-crystal films. These films, which may
be a micron thick are deposited on sapphire substrates. We have
obtained samples with a silicon thickness of about 10 microns,
and have etched them and looked for the photoinduced signal, and
have not found it. Now if we are observing photoinduced
resonances directly from the paramagnetic surface centers, it
should not make any difference if we are using an unusually thin
sample. However, if we are observing an indirect effect of conductivity which involves the bulk of the material, it is understandable that the thin silicon-on-sapphire sample is ineffective.

NMR Relaxation Enhancement by Si Powders

Finally we have explored an indirect method of studying the
surface of silicon by nuclear-magnetic-resonance measurements of
liquids in contact with silicon powders (16). It is well known
that paramagnetic molecules dissolved in liquids drastically
shorten the NMR spin-lattice relaxation of the liquids. Furthermore, in such systems one can have the dynamic polarization effect,
where pumping the electronic paramagnetic levels produces large
enhancements of the NMR signal. It is also possible to produce
these effects by mixing powders of paramagnetic materials with
liquids that do not dissolve them, where the electron-nuclear
interaction only takes place at the surface of the powder grains.
For example, granules of DPPH sieved between 50 and 100 mesh
screens, and mixed with water, give a proton dynamic polarization
enhancement of about -20 (extrapolated to infinite pumping power)
(21). This result was obtained with a dynamic nuclear polarization (DNP) apparatus operating at a low magnetic field (74 gauss).
A corresponding shortening of T_1 also is observed.
    We then wanted to see if similarly sieved crushed Si powder
has enough paramagnetic defects at its surface to cause surface
relaxation and dynamic polarization effects. A number of samples
were prepared using powders prepared from two different kinds of
rather low purity silicon, and these were mixed with various
liquids. Also a set of oxidized silicon powders were prepared.

In all cases, the relaxation time became considerably shorter, but there was no evidence of dynamic polarization. The possibility was considered that the liquid molecules complexed with the surface so as to form a region of much longer molecular correlation time than in the bulk. However, this hypothesis proved quantitatively unreasonable. The conclusion was thus drawn that paramagnetic defects at the silicon surface were responsible for the shortening of T_1, and these defects have a very broad resonance (unlike the g = 2.0055 surface-damage signal) which makes it unobservable by electron paramagnetic resonance and also prevents the dynamic polarization effect.

Further experiments were performed with a single liquid, hexane, in order to study the effects of various processes on the surface relaxation. Three samples of silicon powder are included in Table II, which gives the change in the hexane T_1 due to successive treatments of these powders. All samples were 50-100 mesh size, and were made from Fisher Chemical Co. material. Note that T_1 of the bulk hexane is about 9.5 sec.

From the results listed in Table II we concluded that

Table II
Proton Relaxation Times of Hexane Mixed with Silicon Powders

Sample	T_1 (sec)
No. 1 freshly crushed	.55
treated with HF	.7
slow Si etch	7.5
heated 600°C 1 hr O_2	8.5
heated 1000°C 2 hr O_2	10.5
No. 2 oxidized 1 hr 1050°C	.65
etched in HF	5.2
No. 3 oxidized 2 hr 1000°C	1.5
etched in HF	5.9

oxidation in itself has a minor effect on the silicon surface relaxation. Likewise treatment of freshly crushed silicon with HF has little effect. However, when an appreciable portion of the surface-damage layer is removed, a large increase in T_1 occurs. This can be accomplished either by using an etch that attacks silicon, or by thermal oxidation followed by removal of the oxidized layer with HF. This may mean that the surface paramagnetic centers are associated with damage sites, or that removal of the innumerable irregularities of the damaged surface drastically reduces the effective area of the surface and the total number of paramagnetic sites accessible to the molecules of the liquid. In any case, the electron paramagnetic centers must be located at the outer edge of the oxide layer, whether it is the native oxide or a thermally grown oxide. Thus, in contrast with the ESR P_B

signal, which is due to defects at the $Si-SiO_2$ interface, the NMR relaxation measurements reflect unpaired electrons very close to the oxide surface.

Abstract

Several magnetic resonances have been observed in studies of the surface region of processed silicon. A readily detectable resonance occurs when the surface is damaged by abrasion, and it is removable by etching. Treatment with $HF-HNO_3$ etches creates paramagnetic states which become observable upon optical irradiation. Thermal oxidation causes three resonances to appear at different locations with respect to the $Si-SiO_2$ interface. Indirect evidence for a broad ESR resonance center located near the oxide surface has been inferred from NMR relaxation-time measurements of silicon powder mixed with liquids. The relevance of these results for elucidation of the electrical properties of the important $Si-SiO_2$ structure is noted.

Literature Cited

(1) Deal, B.E., J. Electrochem. Soc. (1974) 121 198C.
(2) Revesz, A.G. and K.H. Zaininger, RCA Rev. (1968) 29 22.
(3) Walters, G.K., J. Phys. Chem. Solids (1960) 14 43.
(4) Fletcher, R.C., W.A. Yager, G.L. Pearson, A.N. Holden, W.T. Read, and F.R. Merritt, Phys. Rev. (1954) 94 1392.
(5) Feher, G., Phys. Rev. (1959) 114 1219.
(6) Chung, M.F. and D. Haneman, J. Appl. Phys. (1966) 37 1879.
(7) Chung, M.F., J. Phys. Chem. Solids (1971) 32 475.
(8) Haneman, D., Phys. Rev. (1968) 170 705.
(9) Lemke, B.P. and D. Haneman, Phys. Rev. Lett. (1975) 35 1379.
(10) Kaplan, D., D. Lepine, Y. Petroff and P. Thirry, Phys. Rev. Lett. (1975) 35 1376.
(11) Kusumoto, H. and M. Shoji, J. Phys. Soc. Japan (1962) 17 1678.
(12) Chan, P. and A. Steinemann, Surface Sci. (1966) 5 267.
(13) Miller, D.J. and D. Haneman, Surface Sci. (1970) 19 45.
(14) Revesz, A.G. and B. Goldstein, Surface Sci. (1969) 14 361.
(15) Nishi, Y., Japan. J. Appl. Phys. (1971) 10 52.
(16) Caplan, P.J., J.N. Helbert, B.E. Wagner and E.H. Poindexter, Surface Sci. (1976) 54 33.
(17) Mizutani, T., O. Ozawa, T. Wada and T. Arizumi, Japan. J. Appl. Phys. (1970) 9 446.
(18) Shiota, I., N. Miyamoto and J. Nishizawa, Surface Sci. (1973) 36 414.
(19) Ruzyllo, J., I. Shiota, N. Miyamoto and J. Nishizawa, J. Electrochem. Soc. (1976) 123 26.
(20) Lepine, D.J., Phys. Rev. (1972) B6 436.
(21) Caplan, P.J., J.N. Helbert and E.H. Poindexter (unpublished).

16

NMR Studies of Oil Shales and Related Materials

F. P. MIKNIS and D. A. NETZEL

Energy Research and Development Administration, Laramie Energy
Research Center, Laramie, Wyo. 82071

The applications of nuclear magnetic resonance techniques in
the areas of biological, chemical and physical sciences has
steadily increased over the last several years. With the avail-
ability of high resolution Fourier Transform NMR spectrometers,
and recent advances in multiple pulse (1) and double resonance
techniques (2) the usefulness and applicability of NMR shows no
signs of slackening. Indeed, the wide variety of papers to be
presented at this symposium attests to this. However, in the area
of fossil fuels, the applications of NMR have been lacking. With
the current interest on energy research and development and the
future prospects of synthetic liquid fuels from oil shale, coal
and tar sands, NMR should become increasingly more useful for the
study of fossil fuels.
    Few NMR studies dealing with oil shale, coal or tar sands
have been published. Of those that have, the majority have been
wide-line NMR studies of coals and coal extracts (3-7). Carbon-
13 wide-line NMR has also been used in the study of coals (8,9).
High resolution proton NMR has been used to study asphaltenes
from Athabasca tar sands (10), and wide-line (11) and pulsed NMR
(12) have been used to estimate potential oil yields of oil
shales. More recently NMR double resonance techniques have been
applied to the study of coal aromaticity (13) and coal lique-
faction studies (14) and multiple pulse NMR techniques have been
applied to determine the chemical environments of protons in
coals (15).
    In this paper some applications of pulsed NMR to the study of
fossil fuels will be discussed. Emphasis will be on oil shales
and to a lesser extent, tar sands and coals. Topics to be discussed are NMR methods for assaying oil shales, temperature dependence of relaxation times in oil shales, and proton-enhanced
nuclear induction spectra of oil shales. The first is related to
resource evaluation, the second to processing and the third to
characterization studies. Throughout the discussion emphasis will
be on the practical utility of such measurements. Hopefully the
topics chosen for discussion will show the broad applicability of
NMR techniques to fossil energy research.

Background

As the topic of this paper deals primarily with oil shales, some qualitative aspects of oil shale are worth mentioning. First, oil shale is a misnomer. The rock is not a shale, but a marlstone, and it does not contain any liquid oil as such. Instead oil shale contains a solid organic material, called kerogen, which decomposes upon heating to form a shale oil. Kerogen, by definition, is not soluble in petroleum solvents. The small percentage of the organic matter in oil shale which is soluble in petroleum solvents is called bitumen.

Oil shales occur worldwide. The largest deposit is the Green River Formation in Colorado, Wyoming and Utah. The richness, or quality of an oil shale is measured as gallons of oil per ton of shale. Generally such assessments are made by the Fischer assay method, although other methods are available (12, 16-19) of which pulsed NMR is one of them. Because the gal/ton estimates are easy to use, all of these methods correlate their results against the Fischer assay methods.

Applications

NMR Assays   A simple but very important application of NMR measurements in oil shales is that of determining the potential oil yield of oil shales. A rapid and practical method for determining oil shale richness is of considerable importance, both for resource evaluation and processing technology. The Fischer assay method (20) has been the traditional method for assaying oil shales, although the method is largely empirical. There is a definite need to relate results of the Fischer assay to the more fundamental parameters of oil shale, such as organic carbon and hydrogen contents (16). NMR techniques should prove useful for the determination of these parameters.

The basic assumption underlying the NMR assay method is that the organic hydrogen content of an oil shale is directly relatable to the shale's potential oil yield. This assumption has been tested on 90 oil shales using wide-line NMR (11), and over 400 oil shales using pulsed NMR (12). In pulsed NMR, the hydrogen contents are measured simply by measuring the amplitude of free induction decay (fid), after the application of a 90° pulse. Interferences may arise due to inorganic hydrogens present in minerals and bound water, etc. Ways to minimize these interferences have been discussed (12).

Figure 1 shows a correlation between oil yields determined by the Fischer assay method, and the fid amplitude, normalized to unit sample weight. The oil shales used in this particular study were taken from a set of oil shales which had been carefully prepared for an ASTM study group. The aim of the study group is to investigate the precision of Fischer assay procedures. The oil yields are the averaged data for six participating laboratories.

The fid amplitude was sampled 20 us after the pulse, averaged 64 times in signal averaging computer and normalized to unit sample weight. The important features of Figure 1 are the excellent linear correlation between NMR data and Fischer assay data. Previous work (12) had established similar linear correlations, but with much greater scatter in data about the regression line. The scatter was attributed largely to imprecision in the Fischer assay method itself. The data also show that the NMR response is independent of particle size.

A significant advantage of the NMR method is its rapidity, compared to the Fischer assay methods. Typically a Fischer assay requires about 100 mins. per sample, whereas the NMR method requires 5 min. or less. The main disadvantage of the NMR method is the small sample sizes (5 g or less) normally used. Therefore, care must be taken to insure that the NMR measurements are made on representative samples.

Temperature Dependence of Relaxation Times   Because processes to convert oil shales to shale oils involve the application of heat, a natural NMR investigation would be the temperature dependence of relaxation times of oil shales. When oil shale is heated to 500°C for an extended period of time, the kerogen thermally degrades into a liquid shale oil which can be further refined into conventional petroleum products. Similarly by heating tar sands, the viscosity of the tar sand bitumen decreases, allowing for potential recovery of usable fuels from tar sand deposits via in-situ techniques.

Spin lattice relaxation times, T_1, of a tar sand core and its bitumen, and an oil shale and organic concentrate are shown on Figure 2. These measurements were made at a resonant frequency of 20 MHz, and using the 180°-90° pulse sequence. For all cases, the spin lattice relaxation times were non-exponential. The T_1's shown on figure 2 are the weighted average relaxation times, determined from the initial slope of the T_1 curves (21). Because of the complexity of the organic material in oil shales and tar sands, the use of average values is justified.

Both the tar sand and the bitumen exhibit a broad minimum at about 330°K, indicating that the average motions in each case are the same. A slightly broader distribution of correlation times is exhibited by the tar sand cores. This would be expected because of the reduced mobility of the bitumen, when confined to the core. Overall, the temperature dependence of T_1 for tar sands resembles that observed in many polymer systems. The T_1 curves for the tar sands follow the expected trends as a function of frequency at 60 MHz, the magnitude of T_1 at the minima was approximately 3 times the value at 20 MHz, and the minima shifted to a higher temperature ($\approx 350°K$).

Oil shales show a more complicated temperature dependency. The low temperature region suggests the possibility that another T_1 minimum may occur below 250°K, presumably due to methyl group

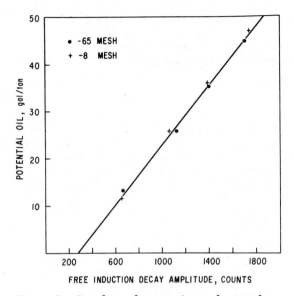

Figure 1. *Correlation between free induction decay amplitude and average oil yields, determined by Fischer assay for ASTM oil shales of different particle size*

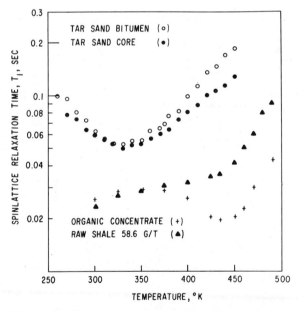

Figure 2. *Temperature dependences of spin lattice relaxation times in oil shale and tar sand*

reorientation. At higher temperatures (>400°K), a definite minimum in the organic concentrate (kerogen) curves is shown, which is
absent for the raw oil shale. Apparently the mineral matrix in
the raw shale prevents the same characteristic motions from occurring.

From a processing viewpoint the temperature range, 250°K to
500°K, is fairly low. The more interesting region for oil shales
and tar sands would be the range, 500°K to 750°K, the range in
which oil shale kerogen decomposes to bitumen which in turn produces the liquid products. Similarly at elevated temperatures,
T_1 or self-diffusion measurements could be made on tar sands, from
which the viscosity of the bitumen in place, might be obtained.
Such a measurement would be of practical utility for in-situ recovery methods. Unfortunately, we are unable to make NMR measurements at these temperatures.

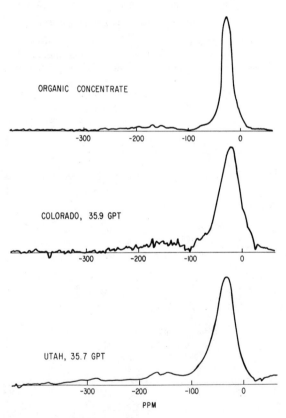

*Figure 3. Proton-enhanced ^{13}C NMR spectra of oil
shales*

Proton Enhanced ^{13}C NMR in Oil Shales   As a final application
of NMR to fossil fuels, we wish to discuss some preliminary work
on the application of double resonance techniques to the study of
oil shales. Recent advances in double resonance techniques (2)
have provided a method, whereby high resolution NMR spectra in
solids may be obtainable. By combining double resonance techniques, with sample spinning at the magic angle, the quality of the
NMR spectra obtained from solids, approaches that of materials in
solution (22).
     Some representative proton enhanced ^{13}C NMR spectra of oil
shales are shown on Figure 3. The spectra were obtained at a
resonance frequency of 46.6 MHz. The frequency scale is calibrated in ppm relative to an external TMS liquid reference.
These spectra are the Fourier transforms of 8000 transients obtained with a repetition rate of 5 sec, a 4 msec mixing time and
for a single contact. At first glance, the proton enhanced
spectra of oil shales are quite disappointing, exhibiting only
a broad band in the aliphatic region, and a broader, barely resolvable band in the aromatic region. However, it should be
stressed that the proton enhanced technique is probably the only
one available capable of a direct measurement of these types of
carbons. Previous work in oil shale kerogen (23) has relied on
rather severe oxidation, reduction and thermal degradation to
arrive at some knowledge of the chemical composition of oil shale
kerogens.
     In general Green River Formation kerogen is a macromolecular
material composed of approximately 5 to 10 percent chain paraffins,
20 to 25 percent cyclic paraffins, 10 to 15 percent aromatics,
and 45 to 60 percent heterocyclics. The overall features of the
proton enhanced spectra support this compositional make up. The
greater width of the aliphatic resonance bands for the raw oil
shales is probably due to differences in molecular mobility of the
kerogen in the mineral matrix. The aliphatic bands of the Colorado and Utah shales show a shift of about 10 ppm relative to each
other. Whether this shift is reflective of differences in chain
lengths, branching, types of cyclic materials, etc between Utah
and Colorado shales is not known. Clearly, there is a definite
need to apply sample spinning techniques to oil shales, before
these questions can be answered.

Summary

     In this paper we have attempted to describe some aspects of
oil shale technology, for which NMR technques might serve a useful
purpose. These ranged from the very simple NMR measurements for
assaying oil shales, to the state-of-the-art proton enhanced
methods, which show definite possibilities for oil shale characterization, and compositional studies. Although oil shales were
emphasized the same types of measurements could be applied in a
similar manner, for similar pupuses in tar sands and coals.

Acknowledgment

The authors wish to thank Dr. Alexander Pines and Mr. David
Wemmer of the Chemistry Department, University of California at
Berkeley for providing us with the proton enhanced ^{13}C spectra
of oil shales.

Literature Cited

1. Vaughan, R. W., Ann. Rev. Mat. Sci., (1974), 4, 21-42.
2. Pines, A., M. G. Gibby, and J. S. Waugh, J. Chem. Phys.,
    (1973), 59, 569-590.
3. Oth, J. F. M., and H. Tschamler, Fuel, (1961), 40, 119-123.
4. Ladner, W. R., and A. E. Stacey, ibid, (1961), 40, 295-305.
5. Sanada, Y., and H. Honda, ibid, (1962), 41, 437-441.
6. Smidt, J., W. van Raegen, and D. W. van Krevelen, ibid,
    (1962), 41, 527-535.
7. Ladner, W. R., and A. E. Stacey, ibid, (1964), 43, 13-21.
8. Retcofsky, H. L., and R. A. Friedel, Anal. Chem., (1971), 43,
    485-487.
9. Retcofsky, H. L., and R. A. Friedel, J. Phys. Chem. (1973),
    77, 68-71.
10. Speight, J. G., Fuel, (1971), 50, 102-112.
11. Decora, A. W., F. R. McDonald, and G. L. Cook, Using Broad-
    Line Nuclear Magnetic Resonance Spectrometry to Estimate
    Potential Oil Yields of Oil Shales. BuMines RI 7523, (1971),
    30 pp.
12. Miknis, F. P., A. W. Decora, and G. L. Cook, Pulsed Nuclear
    Magnetic Resonance Studies of Oil Shales-Estimation of
    Potential Oil Yields. BuMines RI 7984 (1974) 47 pp.
13. VanderHurt, D. L. and H. L. Retcofsky, Fuel, to be published.
14. Retcofsky, H. L., and Schweighardt, Presented at 27th
    Pittsburgh Conference on Analytical Chemistry and Applied
    Spectroscopy, Cleveland, Ohio, March 1-5, 1976.
15. Chow, C., R. G. Pembleton, and B. C. Gerstein, ibid.
16. Cook, E. W., Fuel (1974), 53, 16-20.
17. Smith, J. W., Ind. and Eng. Chem. (1956), 48, 441-444.
18. Biscar, J. P., J. Chromat. (1971), 56, 348-352.
19. Reed, P. R., and P. L. Warren, Quart. Colo. School of Mines,
    (1974), 69, 221-231.
20. Stanfield, K. E., and I. C. Frost, Method of Assaying Oil
    Shale by a Modified Fischer Retort. BuMines RI 3977, (1946)
    11 pp.
21. Woessner, D. E., B. S. Snowden, R. A. McKay and E. T. Strom,
    J. Magn. Resonance (1969) 1, 105-118.
22. Schaefer, J., and E. O. Stejskal, J. Amer. Chem. Soc., (1976),
    98, 1031-1032.
23. Robinson, W. E., in "Organic Geochemistry", Eglington, G. and
    M. T. J. Murphy, Eds., Springer-Verlag, New York, N. Y., 1969,
    Chapter 26.

On the Mobility of Benzene and Cyclohexane Adsorbed on Graphitized Carbon Black

B. BODDENBERG and J. A. MORENO

Institute of Physical Chemistry, Technical University of Hannover, Welfgarten 1, 3000 Hannover, W. Germany

Technical University of Hannover

Highly graphitized carbon blacks are known to possess energetically homogeneous surfaces as is well documented by numerous adsorption studies. This property is explainable with the notion that the faces of the polyhedral (1) particles are the graphite basal planes. There is much controversy in the literature about the structure and the dynamical state of the adsorption layer of molelules in the mono- and submonolayer regions (1,2).

The most controversial subject with this respect seems to be the system benzene / graphitized carbon black. Whereas Pierce and Ewing (3) conclude from adsorption isotherm and heat of adsorption measurements (-20 to 0°C) that localized adsorption is prevailing with the benzene molecules lying flat on the surface, Ross and Olivier, on the other hand, were able to fit adsorption isotherms (0 to 50 C°) with the assumption of two-dimensional real gas behaviour and free rotation of molecules about at least one axis. In view of the high anisotropy of the polarizability tensor of the benzene molecule the latter authors favour a model with the molecules standing up on the surface and rotating freely about the hexad axis. Steele (2) considers the molecules lying flat on the surface as the most probable orientation. The arguments of other authors (4,5,6) are on similar lines. To our knowledge there is up to now no spectroscopic evidence in favour of one of these models. We were able to show from NMR measurements that neither of these models explains the low temperature (<150 K) NMR data.

Using pulse NMR techniques the proton magnetic relaxation times T_1 and T_2 at 16 and 60 MHz have been measured of benzene adsorbed on Graphon (Cabot Corp.) in the temperature range 280 to 80 K. The amounts adsorbed were 0.18, 0.36 and 0.72 mmol/g which correspond to surface coverages $\Theta = 0.5$, 1 and 2, respectively, assuming 40 Å ^{2} to be the area occupied by one adsorbed molecule. In addition, T_1 adn T_2 have been measured for C_6H_6/C_6D_6-mixtures at the same surface coverages in order to separate the intra- and intermolecular contributions to the relaxation rates. Using semiclassical NMR relaxation theory the following model is able to represent the measured T_1 data ($\Theta = 1$) for temperatures between 80 and 150 K within the range of experimental error.

(i) The benzene molecules are adsorbed at fixed sites with their planes oriented perpendicularly to the graphite basal planes,

(ii) the molecules perform reorientational jumps about diad axes to next positions 60° apart at a rate given by $2.3 \cdot 10^{13} \exp (1060/T) s^{-1}$,

(iii) there is a two-demensional ordered structure of benzene molecules with hexagonal symmetry. The distances of the centers of gravity of the molecules are determined by the graphite lattice.

Remarkably, with only slight deviations, the T_1-data are the same for surface coverages 0.5 to 2 which indicates that lateral interactions of the molecules are of importance. For $\Theta = 2$, a second phase of benzene was detected the T_1 data of which are nearly equal to those of solid benzene.

The transversal relaxation function is temperature independent in the temperature range 80 to 150 K, but is of complex shape.

At temperature above 150 K additional, presumably translational motions show up in the relaxation times. The interpretation of these data is being undertaken.

Literature Cited

(1) Ross, S. and Olivier, J.P., "On Physical Adsorption", Interscience, New York, 1964.

(2) Steele, W. A., "The Interaction of Gases with Solids", Pergamon, Oxford, 1974.

(3) Pierce, C. and Ewing, B., J. phys. Chem. (1967) 71, 3408.

(4) Isirikyan, A. A. and Kiselev, A. V., J. phys. Chem. (1961) 65, 601.

(5) Avgul, N. N. and Kiselev, A. V. Chemistry and Physics of Carbon (1970) 6, 1.

(6) Pierotti, R. A. and Smallwood, R. E., J. Coll. Interf. Sci. (1966) 22, 469.

Quantitative Electron Spin Resonance Studies of Chemical Reactions of Metal Impregnated Activated Charcoal

RAYMOND A. MACKAY

Drexel University, Philadelphia, Pa. 19104

EDWARD J. POZIOMEK and RICHARD P. BARRETT

Edgewood Arsenal, Aberdeen Proving Ground, Md. 21010

We recently reported (1) the discovery that the reaction of cyanogen chloride gas with ASC whetlerite, a Cu/Ag/Cr impregnated activated charcoal, produces a strong electron spin resonance (esr) signal. ASC whetlerite (2) is impregnated with salts of Cu(II), Cr(VI) and Ag(I). The removal mechanism of the carbon for CNCl involves copper, with a synergistic effect being exerted by chromium. The presence of silver is apparently unimportant for CNCl activity (1, 3). The esr (derivative) signal consists of two partially overlapping lines. One line has a g value of 2.12 and a half width of 190 gauss, and the other a g value and half width of 2.00 and 425 gauss, respectively. The signals have been shown to arise from monomeric hydrated Cu(II) species on the charcoal surface, resulting from the reaction of the CNCl(1). This esr detection method appears to be quite specific for cyanogen chloride, and preliminary results indicated that the esr signal intensity could be used quantitatively as a measure of the extent of reaction. We report here a more detailed quantitative esr investigation of the reaction of CNCl with ASC whetlerite.

Experimental

ESR Studies. The X-band (9.5 GH_z) derivative esr spectra were obtained on a Varian E-4 spectrometer system employing 100 kHz field modulation and a modulation amplitude of 10 gauss. After mechanical grinding of the sample in a Wig-L-bug for two minutes, samples of about 5 mg of the charcoal powder were placed in glass melting point capillary tubes and packed by dropping down a glass tube of fixed height a preset number of times. Both the sample weight and height in the capillary were recorded. The capillary was placed in a sample holder which positioned the tip containing the sample in the center of the cavity. All samples were run at a microwave power level of 4mW, and at ambient temperature.

Tube Reactions. The reaction of CNCl with charcoal beds under flow conditions was performed as previously described ([1]).

Results and Discussion

Quantitative esr Measurements. At ambient conditions there is no discernable variation of signal shape or intensity with temperature, so careful control is not required. The actual esr signal intensity is given by the area under the absorption curve. If the line width does not change, the peak to peak height of the derivative line is proportional to the intensity. If both the sharp and broad esr lines maintain the same shape and ratio, it should still be possible to use the height of the sharp line as a measure of signal strength. As will be seen, this generally appears to be the case, at least when the exposure is below the maximum capacity. For absolute comparison between different samples, it is necessary to correct for background (no CNCl). This is approximated by subtracting the difference in signal height of the blank, at the peak positions of the sample, from the sample signal height, as shown in Figure 1.

The dependence of signal height (intensity) and reproducibility on the sample size was investigated. By sample size, we refer to sample weight and height (packing density). The effect of small variations in positioning of the capillary in the cavity was concurrently examined. It was found that the signal intensity of samples much larger than 5 mg depended critically on positioning; while smaller samples were difficult to weigh accurately. Aging of the sample, at least over periods of days, had no effect on the signal intensity. In general, reproducibility both for the same sample run at different times in the same day and on different days, as well as for different samples from the same batch of charcoal, was within $\pm$ 5%.

ASC Whetlerite. The esr samples were obtained from charcoals run in CNCl tube tests. The CNCl gas was passed through charcoal beds of various weights (w) in tubes of fixed diameter. Flow rate, relative humidity, and CNCl concentration were held fixed, and the time required for a fixed fraction of the inlet CNCl to break through the bed (t_b) was determined. A plot of t_b vs w is linear at higher bed weights, as given by the modified Wheeler equation

$$t_b = \frac{W_e}{C_0 L} (w - w_c)$$

where C_0 is the initial concentration of CNCl in the gas stream, and L is the flow rate. The slope is taken as a measure of the dynamic capacity of the charcoal W_e in min/gm, and the intercept at t_b = 0 as the critical bed depth (w_c) in grams, as shown in

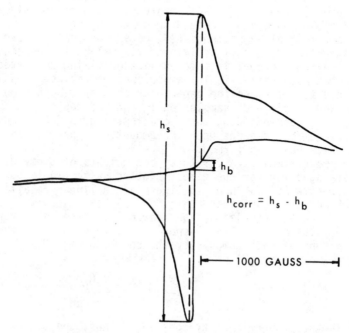

Figure 1. Correction of peak-to-peak height

Figure 2. Samples taken from the upstream end of the tube are referred to as "top of the bed". Samples obtained by first mixing and grinding up the entire tube contents are referred to as "whole bed".

In order to obtain a wide range of charcoal activity, batches of a given lot of ASC whetlerite were sealed in bottles saturated with water vapor for varying lengths of time at high temperatures ("aged"). The aged charcaols gave lower values of t_b for a given bed weight, depending on the aging time.

In our initial experiments (1), we showed that the esr signal strength (S) was directly proportional to the percentage of CNC1 treated charcoal in samples made by mixing untreated and CNC1 treated ASC whetlerite in various weight ratios. However, it remained to determine the dependence of S from both "top of the bed" and "whole bed" samples on t_b and bed depth. In particular, t_b depends upon the mechanism(s) of a CNC1 removal while S presumably depends upon the amount of a particular copper species which has been formed.

The break times (t_b) for total bed weights of 5 gms of ASC whetlerite aged for various times, plotted vs the whole bed esr signal strength ($\bar{S}$) is shown in Figure 3. A linear relationship between $\bar{S}$ and t_b is obtained.

If the copper species on the charcoal surface is responsible for the removal of CNC1, the $\bar{S}$ should be proportional to the average amount of CNC1 removed by the bed (m grams CNC1 removed per w grams of charcoal) or

$$\bar{S} = k \left(\frac{m}{w}\right) = K \left(1 - \frac{w_c}{w}\right)W_e = \frac{kLC_o}{w} t_b$$

where k is the proportionality constant. Thus, the linear relationship between $\bar{S}$ and t_b is consistent with this interpretation.

Various bed weights of the impregnated charcoal were tube tested with CNC1 to the break point, and the esr signal measured for top of the bed samples. A plot of S_t vs t_b for two different charcoals of about the same capacity and somewhat different critical bed depths is shown as Figure 4. All such curves exhibit the same general features. The value of S_t rises with increasing bed weights (or t_b), and levels off at high t_b.

Based on the previous result, S_t should be proportional to the capacity W at w = o. Assuming that $\int_0^w Wdw \approx \frac{Ww}{f}$ where f is a constant ($\sim$2), it follows from the Wheeler equation that

$$\frac{S_t}{S_t^{max}} \cong \frac{at_b}{1 + a\,t_b}$$

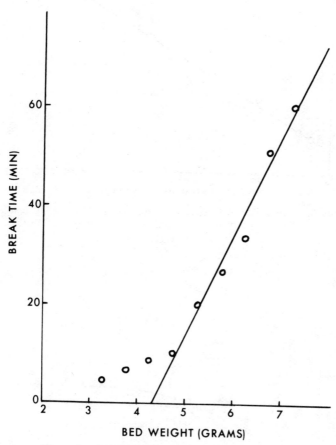

Figure 2. *CNCl breakthrough time vs. bed weight*

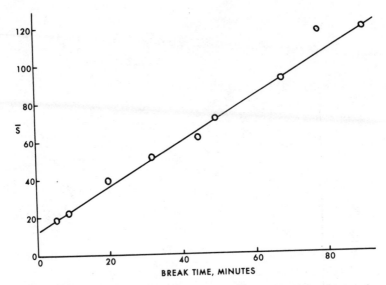

Figure 3. Whole bed ESR signal strength ($\overline{S}$) vs. CNCl break time for five-gram samples, aged for various times

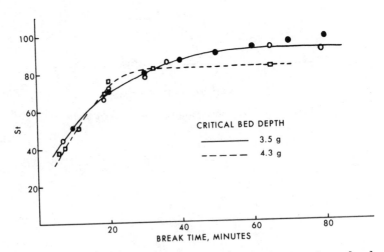

Figure 4. ESR signal strength (S) from top of the bed samples vs. break time

where $a = LC_0/W_e w_c$ and $S_t^{max} = kW_e$. The ratio of S_t^{max} is 1.4, while it should be 1.0. The ratio of $\underline{a}$ values is 0.7, compared with the ratio of critical bed depths (corrected for W_e ratio of 1.4) of 0.9. Thus, while $\bar{S}$ can be related to t_b, S_t can be related to w_c only at short break times where $at_b<1$). In this case,

$$S_t \cong \frac{kLC_0}{w_c} \qquad t_b = \frac{w}{w_c} \quad \bar{S}$$

However, the values of S are small for $w \sim w_c$, and thus this relationship is not expected to be very useful. At long times the value of S_t^{max} is obtained, but this is only a measure of W_e. To obtain W_e, the value of $\bar{S}$ at longer bed weights ($w >> w_c$), and thus higher t_b, can be used.

An examination of the bed depth profile provides further insight into these results. An 8.5 gram bed weight of ASC whetlerite was tube tested with CNCl to break, and samples withdrawn in sections proceeding along the bed. A plot of the esr signal strength (S) as a function of total bed weight (position along the tube) is shown in Figure 5 for flow rates of 1.0 and 1.63 l/min. The origin represents the upstream end of the tube. Although the general shape of the curve is in accord with that expected for a bed profile, a few salient features may be noted. First, the value of S at the origin is below the maximum indicating a decrease of signal by "excess" CNCl. Second, the curves for both flow rates have approximately the same value of S at the origin, indicating that the esr signal from a top of the bed sample depends on the total amount of CNCl to which it has been exposed. An examination of the esr curves indicate that the signal is not being destroyed, but rather that the relative proportions of the sharp and broad lines are changing. When the total intensity of the esr signal is plotted vs bed depth, the "drop" in S_t is eliminated. Based on these and our earlier results ([1]), the following mechanism is proposed. The structures given are meant to be schematic only:

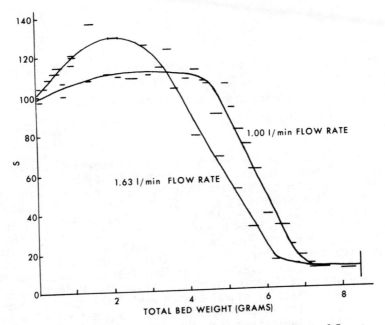

Figure 5. *ESR signal strength (S) vs. total bed weight for different CNCl flow rates*

Equations (i) and (ii) represent the initial hydrolysis of ClCN, while equation (iii) represents the conversion of the sharp to the broad esr line species by continued exposure to ClCN.

It has also been observed in static tests that a sufficiently long exposure to CNCl will completely destroy all traces of the esr signal. Since the sharp line is first converted to the broad, there is a fourth reaction of CNCl with the (schematic) species CN(OH)(CN) which converts it to a diamagnetic one also. This mechanism predicts that a maximum of 3 moles of CNCl can be removed per mole of active copper impregnant.

It is therefore apparent that esr measurements can be used to characterize the activity or history of an impregnated charcoal with respect to CNCl exposure. In addition, it can be a valuable adjunct to such studies in examining the nature of the active species and the mechanism(s) of removal. Under the proper conditions, the esr signal strength of a total bed sample of CNCl treated charcoal can be used as a measure of capacity, while the relative ratio of the sharp and broad lines can be related to the total ClCN that sample has seen.

Summary. A set of experimental conditions has been determined for obtaining quantitative esr data on impregnated charcoals. It has been demonstrated that esr measurement can be used to characterize the ClCN activity and history of the charcoal. We have proposed a mechanism and some details concerning the nature of the active species and reaction products.

Literature Cited

(1) E. J. Poziomek, R. A. Mackay and R. P. Barrett, "Electron Spin Resonance Studies with Copper/Silver/Chromium Impregnated Charcoals", Carbon, 13, 259 (1975).
(2) R. J. Grafenstetter and F. E. Blacet, "Impregnation of Charcoal", Ch 4, in Summary Technical Report of Division 10, National Defense Research Committee, Military Problems with Aerosols and Nonpersistent Gases, Vol 1 (1946), pb 158505.
(3) Ibid, Ch 7, pp 161-164, "Mechanism of Chemical Removal of Gases", J. William Zafar.

19

NMR Studies of ^{3}He on a Porous Heterogeneous Adsorbent at Low Temperatures

DOUGLAS F. BREWER

University of Sussex, Sussex, England BN1 9RH

Although the work I shall describe in this paper has produced many interesting results concerning the interaction between ^{3}He atoms and a substrate, the original reason for starting them was quite different. The primary aim was to investigate the effects of size and dimensionality in quantum fluids of ^{3}He and ^{4}He at temperatures from 50mK or less, up to about 3K. Since size effects become important at around 100Å or so, depending on temperature, it was necessary to work with He systems of about these dimensions.

An obvious system is the adsorbed film, which can readily be formed in thicknesses of less than one monolayer (~ 3Å) up to 100Å or more. If experiments are to be done which need reasonably large samples of He (of order 1 cm^3, for example, in a heat capacity experiment) then a large surface area for adsorption must be obtained - or, more exactly, a large surface-to-volume ratio. One adsorbent which is highly suitable for this purpose is Vycor porous glass, with an average pore diameter of around 70Å, and nearly all the experiments I shall discuss were done with this material as substrate. It has a surface-to-volume ratio of about 6×10^6cm^{-1} and hence a very large fraction of the adsorbent and adsorbate atoms lie close to their interface.

For the original purposes of the experiments, the interface interactions are a nuisance: ideally, one would like to work with a fluid system whose size is limited by rigid walls (or a free surface) which do not themselves exert a force on the fluid but simply impose boundary conditions on the wave function describing its behaviour. This is not exactly possible, and for practical systems the observed behaviour is determined not only by pure size effects but also by the interfacial forces. The separation of these presents difficult problems, but a first step is clearly the investigation of the helium-Vycor composite system when the wall interactions are by far the dominant ones, i.e. when only one monolayer or less of He is adsorbed: hence the relevance to this symposium. Before going on to describe these experiments, I shall first say something about the porous glass adsorbent, and also about the original aims of the research.

Size Effects in Liquid Helium

It is a matter of great luck that the only two substances that remain liquid down to the absolute zero obey different statistics (^{3}He is a fermion and ^{4}He a boson), allowing us to study the differing effects of quantum statistics on fluids at the lowest temperatures. Liquid ^{4}He, whose normal boiling point is 4.2K, undergoes a transition to an ordered superfluid state at 2.17K (the lambda point). This is generally thought to be a Bose-Einstein condensation which in a non-interacting boson gas is a purely statistical effect and in liquid ^{4}He is modified slightly by the atomic interactions.

Liquid ^{3}He at sufficiently low temperatures behaves as a fluid obeying Fermi-Dirac statistics, and does not undergo a purely statistical transition at these temperatures. At very much lower temperatures, of order 1 millikelvin, it does undergo a transition resulting from pair correlation effects between nuclear spins, of the same type which lead to superconductivity of electrons in metals which also form a Fermi fluid. Like ^{4}He, the transition is to an ordered superfluid state characterised by an order parameter Ψ which is the wave function describing the superfluid. Ψ is not allowed to vary significantly over a distance called the coherence length, ξ, which is temperature dependent. In superfluid ^{3}He this is probably about 100Å and in ^{4}He it is about 2Å at the lowest temperatures, but it increases rapidly near the transition temperature where it becomes infinite. Size effects occur when the system size becomes of the order of ξ.

A good deal is known about the behaviour of adsorbed ^{4}He in film thicknesses between 3Å and about 100Å. For example, the specific heat, which in bulk liquid has a logarithmic infinity at the lambda point, becomes smeared out, with a maximum occurring at lower temperatures for thinner films. (1) This striking behaviour is shown schematically in Figure 1. Superfluidity no longer starts at the temperature T_c of the maximum of the specific heat anomaly, but at a still lower temperature T_0, also indicated in Figure 1. The reason for these effects, and the nature of the state between T_0 and T_c, is not understood, and the various proposals that have been put forward are controversial.

Two other effects in He adsorbed on porous glass are worth pointing out in this brief survey. One is that the specific heat at the lowest temperatures rises above the bulk value when a free surface is present, due probably to surface excitations (2). The second is the

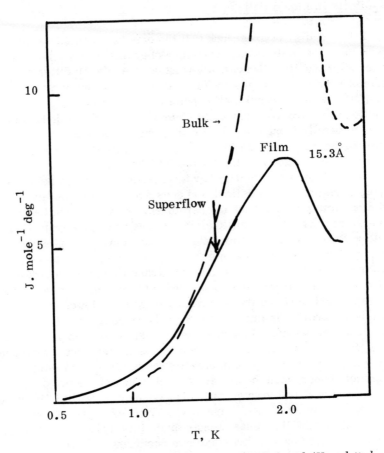

Figure 1. Specific heat of a ⁴He film and bulk liquid ⁴He, plotted against temperature

anomalous behaviour of the dielectric constant, which has been measured as a function of temperature and pressure in an attempt to correlate it with density observations. It has been found impossible to do this in a consistent way and it may be necessary to invoke the existence of induced dipole moments in the first layer of He atoms adsorbed at the solid surface. Boundary effects have also been observed with bare Vycor porous glass which has a very large temperature variation of dielectric constant (as has also been observed in solid glasses, though not necessarily for the same reason) (3).

In the first monolayer, specific heat measurements (4) have been made for both ^{3}He and ^{4}He, and show a T^2 temperature dependence which is characteristic of a two-dimensional Debye solid, as shown in Figure 2. These observations, together with several others including the NMR experiments which I shall talk about later, show with virtual certainty that the atoms of the first monolayer adsorbed in Vycor are localised, with a high density corresponding to solid at several hundred atmospheres pressure.

The Substrate

Vycor is a high silica content glass manufactured by Corning, the variety used in these experiments being porous Vycor 7930, described originally by Nordberg (5). It is made by heat treating an alkali-borosilicate glass at around 600°C for a long period during which phase separation occurs, then leaching out the boron-rich phases with acid to leave a high porosity glass containing 96% SiO_2, 3% B_2O_3, and small quantities of Na_2O_3, Al_2O_3 and other oxides. The structure possesses short range order like a supercooled liquid, with Si and O atoms in a ring configuration and the boron impurities at the centre of the rings.

Various methods can be used to determine surface areas and porosities giving results which do not differ greatly although the differences can be outside the experimental precision. In the work to be described, the area A was determined by BET analysis of a nitrogen adsorption isotherm at its normal boiling point and the total pore volume V by the amount of nitrogen adsorbed at saturation. A uniform cylindrical pore model of the structure then gives the pore diameter as $d = 4V/A$. Application of such a model has the virtue of enabling these quantities to be evaluated consistently and quite precisely, but they should be regarded primarily as parameters whose values as derived from various experiments can be compared in order to investigate the validity of the model and as a help in analysing and interpreting results. Typical results of applying this approach to

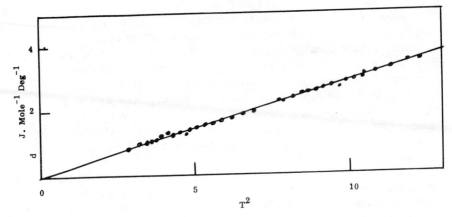

Figure 2.   Specific heat of a pure monolayer ^{3}He film as a function of (temperature)^2

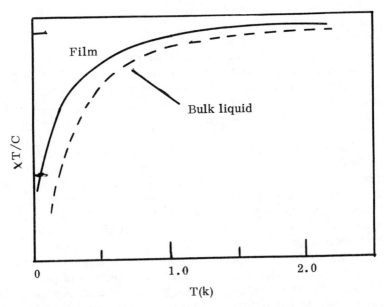

Figure 3.   A plot of $\chi T/C$ vs. T showing that Curie's Law is obeyed above about $1°K$, but the film susceptibility is enhanced at lower temperatures. The origin of coordinates is the point $(0, 0.2)$.

Vycor are:

$$
\begin{aligned}
\text{Surface area} &= 135\text{m}^2\text{gm}^{-1}\\
\text{Volumetric pore diameter} &= 62.4\text{Å}\\
\text{Porosity} &= 32\%
\end{aligned}
$$

Different samples of Vycor 7930 may give slightly different values: for example, one sample used gave a pore diameter of 71Å , which probably represents a real, though small, difference in pore structure. Electron micrographs show that there may be variations in pore diameter of a factor of two, although the pore size distribution is thought to be quite narrow.

The Nuclear Magnetic Susceptibility

NMR measurements give us valuable information on the motion and disposition of the atoms. In examining the properties of the adsorbate as a degenerate quantum system and the way in which it is affected by substrate interactions, the nuclear magnetic susceptibility χ is a particularly important quantity. It has been measured both by continuous wave and by pulsed NMR techniques for bulk liquid and for ^{3}He films over a wide range of temperature. Figure 3 shows the striking result found for the liquid at the saturated vapour pressure at temperatures below 2K, plotted as χT versus T (6). In these measurements, χ cannot be experimentally determined in absolute measure, and the results have been normalised to

$$
\frac{\chi\,T}{C} = 1
$$

in the higher temperature region where χT is found to be temperature independent; C is a normalising constant. In this region the liquid, although mobile and with significant interaction forces, obeys the classical Curie law (χT = constant) as for a system of free spins. Although χ cannot be measured absolutely, careful comparison measurements have shown that the value of the constant C is in fact the free spin-half value. In addition, and most surprisingly, the value of χT in this region is independent of the density of the liquid; and it also remains the same when solidification takes place when the exchange interaction of the solid is small.

At the lower temperatures χT/C falls away from unity, and eventually we find that χ becomes temperature independent (Pauli paramgnetism) as the absolute zero is approached (provided, as mentioned below, that the temperature is not less than a few millikelvin). This general behaviour is exactly what one would expect of an ideal Fermi gas and the curve does indeed fit extremely well if a Fermi degeneracy temperature $T^* = 0.45K$ is used. The effect of spin-dependent interaction is thus shown to be strong, since an ideal gas with the density of liquid ^{3}He would have a much higher degeneracy temperature—5K. These interactions can be taken account of in the Fermi fluid theory of Landau, which is now fully accepted as a phenomenological model of the liquid in the highly degenerate region, until the superconducting type transition occurs at around 1mK. In solid ^{3}He, the exchange interaction is very small - less than 1mK - and the dipole-dipole interaction is $\sim 10^{-7}$K. Down to temperatures of a few millikelvin, therefore, the nuclear susceptibility follows a classical Curie law, as mentioned above.

We now examine the results obtained with ^{3}He films adsorbed on porous Vycor glass, in the light of the above observations. When only a few atoms are adsorbed they are sufficiently tightly bound to the substrate to be localised, and again for temperatures between 0.3K and about 2K they obey a Curie Law, $\chi T =$ constant. This has been experimentally observed for coverages between 1/3 monolayer and 1 monolayer, and must certainly hold for lower coverages where experimental difficulties have so far prevented direct observation.

This ideal classical behaviour indicating the essentially non-interacting magnetic behaviour of adsorbed ^{3}He atoms in the monolayer, is shown in another way in Figure 4, where the total susceptibility signal (total magnetisation) in a CW measurement at frequency 6.24MHz and temperature 0.405K is plotted as a function of the amount of ^{3}He adsorbed (7) (1). The linearity of the graph in the region up to one monolayer coverage suggests very strongly that the addition of each spin does not affect the susceptibility of those already there, which is characteristic of a non- interacting system. This is convincing evidence of the localised nature of the atoms in the first complete layer, especially since, as Figure 4 also shows, higher coverages exhibit progressive dengeneracy as the atoms interact more strongly. The formation of the second and higher layers is accompanied by a smaller susceptibility for each added spin, which can be interpreted as a Fermi degeneracy effect. When the pores are very nearly full, the graph is linear again but with a smaller slope than for sub-monolayer films. When the pores are completely full, the additional signal is provided by the bulk liquid which then surrounds the porous glass in the cell, and the slope of the graph here therefore corresponds to degenerate bulk liquid and gives a susceptibility calibration which shows that the outer layers of the adsorbate behave magnetically like bulk liquid. The extrapolation of the high covering linear region back to zero cooling gives a finite intercept on the susceptibility axis which shows that the layers near the wall retain their enhanced susceptibility even in the presence of the overlayers.

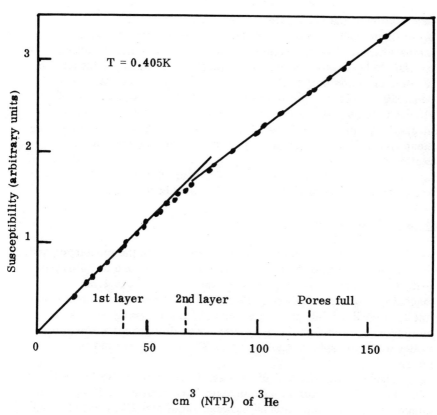

Figure 4.   *The total susceptibility of an adsorbed ³He film as a function of coverage at 0.405° K*

These observations demonstrate that the nuclear magnetic
susceptibility of even a rather thick ^{3}He film is dominated by the
substrate interaction which leads to a very marked interfacial
enhancement of the susceptibility.

It is important to note that the substrate-enhanced susceptibility,
which in the temperature region above ~0.4K can be attributed to
localisation of the adsorbate atoms at the wall which then contribute a
large Curie susceptibility, becomes relatively much greater at lower
temperatures. Even in gaps as large as 6μ the average susceptibility
is found to be substantially greater than in bulk liquid, although the
fraction of atoms next to the wall is minute (8). The wall effects
penetrate deeply into the liquid in a way which is not understood,
although the suggestion has been made that the enhanced wall
magnetisation may be carried into the bulk liquid because of the long
mean free path of the ^{3}He quasiparticles emitted from the wall
region (9).

In superfluid ^{3}He below 2.65mK, much more complicated effects
are found which we shall not discuss further here.

NMR Relaxation Times

The nuclear susceptibility measurements in the monolayer and
sub-monolayer region described above give us the useful information
that in the temperature region studied - about 0.4K to 2.5K - the
magnetic interaction between ^{3}He nuclei has a negligible effect. In
itself, it does not tell us a great deal about the arrangement and
motion of the atoms on the substrate. For this purpose, the spin-
lattice and spin-spin relaxation times (T_1 and T_2) are more
informative.

There is a great deal of data available on these two quantities for
^{3}He adsorbed on Vycor in submonolayer and multilayer films of various
thicknesses, for temperatures between 40mK and 2K or so. In many
cases, several different interpretations are possible of the physical
processes involved and further careful experimentation is needed
before the situation is clarified. One important experimental effect is,
however, clear, and that is the necessity for careful annealing of the
adsorbate in order to obtain reproducible results. If the films are
adsorbed at temperatures below 4K, or even at 4K and then
immediately cooled, the results are not reproducible. In general, T_1
is much shorter and T_2 is longer for such unannealed films. This is
presumably because the films form in clumps on one part of the
adsorbent near the gas inlet tube, but spread out into an equilibrium
configuration given enough time at a high enough temperature. The
effect presumably depends to some extent on the design details of the

apparatus, but it must clearly be looked for very carefully, and the possible need for annealing must always be investigated. All the results presented below were taken on films fully annealed at 4K for at least 15 minutes, and were unchanged after further annealing at this temperature. It is not clear how much the annealing process depends on surface diffusion, and how much on equilibration through the vapour phase.

Figure 5 shows how T_1 and T_2 evolve during the build-up of a ³He film on Vycor from a fractional coverage θ of 1/4, up to a full monolayer, at a temperature of 0.405K (full monolayer coverage corresponds to 38.6 cm^3 NTP) (10). All of the T_1 measurements were taken with a CW technique at 6.24MHz, and were made by the standard method of saturating the resonance, then lowering the rf level well below saturation and measuring the time constant for recovery of the signal. All recoveries observed in this way were exponential. For a reason which will be explained later, these values of T_1 will be referred to as the "long component". The T_2 measurements were taken in two different ways. With CW at 6.24MHz, the line width was always greater than the inhomogeneity of the magnet (which was ~0.16G) for $\theta < 1$. In these cases, T_2 was taken as the reciprocal of the difference between the observed line width and the magnet width. For the complete monolayer, a different method was used, explained in reference (10), involving measurement of T_1 and the saturation factor $Z = (1 + \gamma^2 H_1^2 T_1 T_2)^{-1}$ where γ is the gyromagnetic ratio and H_1 the amplitude of the rf field. In addition, T_2 was determined with 90°-180° pulse sequences and observation of spin echoes, at 2.14MHz. It is notable that T_2 recoveries are always experimental. As shown in Figure 5, agreement between the two methods is very good.

Two striking features of Figure 5 are the very rapid decrease of T_1 at around $\frac{1}{2}$ monolayer from about 200 seconds to 0.4 seconds while T_2 increases only slowly, and the fact that $T_1 \gg T_2$ at low coverages. For $\theta \simeq 1/4$, $T_1/T_2 \simeq 4 \times 10^5$, and even at $\theta = 1$, $T_1/T_2 \simeq 20$. It is clear that a significant change is taking place in the state of the film between low and high coverages.

If these data can be interpreted in terms of standard BPP theory, it can be inferred that the large T_1/T_2 ratio indicates that the adsorbate corresponds to an immobile, rigid lattice at low coverages, but that the degree of mobility increases when $\theta \lesssim \frac{1}{2}$. It is possible that the initial formation of the film is in rigid-lattice islands on the lowest energy adsorption sites, and that as the coverage increases a more mobile film is formed between the islands. Such an interpretation is consistent with Roy and Halsey's explanation (11) of the specific heat of helium adsorbed on argon-coated sintered copper.

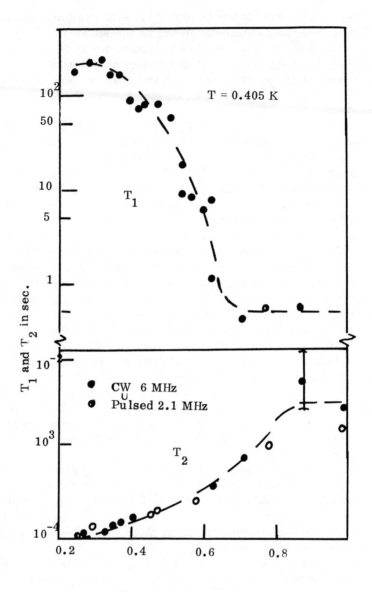

Figure 5. T_1 and T_2 at 0.405°K as a fraction of coverage

Even for monolayer films we find that $T_2 \ll T_1$, and hence the adsorbate in all monolayer and submonolayer coverages is unlikely to be liquid-like or gas-like. It should be noted that relaxation times measured on Zeolite ($\underline{12}$) have exactly the opposite behaviour in the sense that T_1 increases and T_2 decreases as the coverage increases. Zeolite is, of course, a very different structure from Vycor, with much smaller pores of diameter $\sim 9\text{Å}$.

A more precise measurement of the recovery of the magnetisation particularly the initial part, can be made by spin-echo techniques. Figure 6 shows the signal amplitude as a function of time taken in this way, for various frequencies using one monolayer at 0.3K. Clearly the recovery is not given by one single exponential term, although the CW method for measuring T_1, which looks at the later stages of recovery, did appear to give a simple exponential, with a characteristic time which we referred to above as the "long component."

Various reasons can be given for non-exponential behaviour. The most obvious one is the heterogeneity of the adsorbent, which leads to different spin-lattice interactions at different parts of the surface, and hence to a range of relaxation times. A second possibility arises from comparison with solid ^3He, where a similar effect is found. In that case, as discussed in detail by Guyer et al ($\underline{13}$), it is due to the existence of an additional thermal reservoir (the exchange bath) through which the energy of the Zeeman system passes on its way to the lattice. The overall process is then determined by the Zeeman-exchange relaxation time and by the exchange-lattice time, and should be given by the sum of two exponentials rather than of many. Our data is not sufficiently accurate to discriminate between the two explanations, although the second is supported to some extent by the observation that the character of the recovery is different if the system is first saturated for a long time. For example, with 3/4 layer coverage a sequence of 20 pulses separated by 20 msec removes the initial fast component and the magnetisation recovers exponentially with a long time constant. A third possible explanation of non-exponential signals has been put forward recently by Mullin, Creswell and Cowan ($\underline{14}$), who point out the possibility of anisotropy with respect to field direction in the case of a randomly oriented adsorbent surface such as that provided by Vycor. It may be that all of these effects are present to some degree.

From the graphs of Figure 6 one can extract, for each frequency, a "short component" of T_1 for the initial recovery. Figure 7 shows that this varies linearly with frequency, as also observed with zeolite ($\underline{12}$). This may be a characteristic of surface relaxation, due to anisotropy ($\underline{14}$) or heterogeneity of the surface which leads to a range of correlation times (see also below).

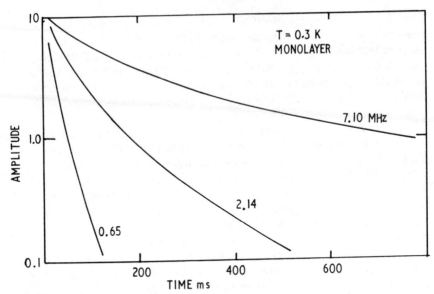

Figure 6.   Spin-echo amplitude as a function of time for a monolayer of ^3He at 0.3°K,
showing nonexponential decay

The temperature dependence of the relaxation times reveals
another feature which is similar to the case of solid ^3He (16)(17),
namely (Figure 8) a minimum in the short component of T_1 as a
function of temperature for 2/3 monolayer coverage (15). According
to BPP theory, the minimum occurs when the diffusion time τ is
around the inverse of the Larmor frequency ω_0. In solid ^3He the
Zeeman energy relaxes direct to the lattice at high temperatures,
whereas on the low temperature plateau T_1 is determined by the
Zeeman-exchange bath relaxation mentioned above, taking place by
tunnelling or exchange. Although the minima are not very clearly
defined, one can get a rough estimate of an activation energy ϵ by
fitting the temperatures where the minima occur to the equation

$$\tau = \tau_0 \exp(\epsilon/\tau)$$

giving $\epsilon \simeq 30K$ and $\tau_0 \simeq 10^{-13}$ sec. The latter may be interpreted
as the average period of vibration of a ^3He atom in the adsorbed
layer.

A detailed analysis of these and similar experiments is still
taking place, with the assumption that the very shallow minimum is a
result of a wide range of correlation times, with a composite spectral
density of Lorentzian form. Preliminary indications are that this can
explain the linear variation of T_1 with frequency (Figure 7), and a
large ratio T_1/T_2 (Figure 5); it gives a range of correlation times
varying from 10^{-5} sec to 10^{-9} sec.

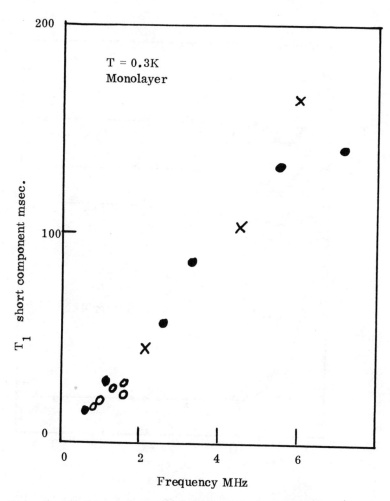

Figure 7. The "short component" of T_1 for a monolayer of ³He at 0.3°K, plotted against frequency. Three sample chambers were used.

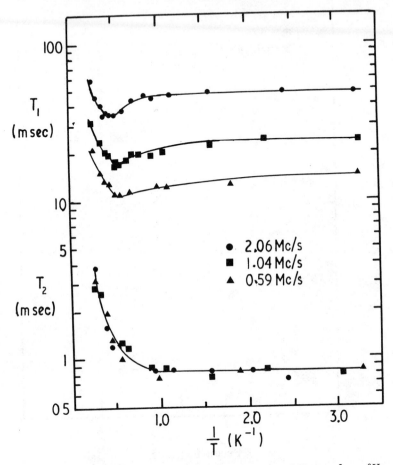

Figure 8.   Minima in the spin-lattice relaxation of a 2/3 monolayer ^{3}He film for different frequencies, as a function of T^{-1}

The above brief summary has necessarily omitted a great deal of detail of the many observations that have been made of ^{3}He adsorbed on Vycor, and also the use of other adsorbents. Although the data are often difficult to correlate and to analyse in terms of more usual bulk systems, some progress is being made in understanding them, and the techniques are a useful tool for characterising the surfaces.

This work was supported in part by the Science Research Council and by the U.S. Army through its European Research Office (Grant DA-ERO-124-74-G0046).

Literature Cited

(1)   Brewer, D.F., J. Low Temp. Phys. (1970) 3 205.

(2)   Brewer, D.F., Symonds, A.J. and Thomson, A.L., Phys. Rev. Letters (1965) 15 182.

(3)   Baker, J.C., D.Phil. Thesis, University of Sussex (1976), unpublished.

(4)   Brewer, D.F., Evenson, A., and Thomson, A.L., J. Low Temp. Phys. (1970) 3 603.

(5)   Nordberg, M. E., J. Am. Ceramic Soc. (1944) 27 299.

(6)   Thompson, J. R., Ramm, H., Jarvis, J.F., and Meyer, H. J. Low Temp. Phys. (1970) 2 521.

(7)   Brewer, D. F., Creswell, D. J. and Thomson, A.L., Proc. 12th Int. Conf. on Low Temp. Phys. Ed. E. Kanda, p.157, Academic Press of Japan, Tokyo, 1970

(8)   Ahonen, A. I., Kodama, T., Krus ius, M., Paalanen, M.A., Richardson, R.C., Schoepe, W. and Takano, Y., preprint (1976)

(9)   Creswell, D. J. and Brewer, D.F., "Liquid and Solid Helium" p.329, John Wiley and Sons, New York 1974.

(10)   Creswell, D.J., Brewer D.F., and Thomson, A.L., Phys. Rev. Letters (1972) 29 1144.

(11)   Roy, N.N., and Halsey, G.D., J. Low Temp. Phys. (1970) 4 231

(12)   Monod, P. and Cowen, J.A., private communication.

(13)   Guyer, R.A., Rev. Mod. Phys. (1971) 43 532.

(14)   Mullin, W.J., Creswell, D. J. and Cowan, B. (1976). J. Low Temp. Phys., to be published.

(15)   Thomson, A.L., Brewer, D.F., and Goto, Y., Proc. 14th Int. Conf. on Low Temp. Phys. 1, 463, North Holland 1975.

(16)   Richardson, R.C., Hunt, E, and Meyer, H. Phys. Rev. (1965) 138 A1326

(17)   Richards, M.G., Hatton J., and Giffard, R.P., Phys. Rev. (1965) 139 A91

20

Magnetic Resonance Studies of Thorium Oxide Catalysts

WALLACE S. BREY, BENJAMIN W. MARTIN, BURTON H. DAVIS, and
RICHARD B. GAMMAGE

Department of Chemistry, University of Florida, Gainesville, Fla. 32611

Thorium oxide, prepared in suitable form with a high surface area, has been found to be active as a heterogeneous catalyst for a variety of vapor phase reactions. For example,
when an alcohol is passed over the oxide, it may be dehydrated
to an olefin or dehydrogenated to a ketone or aldehyde. The
product composition has varied with the investigator, and thus
the course of the reaction must be related to the history of a
particular catalyst sample (1,2,3). Various samples of the
oxide do indeed have different selectivities for the two competing reactions, and thus thoria, in addition to being a catalyst
of practical importance, presents an interesting system with
which to seek answers to questions concerning the basis for
selectivity of a heterogenous catalyst.

Certain aspects of the alcohol decomposition reaction are
particularly outstanding. Lundeen and Van Hoozer (4) reported
that thoria selectively dehydrates 2-alkanols to 1-alkenes,
rather than to the thermodynamically favored 2-alkenes. Results
of J. W. Legg, working in our laboratory, indicate that butanol
is dehydrated almost exclusively to butene-1 over thoria catalysts prepared by heating thorium hydroxide which had been
precipitated by addition of ammonia to a solution of thorium
nitrate (5). More recent studies by Davis demonstrate that this
kind of selectivity, as measured in the dehydration of 2-octanol,
is dependent upon the nature of the catalyst used, varying from
a small preference for 2-octene formation on some samples to a
strong preference for 1-octene on others (6,7). It appears
likely that the surface reaction is a concerted one, but the
manner in which the detailed surface structure determines
reaction stereoselectivity is not understood in any detail. Work
by B.H. Davis (7) has demonstrated that alcohol dehydrogenation
can be facilitated by pretreatment of the catalyst is an atmosphere of oxygen and that alcohol dehydration can be enhanced by
pretreatment in an atmosphere of hydrogen. This result is
certainly suggestive that non-stoichiometric active sites can be
generated on the surface.

In addition to the direct study of the catalytic reactions and use of other methods such as infrared spectroscopy and dielectric relaxation measurements, we have extensively applied magnetic resonance methods in an effort to learn something about the structure of active thoria and about its interactions with substrate molecules. Some of the earlier results will be briefly reviewed and some newer ones will be described in more detail.

Preparation of Active Oxide

Thorium oxide may be prepared from a variety of sources and its properties depend critically upon the preparation procedure. Heating of the oxalate, the nitrate, or the hydroxide has been employed in this work. Material from decomposition of the oxalate in vacuum is contaminated by carbon deposits and therefore the heating was carried out in a stream of air at 600° for six hours. Other materials were activated in vacuum at 500° to 600° C.

The hydroxide used as a source of the oxide was prepared by precipitation by addition of ammonia to a solution of one of the thorium salts. The surface area, an easily measurable property, obtained by the BET analysis of adsorption of nitrogen at liquid nitrogen temperature, is very sensitive to the exact details of the precipitation procedure, including the initial concentration of the thorium salt solution, the rate and condition of addition of ammonia, the pH reached, amount of washing, and so on.

One procedure which consistently produced oxide of area of the order of 100 square meters per gram involved final precipitation of the hydroxide from a colloidal suspension, following an initial precipitation. To a thorium salt solution of concentration about 50 grams per liter, is added ammonia, rapidly with stirring, until a pH of 7 is reached. The precipitate is filtered off, washed with an amount of water equal to one-half the volume of the initial solution, and then warmed until it liquefies to form a bluish opalescent liquid. Ammonia is again added to a pH of about 7.2, and the resulting solid is filtered off, dried at 110° and activated by heating in vacuum at 600° for six hours.

NMR Spectra of Adsorbed Water and Alcohols

Wide-line NMR spectra were obtained on a Varian DP-60 spectrometer using 80 Hz modulation. The samples were contained in 5 mm diameter sample tubes attached to vacuum stopcocks and ground joints, which permitted connection to a vacuum line.

Early results of Lawson (8,9) compared T_1 and T_2 values, as estimated from linewidths and saturation behavior, for water on thoria made from the hydroxide with those for water on thoria made from the oxalate, using samples with surface areas in the range of 10 to 30 square meters per gram. For both types of

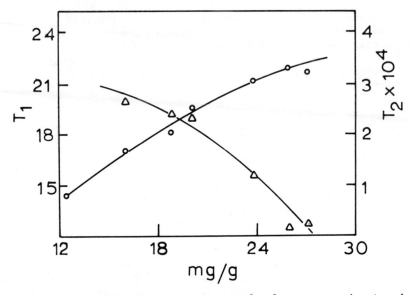

Figure 1. *Relaxation times, in seconds, of methanol protons as a function of the amount of alcohol adsorbed on thorium oxide. The oxide was prepared from precipitated hydroxide and had a surface area of 14.1 m^2/g. Triangles, T_1; circles, T_2.*

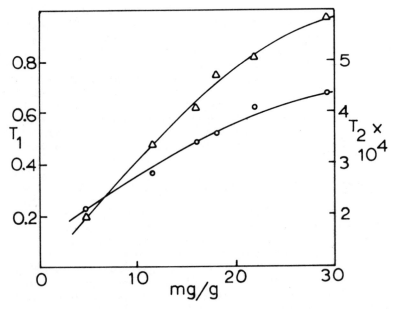

Figure 2. *Relaxation times, in seconds, of water on thorium oxide prepared from the oxalate. Surface area, 12.5 m^2/g. Triangles, T_1; circles, T_2.*

oxide T_2 increased with increasing coverage of the surface, the values being about 1 to 5×10^{-4} seconds. In some of the samples, particularly those activated at 500° C, adsorption-desorption cycles showed marked hysteresis in the linewidth, with lines broader during desorption. There were also long-time trends: following adsorption of water, the linewidth increased for a period of weeks to months. These observations are consistent with the adsorption-desorption hysteresis and the slow approach to equilibrium in the thoria-water system found by several earlier workers making careful measurements of adsorption isotherms (10, 11, 12).

Values of T_1 for water on various samples are either substantially independent of coverage of the surface or increase with increasing surface coverage. This is in contrast to the behavior of methanol, shown in Figure 1, for which T_1 averaged over both kinds of hydrogen, decreased with increasing coverage, as is expected for motional dipole-dipole relaxation when the correlation time is longer than that corresponding to a minimum T_1 value. For hydroxide-derived samples, the T_1 values for water have magnitudes of 1 to 10 seconds; those for the oxalate-derived samples, which show a striking rise with increasing coverage, as in Figure 2, are only about one-tenth as long. However, as the temperature decreases, values of T_1 increase, as shown in Figure 3, suggesting that the system is still on the long-correlation-time side of the T_1 minimum.

The unexpected increase in T_1 with increasing surface coverage suggests another mechanism of relaxation. One possibility involves a complication from transfer between two phases, and another is a relaxation contribution from paramagnetic impurities in the surface, which would be felt most strongly by the material first adsorbed.

It was also observed that molecules of methanol, ethanol, or butylamine adsorbed on these samples of thoria gave a rather interesting change in spectrum with decrease in temperature. The peaks broadened around zero degrees and could be partially resolved over some range of temperature into a broad component and a narrow component. With further decrease in temperature, the former component, attributable to the hydrogens in the adsorbed polar group, became too broad to observe and the narrow component persisted to temperatures of -100° C or below.

The hydrogen resonance of one of the catalysts prepared by precipitation of the hydroxide from colloidal solution, as described above, showed the presence of both broad and narrow lines, as indicated in Figure 4(a). The narrow line was not detected, however, with the lowest modulation amplitude which could be set on the instrument. Deuterium oxide was adsorbed on the sample and then pumped off after one hour. This process was repeated a number of times over a 48-hour period. Following the exchange, the broad proton peak had disappeared entirely, and what remained was a narrow peak, so narrow that now it could be detected at the lowest modulation setting. The sample was then allowed to stand for a week under saturation pressure of D_2O and several further saturation and evacuation cycles in D_2O were performed. The amplitude of the narrow peak remained substantially constant. Back-exchange of the D_2O with H_2O restored the broad peak and unexpectedly broadened the narrow peak. Thus, we are left with the surprising situation that the hydrogens responsible for the narrow peak are not exchangeable with D_2O, but are close enough to exchangeable water to have their linewidth affected by the presence of H or D in the other water.

ESR Signals of Oxide Samples

The relaxation behavior of water adsorbed on the oxalate-derived oxide suggested the presence of paramagnetic impurities in the adsorbent surface. Accordingly, we examined the ESR spectra of a number of thoria samples, using a Varian 4502-14 X-band spectrometer with a 12-inch magnet, 100 kHz modulation and dual sample cavity (13). Oxide made by directly heating a commercial sample of hydrated thorium nitrate showed no ESR signal when heated at any temperature, but oxide made by thermal decomposition of hydroxide or of oxalate precipitated from an aqueous solution of the same material showed spectra such as those in Figures 5 and 6. Although the hydroxide from nitrate solution showed a well-developed spectrum after drying at 150°, the oxalate from the same solution showed no absorption when similarly treated. Heating the oxide from the hydroxide caused some changes in detail in the spectrum but no substantial overall alteration. The oxide from oxalate developed a sharp central peak at g = 1.994 when heated in air or oxygen at 350°, but this peak disappeared on heating in vacuum at that temperature, to be replaced by other absorptions at higher and lower fields. The spectrum is substantially modified by addition of water vapor, but subsequent exposure to the atmosphere restores the central peak, leaving only weak absorptions at the high and low field positions. Figure 7 shows the results of heating material of both the hydroxide and oxalate types to high temperatures; it is striking in view of the low-temperature differences in behavior that the spectra become almost identical. Heating to somewhere between 800° and 1000° C removes all but slight traces of the ESR signals from all thoria samples.

We have not been able to correlate the intensity or nature of ESR signals with the relaxation behavior of adsorbed water. It is, however, surprising that so wide a variety of such signals can be observed. The modifications of the oxalate-derived oxide spectra by oxygen and by water indicate that most of the resonances observed arise from surface sites. These sites may be stabilized by the presence of impurities or the resonances themselves may have contributions from an impurity. However, addition of a wide variety of different species to the thorium

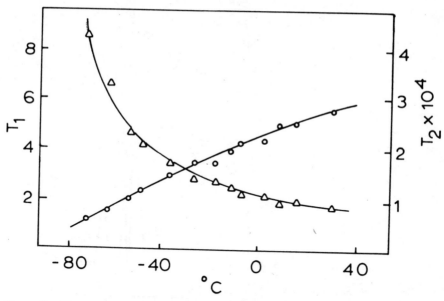

Figure 3. The temperature dependence of the relaxation times, in seconds, of water on thorium oxide of area 31.1 m²/g. Water coverage, 16.7 mg/g. Triangles, T_1; circles, T_2.

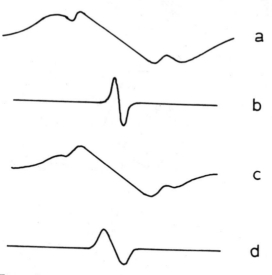

Figure 4. Spectrum of water on thorium oxide during exchange with D_2O. The oxide, surface area 110 m²/g, was precipitated from colloidal dispersion. (a) Initial spectrum after evacuation at 400°C. (b) After exchange with D_2O. (c) After reexchange with H_2O. (d) After second exchange with D_2O. Spectra (a) and (c) were obtained with higher modulation amplitude than (b) and (d).

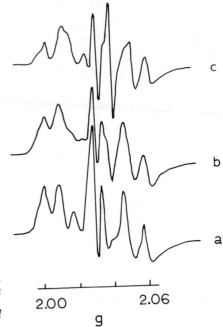

Figure 5. ESR spectra of thorium hydroxide and its dehydration products, taken at room temperature. (a) Dried 24 hr at 150° (b) Heated at 350°C for 2 hr. (c) Heated at 450°C for 4 hr.

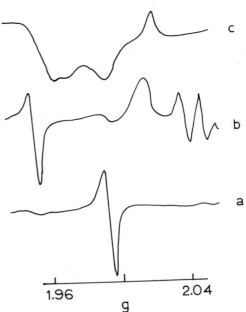

Figure 6. ESR spectrum of oxides from thorium oxalate. (a) Heated at 350°C in oxygen. (b) Treatment (a) followed by heating in vacuum at 350°C. (c) Treatment (b) followed by addition of water vapor at 25°.

solution before precipitation--almost every possible transition metal which might be responsible for the absorptions was employed-- failed to produce spectra resembling those observed.

Failure of the directly-decomposed nitrate to yield an observable spectrum is further evidence that specific defects characteristic of a given form of the oxide are associated with the resonances, and also that the presence of oxides of nitrogen is not responsible for the spectra, despite the fact that noticeable quantities of NO_2 were given off in heating the hydroxide and the oxalate to 600°, although it must be recognized that failure to detect the resonance of a paramagnetic species may depend upon circumstances of its local environment rather than upon its absence from the sample.

Irradiated Samples

In a further attempt to identify the sources of the absorptions, a sample of very pure thoria was obtained from AERE, Harwell, England. Spectroscopic analysis at Oak Ridge National Laboratory showed it to contain 14 ppm Fe, 19 ppm Cr, and 26 ppm Cu as the principal impurities. The material as received had no ESR spectrum, but when it was dissolved, reprecipitated, and ignited to 650° C, giving samples of oxide with surface areas of from 45 to 90 square meters per gram, two ESR peaks appeared. One of these, with a g value of 1.9676, is evident in all low-temperature spectra of materials from this source and is assigned to the Cr impurity (13).

A number of x-ray and gamma-ray irradiation studies of these samples were carried out in order to see whether electrons could be trapped at impurities or defects and thus give ESR patterns (14). It was found that the second peak in the spectrum, having a g-value of just under 2.0, was enhanced in intensity by gamma radiation. Another peak, with g = 2.003 appears after gamma irradiation at room temperature; its ease of saturation suggests that its source is trapped electrons. Both it and other nearby peaks disappear on adsorption of CO_2 on the surface, to be replaced by an anisotropic spectrum with g-values of 2.0048, 2.0089, and 2.0126, which are too large for the expected CO_2^-, at least as found on MgO (15) or in irradiated sodium formate.

Gamma irradiation at liquid nitrogen temperatures resulted in a spectrum with additional features as illustrated in Figure 8. The patterns were sorted out by saturation and pulse annealing experiments. Species B, with an isotropic g value of 2.0059 and species C, with g values of 2.0102, 2.0016, and 2.0128 appear to be types of V-centers. The g values of 1.9978 and 2.0000 for species D, with axial symmetry, along with its ease of saturation indicate an F-center; signal D remained fairly strong up to room temperature.

Another sample of the thoria was irradiated with neutrons, which should create atomic displacements as well as dislodge

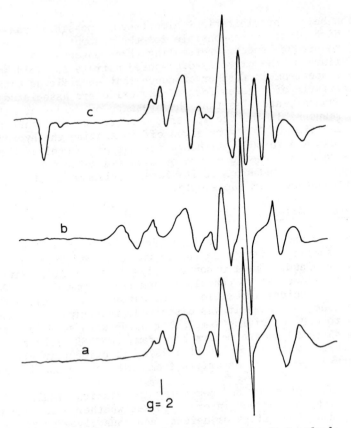

g= 2

Figure 7. ESR spectra of thorium oxides. (a) Hydroxide de-
composed at 550°C in air. (b) Oxalate heated at 600°C in oxygen
for 15 hr. (area of 85 m²/g) followed by heating in water vapor
for 24 hr (area of 28 m²/g). (c) Oxalate heated at 600°C in oxygen for 15 hr followed by heating in vacuum at 800°C for 15 hr
(area 68 m²/g).

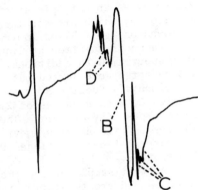

Figure 8. Spectrum at 100°C of thoria gamma-
irradiated at liquid nitrogen temperature. Letters refer to signals mentioned in the text.

electrons. The sample, examined in air at room temperature four months after irradiation showed at least three sharp peaks, superimposed on a broad absorption. The spectrum was not much changed by heating in vacuum; lowering the temperature of the sample to 100 K had the effect of making visible the Cr resonance. A sharp peak with a g-value of 2.0018 was assigned to an F-center; since it was unaffected by the atmosphere, it is a bulk rather than surface defect. On warming the sample to 100° C and then to 200° C, the F-center resonance gives way to an anisotropic signal with an average g value near 1.998, which resembles the F_2 center in MgO, and then this signal is replaced by one at 2.0034 with the saturation behavior of a V center. These results clearly indicate the possibility of observing bulk defects in the thoria and demonstrate that they are quite different in characteristics from the peaks associated primarily with surface defects in samples not neutron-irradiated.

Adsorption of Carbon Monoxide

When CO was admitted to a sample of active oxide prepared from the AERE thoria, there was formed a species having an ESR signal similar to the signal reported for CO on magnesium oxide by Lunsford (16). Within a few minutes, this signal was replaced by another spectrum, shown in Figure 9. These signals increase in intensity over a period of several hundred hours and change reversibly between spectrometer temperatures of 310 K and 100 K. They are associated with strongly bound CO, since they continue to grow after the CO in the gas phase has been pumped out. Thus we interpret the slow process as a migration of a defect from the interior of the solid to react with an adsorbate molecule by donating an unpaired electron to the antibonding CO orbital, leading to a nearly neutral complex.

The adsorption to CO on MgO is reversible, for the molecules can be pumped off unchanged at higher temperatures. For the CO-ThO_2 system, however, raising the temperature while pumping leads to a series of ESR patterns which indicate that the adsorbate reacts with the surface to form CO_2^-, and at still higher temperatures there is a disproportionation reaction which leaves a deposit of carbon on the surface, evident by its characteristic broad ESR signal.

ESR Spectra of Adsorbed Nitric Oxide

Nitric oxide has a diamagnetic ground state, $^2\Pi_{1/2}$, but the first excited state, $^2\Pi_{3/2}$, lies only 0.015 eV higher and thus accounts for the paramagnetism of nitric oxide at room temperature. Under the influence of the field of the adsorbent, the degeneracy of the ground state of the molecule is lifted. The adsorption of nitric oxide on other catalyst surfaces has been investigated (17), and it seems profitable to make comparisons with adsorption on

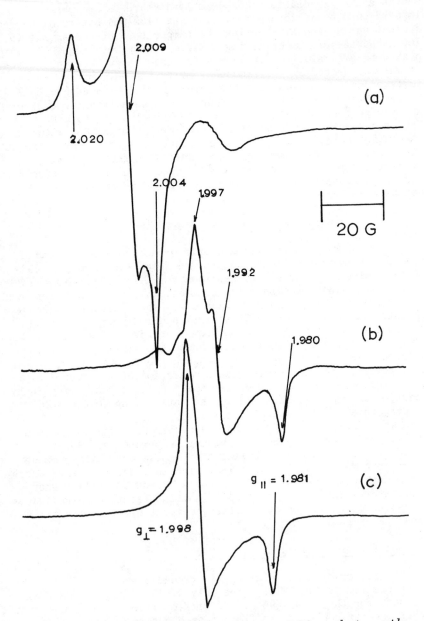

Figure 9. ESR spectra of products of adsorption of CO on thorium oxide. (a) Spectrum at 100°K five minutes after admission of 100 torr of CO at room temperature. (b) Same system at 100°K, after 240 hr. at room temperature. (c) Same sample as in (b), but with spectrum obtained at 298°K.

ThO_2.

All ESR spectra were obtained at liquid nitrogen temperature in a Varian E-3 spectrometer. The samples were contained in quartz tubes fitted with glass vacuum stopcocks and ground joints for attachment to a vacuum line for activation and addition of nitric oxide. The nitric oxide used was purified by passage through a 40 cm column of 60-200 mesh silica gel powder in a tube immersed in dry ice-acetone. An attempt was made to determine an adsorption isotherm for NO at -142° C, but the amount adsorbed was constant at 41 ± 2 milligrams NO per gram ThO_2 over the range of pressure from the lowest measurable in the apparatus up to P/P_0 equal to 0.25, indicating ready saturation of a few active sites at very low pressure.

Three samples of thoria were employed in the study. In each case, the material was prepared by dissolving a salt of thorium in water at a concentration of 50 grams per liter, and then carrying out two successive precipitations, the second being from a colloidal suspension, as described above. As we have indicated earlier, many samples of thorium oxide exhibit rather complex ESR spectra. Of the three salts used in this study--the nitrate, bromide, and iodide--only the last yielded an oxide with an ESR spectrum clean enough to allow the spectrum of the adsorbed NO to be analyzed unambiguously, although once this spectrum was in hand, it was possible to see that the NO plus thoria spectra of the other samples are indeed superpositions of the same NO spectrum on the particular oxide background. The surface areas of the three oxides were 112, 74, and 104 square meters per gram, respectively, and the ESR spectra are shown in Figures 10-12. Although some of the oxide ESR signals are associated with surface sites, it was not possible clearly to observe the loss of any part of the spectrum as the result of NO adsorption.

The spectra obtained are powder spectra, characteristic of randomly oriented nitric oxide molecules, and correspond very closely to the case of axial symmetry. The g values of 1.995 and 1.930 are quite consistent with the values previously reported for adsorbed NO and distinctively lower than those assigned to the possible species NO_2 or NO_2^{-2}. Only the perpendicular component of the hyperfine splitting is observed, and the value of 28.5 ± 1 gauss is also consistent with literature values and characteristic of NO rather than of the other possible species, for which the magnitudes are roughly double this value.

For the hyperfine interaction with the ^{14}N nuclear spin, the single splitting observed appears to involve only one of the two components of the apparent $g_\perp$ absorption, either the g_{xx} or the g_{yy} component. This is a consequence of the fact that the g and A tensors each have a separate principal axis system (18) and results in a lower observed intensity of the $\Delta M_I = \pm 1$ components. This situation may be interpreted as resulting from the unpaired electron being in an almost pure p-type orbital in agreement with the conclusions drawn by Lunsford for cases of

nitric oxide adsorbed on magnesium oxide ([17]), zeolites ([19]), and
zinc oxide and sulfide ([20]).

Analysis of the g-factors can be based on the equations
derived by Künzig et al. for the superoxide ion, O_2^- ([21], [22]),
and modified by Gardner and Weinberger to account for the change
in sign of the spin-orbit coupling constant ([23]). First, the
crystal field splitting, Δ, may be calculated using

$$g_{zz} = g_e - 2\ell\zeta/(\zeta^2 + \Delta^2)^{1/2}$$

where g_e is the free electron g-value, 2.0023, ℓ is an effective
orbital angular momentum quantum number, equal to unity for the
free molecule, and taken to be 1 here, ζ is the spin-orbit coupling constant, here taken to be 0.015 eV ([24]), and Δ is the
splitting between the $2p\pi_x^*$ and $2p\pi_y^*$ levels, whose degeneracy is
removed under the influence of the crystalline electric field
and the applied magnetic field. Using the calculated value of
Δ, excitation energy E may be calculated employing

$$g_{xx} = g_e\Delta/\delta - \zeta/E(\Delta/\delta - \zeta/\delta - 1)$$

where $\delta = (\zeta^2 + \Delta^2)^{1/2}$.

From this analysis, Δ is found to be 0.41 eV, which is a
value intermediate between that of 0.14 found for Zeolite-X ([23]),
and the value of 0.60 observed for decationated Y-Zeolite. The
magnitude of E for thorium oxide is 5.09 eV, and the decrease in
E from 5.45 eV for the free molecule is in agreement with the
decrease observed for adsorption on zeolites.

It is striking that no evidence is found for the formation
of NO_2, which has been observed on other catalyst surfaces with
available oxygen ([25], [26]).

An interesting aspect of this study was the observation that
the ESR signal disappeared if the sample was allowed to warm to
room temperature between the stage of NO adsorption and placing
the sample in the spectrometer. Spot tests applied to the surface of the oxide after it had been in contact with NO at room
temperature and was then degassed indicated the presence of
nitrite ion. This suggested the possibility that NO_2 or NO_2^-
might have been formed. However, as mentioned above, both these
species have hyperfine splitting values considerably larger than
those observed here, as well as at least two out of the three
g-tensor components of 2.00 or larger. No change in the adsorbate spectrum with time could be observed; it simply decreased in
intensity.

Shaking the solid with distilled water and testing the
liquid alone failed to give a positive test, implying that the NO
is bonded to the thoria to give an insoluble substance, possibly
a trivalent thorium compound, such as Th(0)-0-NO, from which
nitrite ion is liberated by treatment with acid.

Figure 10. *ESR spectrum of nitric oxide on oxide made from hydroxide precipitated from thorium nitrate solution*

Figure 11. *ESR spectrum of nitric oxide on oxide made from hydroxide precipitated from thorium bromide solution*

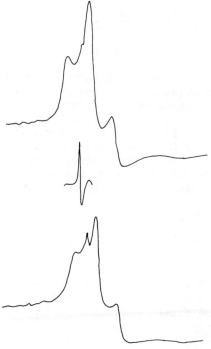

Figure 12. *ESR spectrum of nitric oxide on oxide made from hydroxide*

Reaction of Vanadium Tetrachloride with Surface Hydroxyls

As yet another way of looking at the surface structure of thoria, we have applied the method of Chien (27), in which vanadium tetrachloride is allowed to react with hydroxyl groups on the surface of an oxide. This affords an excellent paramagnetic probe, because the unreacted VCl_4 has a broad, undetectable spin resonance signal, but the product of the reaction has a characteristic eight-line spectrum. Furthermore, other oxidation states of vanadium should not confuse the interpretation, since their absorptions show the effect of fine structure.

For the oxides used by Chien, he was able to follow the stoichiometry of the reaction by conventional chemical determination of the hydrogen chloride liberated in the reaction. With thorium hydroxide, however, we encountered the complication that the amount of HCl released was only a fraction of that which must have been formed, indicating strong retention of the reaction product by the oxide surface. In addition to the spectral intensity, which permits a good evaluation of the amount of VCl_4 reacted with the surface, an additional parameter obtainable from the ESR results is the magnitude of the "superexchange" interaction between adjoining paramagnetic species. A computer program has been written to simulate the ESR lineshape, treating the processes leading to collapse of the hyperfine multiplet according to the Anderson-Kubo theory (28, 29) and Jensen's extension of the stochastic Liouville equation for isotropic rotational diffusion (30). The exchange parameter was then estimated by comparison with the observed spectra.

The hydroxyl content of the oxide surface was assumed to be equal to the loss in weight on heating the sample to constant weight at 800°, and the total amount of vanadium held by the oxide was checked by atomic absorption spectroscopy. The oxide had initially about 4 OH groups per nm^2; increments of water were added to increase this concentration, and it appears that the capacity of the surface is about 18 OH groups per nm^2. Analysis of the exchange interactions indicated that each vanadium bonds to the surface through two hydroxyl groups. Even when the number of hydroxyl groups on the surface is relatively low, the majority of them seem to be close enough to allow this type of bonding.

Summary

Each of the magnetic resonance techniques applied to the study of high-area thoria has given some insight into the complexities of structure of this material, but has raised as many questions as have been answered. Thus there still remains extensive additional work to be done before the details of the manifold structures in which this material is encountered can be fully understood.

Literature Cited

1. Balandin, A. A., Tolstopyatova, A. A., and Dudkiz, Z., Kinet. Katal. (1961) 2, 273.
2. Winfield, M. E., in "Catalysis," vol. VII, P. H. Emmett, ed., Reinhold, New York, 1960.
3. Hoover, G. I., and Rideal, E. K., J. Amer. Chem. Soc. (1927) 27, 104.
4. Lundeen, A. J., and Van Hoozer, R., J. Amer. Chem. Soc. (1963) 85, 2180.
5. Legg, J. W., "Catalytic Reactions of Aliphatic Alcohols Over Thorium Oxide," Doctoral Dissertation, University of Florida, Gainesville, 1964.
6. Davis, B. H., J. Org. Chem. (1972) 37, 1240.
7. Davis, B. H., and Brey, W. S., J. Catal. (1972) 25, 81.
8. Brey, W. S., and Lawson, K. D., J. Phys. Chem. (1964) 68, 1474.
9. Lawson, K. D., "Relaxation Studies of Adsorption on Thorium Oxide," Doctoral Dissertation, University of Florida, Gainesville, 1963.
10. Draper, A. L., and Milligan, W. O., "Structure and Surface Chemistry of Thorium Oxide," Report, The Rice Institute, Houston, 1959.
11. Fuller, E. L., Holmes, H. F., and Secoy, C. H., J. Phys. Chem. (1966) 70, 1633.
12. Gammage, R. B., Brey, W. S., and Davis, B. H., J. Colloid Interface Sci. (1970) 32, 256.
13. Brey, W. S., Gammage, R. B., and Virmani, Y. P., in "Physics of Electronic Ceramics," L. L. Hench and D. B. Dove, Eds., A413, Dekker, New York, 1971.
14. Muha, G. M., J. Phys. Chem. (1966) 70, 1390.
15. Lunsford, J. H., and Jayne, J. P., J. Phys. Chem. (1965) 69, 2182.
16. Lunsford, J. H., and Jayne, J. P., J. Chem. Phys. (1966) 44, 1492.
17. Lunsford, J. H., J. Chem. Phys. (1967) 46, 4347.
18. Ammeter, J., private communication.
19. Lunsford, J. H., J. Phys. Chem. (1968) 72, 4163.
20. Lunsford, J. H., J. Phys. Chem. (1968) 72, 2141.
21. Känzig, W., and Cohen, M. H., Phys. Rev. Letters (1959) 3, 509.
22. Zeller, H. R., and Känzig, W., Helv. Phys. Acta (1967) 40, 845.
23. Gardner, C. L., and Weinberger, M. A., Can. J. Chem. (1970) 48, 1317.
24. Herzberg, G., "Spectra of Diatomic Molecules," p. 462, Van Nostrand, New York, 1965.
25. Pietrzak, T. M., and Wood, D. E., J. Chem. Phys. (1970) 53, 2454.

26. Iyengar, R. D., et al., Surface Science (1969) 13, 251.
27. Chien, J. C. W., J. Amer. Chem. Soc. (1971) 93, 4675.
28. Anderson, P. W., J. Phys. Soc. Japan (1954) 9, 316.
29. Kubo, R., J. Math. Phys. (1963) 4, 174.
30. Jensen, S. K., in "Electron Spin Relaxation in Liquids," L.
    T. Muus and P. W. Atkins, Eds., pp 71-88, Plenum, New York,
    1972.

An NMR Study of Surface Complexes under the Conditions of Fast Exchange between Physisorbed and Chemisorbed Molecules

V. YU. BOROVKOV, G. M. ZHIDOMIROV, and V. B. KAZANSKY

Zelinsky Institute of Organic Chemistry of Academy of Sciences USSR, Moscow, U.S.S.R.

The phenomenon of fast exchange between bound and free solvent molecules is widely used for the investigation of complexformation in solutions by high resolution NMR.

It was determined /1,2/, that on the surfaces of solids such as silica, silica-alumina and alumina, a fast exchange can take place between different kinds of adsorbed species. This exchange leads to an effective averaging of the NMR spectra of physically adsorbed molecules and those molecules coordinated by active surface sites. Hence, the parameters of the observed spectra contain information about the interaction of the adsorbate with the surface sites. The existence of such exchange leads also to the considerable increase in the NMR sensitivity for detection of the complex formation. This fact will be made obvious by the experimental results discussed in this paper.

The use of isolated transition metal ions as the active surface sites allows one to considerably increase the efficacy of NMR for the study of surface phenomena. In addition to line broadening, large paramagnetic shifts appear in the NMR spectra of the adsorbates, and the latter especially gives information concerning the interaction between coordinated molecules and transition metal ions.

Experimental

In our work we used samples of Aerosil, which contained on the surface coordinatively unsaturated ions of Co^{2+} and Ni^{2+}, and Y- zeolites in which the Na^+ ions were partially substituted with bivalent cobalt and nickel ions. The method of sample preparation and the conditions of their pretreatment were described in detail in /3/. The NMR measurements were performed using a Soviet-made NMR spectrometer having an operating frequency of 60 MHz. The main technical data and the process of simultaneous measurements of NMR spectra and adsorbtion isothermes for the adsorbates were also described in /3/.

Before measurements the investigated substances were purified by the freeze-pump-thaw method. Molecular hydrogen was purified by diffusion through a glowing palladium capillary.

Experimental Results

The criterion for the detection of interaction of adsorbed molecules with paramagnetic surface ions is the existence of large paramagnetic shifts in the NMR spectra of the adsorbates and the dependence of these shifts on the surface coverage /1, 3/.

Fig. 1 shows the PMR spectra of propylene adsorbed on nickel-containing Aerosil. Line shifts, on the order of several tens of p.p.m., and their dependence on the equilibrium pressure in the system greatly testify to the complex formation of the propylene molecule with Ni^{2+} ion.

We have investigated the adsorption of many different molecules - benzene, unsaturated hydrocarbons (ethylene, propylene, butene, cyclohexene), alcohols (methanol, ethanol), saturated hydrocarbons (methane, isopentane, n-hexane, cyclohexane, cis and trans-1,4--dimethylcyclohexane) and molecular hydrogen - on adsorbents which have contained cobalt and nickel ions. The results of these investigations are given in table 1, in which the plus or minus sign corresponds to the existance or absence of large paramagnetic shifts in the NMR spectra of adsorbed species.

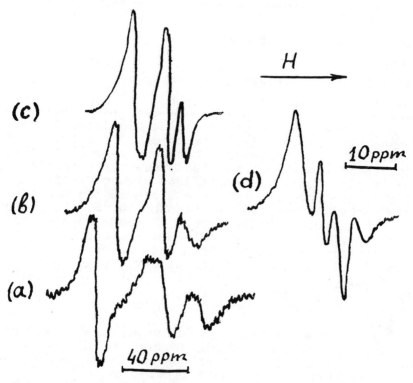

Figure 1. NMR spectra of propylene adsorbed on Ni^{2+} containing Aerosil at equilibrium pressures of 3, 6, 15, and 150 Torr, respectively. All spectra were recorded at $+10°C$.

Table 1.

adsorbate \ adsorbent	Co^{2+} on Aerosil	Ni^{2+} on Aerosil	CoNaY NiNaY
Benzene	+	−	
Unsaturated hydrocarbons	+	+	
Alcohols		+	
Saturated hydrocarbons	+	−	
Hydrogen	+	+	+

The PMR spectra of each kind of adsorbed substance has characteristic features. For example, PMR
spectra of unsaturated hydrocarbons /4/, as a rule,
consist of several lines. The lines corresponding to
the vinyl protons are shifted to higher fields relative to the position of these lines in the spectra of
physisorbed molecules. On the other hand, the protons
on a methyl or methylene group adjacent to a double
bond are shifted to lower fields. Protons at greater
distance from the double bond show no change. Sometimes in PMR spectra one can detect the nonequivalen-
cy of the protons located near a double bond. This is
the reason for the appearance of four lines in the
spectra of propylene adsorbed on nickel-containing
Aerosil (see fig. 1 d).

PMR spectra of alcohols also consist of several
lines. Lines of CH_2 and CH_3 groups are shifted to
lower field, and the position of the lines for the
hydroxyl group changes very little.

Adsorption of all saturated hydrocarbons on cobalt supported Aerosil gave a single symmetrical line
shifted to higher fields. The observed shifts ranged
over several tens of p.p.m.

Hydrogen adsorption on Aerosil containing cobalt
or nickel was studied at liquid nitrogen temperature.
It is interesting to note that for the cobalt-containing samples, the line for adsorbed hydrogen was
shifted to higher fields, but for nickel samples the
shift was in the opposite direction. Upon adsorption
of hydrogen on CoNaY and NiNaY zeolites at temperatures below 0°C, one observes a line shifted to lower
field for both adsorbents. The shift values are weakly dependent on surface coverage and show an exponential increase with a decrease in temperature.

Using the dependence line shifts on coverage,
all complexes which we detected by means of PMR, can

be divided into three different groups.

The first one includes the complexes of benzene, unsaturated hydrocarbons and alcohols, for which is observed a linear increase of shift on decrease in coverage (Fig. 2, 1). The complexes of saturated hydrocarbons with Co^{2+} ions belong to the second group. For all saturated species investigated by us the dependence of the shifts on coverage was similar to that represented on fig. 2, curve 2. In the third group we place the complexes of hydrogen with Co^{2+} and Ni^{2+} ions in Y zeolite, for which there is almost no dependence of the shifts on the amount of the hydrogen adsorption.

A study of ^{13}C NMR spectra of these same adsorbed molecules would allow us to obtain important information about the adsorbate-adsorbent interaction. Unfortunately, the small natural abundance of ^{13}C, and the small magnetic moment of this nucleus, hamper the application of ^{13}C NMR for investigating surface phenomena. From our point of view, the resonance of the ^{19}F nucleus is more convenient for this purpose. The scale of shifts for this nucleus is larger than for ^{13}C and its large magnetic moment provides large sensitivity. It allows one to obtain well resolved NMR spectra of adsorbed species without any difficulties.

At this time we are begining the systematic investigation of adsorption using ^{19}F NMR. For example, fig. 3 shows the NMR spectra of pentafluorobenzene adsorbed on pure Aerosil and Aerosil containing Co^{2+} ions. One can see that the lines for the different kinds of F atoms in the molecule have different paramagnetic shifts. We have already investigated the set of fluorine substituted benzenes C_6F_6, C_6F_5H and C_6FH_5 on cobalt-containing Aerosil. The main feature of the spectra of these compounds is a shift of the ^{19}F lines to lower fields. This is in contrast to the direction of the shift in the ^1H NMR spectra of C_6H_6 adsorbed on the same system.

Theory

The theoretical description of the NMR spectra for the molecules undergoing exchange between the sites of physisorption and chemisorption deals with the problem of the averaging of the NMR spectra by dynemic motion of the molecules. As it will be shown below the methods developed in general theory /8/ can be applied to the description of the NMR spectra of adsorbed molecules.

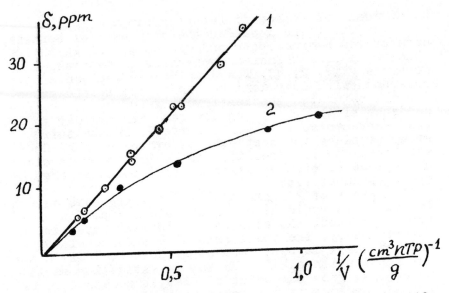

Figure 2. *Dependence of the NMR line shifts upon the amount adsorbed for (1) benzene and (2) cyclohexane on a Co⁺²-containing Aerosil. The spectra were recorded at room temperature.*

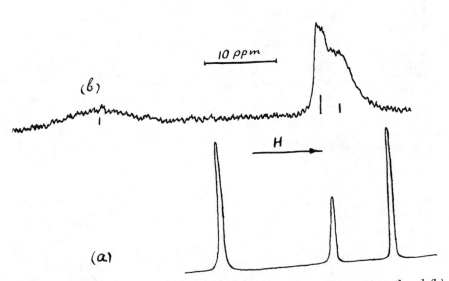

Figure 3. *NMR spectra of pentafluorobenzene adsorbed on (a) pure Aerosil and (b) Co²⁺-containing Aerosil. The spectra were recorded at room temperature.*

(δ)   In the case of fast exchange the observed shift

$$\delta = P_c \cdot \delta_c \qquad (1)$$

where $P_c = \dfrac{V_c}{V}$ - the mole fraction of chemisorbed molecules,

$V = V_c + V_p$ the full number of adsorbed molecules, δ_c - the line shift in NMR spectra of chemisorbed molecule relative to it's position for physisorbed molecule.

The amount of chemisorbed molecules is given by the equelibrium

$$M + \square_c \underset{K_{-1}}{\overset{K_1}{\rightleftarrows}} [M]_c \qquad (2)$$

$$K \cdot V_p \cdot N_c' = V_c \; ; \quad K = K_1/K_{-1} , \qquad (3)$$

here N_c' is the number of empty sites for chemisorption.

For the case $P_c \ll P_p$ ($P_p = \dfrac{V_p}{V}$) from (2) and (3) it follows

$$P_c = \dfrac{K N_c}{1 + KV} \qquad (4)$$

(4) gives the possibility of considering the quantity as a function of N_c, the full number of chemisorption sites ($N_c = N_c' + V_c$), and the number of adsorbed molecules V.

Depending on the value of KV, two limiting situations can take place:

1) "strong" complex: KV$\gg$1

$$\delta = \dfrac{N_c}{V} \delta_c \qquad (5)$$

Such a situation occures for the complexes of benzene, unsaturated hydrocarbons and alcohols, which we earlier referred to as the first class of compounds.

2) "weak" complex: KV$\ll$1.

$$\delta = N_c K \delta_c = N_c K_o \, e^{\Delta H/kT} \cdot \delta_c \qquad (6)$$

This kind of complex corresponds to the complexes of molecular hydrogen with Co^{2+} and Ni^{2+} ions in zeolites.

It should be noted, that the exchange in the system may occur either by a "dissociative" mechanism (scheme /2/) or by a mechanism of substitution:

$$M' + [M]_c \underset{K_1'}{\rightleftarrows} [M']_c + M \qquad (7)$$

This explains, in particular, the existance of fast exchange in the case of rather strong complexes (with large values of ΔH) since the rate of exchange will be determined by the activation energy of this process. The experimental results show that the activation energy of the substitutional process may be rather small and such situations may occur rather often. An example of this is the existence of fast exchange between molecules of alcohol physically adsorbed and coordinated by Ni^{2+} ions.

For the theoretical description of NMR spectra of adsorbates at any rates of exchange, we shall use as a model one which considers the migration of the adsorbed molecules between the sites of physical adsorption (p) and chemisaorption (c) /9, 10/. We assume the number of chemisorption sites is much smaller than the number for physisorption.

For the life times of the molecules in the state (p) and (c) let us assign $\tau_p = \dfrac{\tau_o}{P_p}$ and $\tau_c = \dfrac{\tau_o}{P_c}$, respectively. The value of τ_o characterises the time scale, depending upon the exchange kinetics

$$\tau_o = \frac{1}{K_{-1} + K'_1 V} \tag{8}$$

It is natural to propose that owing to the fast migrations of the molecules the NMR line shape for p-molecules is Lorenzian $(\ell/\omega/)$ with line width $\Delta\omega_p$. On the contrary, the line shape of the coordinated c-molecules should be nonhomogeniously-broadened line of complicate form $f(\omega)$.

So the task arises of averaging the Lorenzian line and the nonhomogeniously-broadened line.

1. The first method /9/ for the calculation of the resulting line shape is based on the consideration of the adiabatic averaging of the spectral contour $\varphi(\omega)$ due to random noncorrelated jumpwise changes of frequencies ω with mean life time τ_o in any state with a given ω inside of the contour $\varphi(\omega)$.

For the system under discussion such a contour is

$$\varphi(\omega) = P_p \ell(\omega) + P_c f(\omega) \tag{9}$$

For $f(\omega)$ it is convenient to consider an aggregate of Lorenzian lines with line widths $\Delta\omega_c$ and with distribution function $g(\omega)$ $(\int g(\omega)d\omega = 1)$. The relaxation function $X(\tau)$ for the system of particles with spin 1/2 averaged over all exchange process possibilities in accordance with the model proposed above can be found from the equation:

$$X(\tau) = e^{-\tau/\tau_o} R(\tau,0) + 1/\tau_o \cdot \int_0^\tau R(\tau,t) e^{-\frac{\tau-t}{\tau_o}} X(t)dt \tag{10}$$

$$R(\tau,t) = \int_{-\infty}^{\infty} e^{i\omega(\tau-t)} \varphi(\omega) d\omega$$

2. A second approach /10/ is based on the application of modified Bloch's equations /11/. This approach preserves all the main features of model I, but for practical calculations it sometimes proves to be preferable. The modified Bloch's equations for the chemical exchange in this case are:

$$\dot{M}_p = \sum_n P_{np} M_n - \left(\sum_n P_{np}\right) M_p + i(\omega_p + i\Delta\omega_p - \omega) M_p + i Y P_p$$

$$\dot{M}_n = P_{pn} M_p + \sum P_{mn} M_m - \left(\sum_m P_{nm}\right) M_n - P_{np} M_n +$$
$$+ i(\omega_n + i\Delta\omega_n - \omega) M_n + i Y P_c \cdot f_n \qquad (11)$$

Here it is accepted that the form of the line in the state c is the superposition of N Lorentzian lines with resonance frequencies ω_n and line width $\Delta\omega_n$; f_n is the relative intensity of the lines, M_p and M_n are magnetic moments; $Y = \gamma H_1 M_0$, H_1 is the magnetic component of the radiofrequency field, M_0 is the full magnetic moment;

$$P_{np} = N_p'/\tau_c N' \ ; \quad P_{pn} = f_n N_c'/\tau_p N' \ ;$$

$$P_{nm} = f_m N_c'/\tau_c N' \ ; \quad N' = N_p' + N_c'$$

Equation (11) can be generalized to the direct consideration of the averaging influence of the ion's spin-lattice relaxation (T_{1e}) on the NMR spectra of the adsorbed molecules. It is important as these ions can be the sites of chemisorption. In /10/ the general equations for NMR line shape were derived for arbitrary values of τ_p, τ_c and T_{1e}, and the method for calculation was given. As an example, the line shape of an NMR spectrum at different rates of exchange is represented in fig. 4 for the case

$$g(\omega) = \begin{cases} 0 & \omega \le \omega_1 \ ; \ \omega \geqslant \omega_2 \\ 1/\omega_2 - \omega_1 & \omega_1 < \omega < \omega_2 \end{cases}$$

and

$$\tau_p = \tau_c \ ; \quad P_p = P_c \ ; \quad \omega_2 - \omega_1 = \Delta\omega$$

Discussion

1. Investigation of the dependence of NMR spectra parameters on temperature and surface coverage of adsorbed species gives, in principle, the opportunity for the determination of the equelibrium constants K and heats of complex formation H. Fig. 5, shows the dependence $\ln \sigma$ vs $1/T$ for the complexes of hydrogen in CoNaY and NiNaY zeolites, a situation in which "weak" complexes form (6). The heat of complex formation, 4 $\frac{kcal}{mol}$, corresponds to the difference between the heats of H_2 adsorption on Co^{2+} and Ni^{2+} ions and that for physisorption alone.

Complexes of saturated hydrocarbons with Co^{2+} ions, referred to as the second group of compounds,

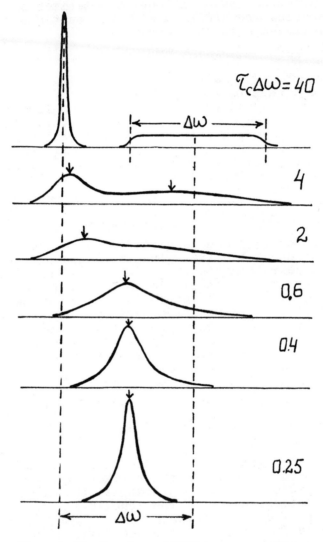

Figure 4. *The line shape of NMR spectrum at different rates of exchange between physisorbed and chemisorbed species*

correspond to complexes of "intermediate" strength, and expression (4) is the one to use for them. In this case it is convinient to analyse the dependence of σ vs ΔV. Measurements performed at different temperatures showed that the heat of such complex formation exceeded the heat of physisorption by 2-3 $\frac{kcal}{mol}$.

2. To interpret the line shifts in NMR spectra caused by complexation of adsorbed molecules with paramagnetic ions, one should take into account that these shifts can be produced by both contact and pseudocontact hyperfine interactions of the molecule's nuclei with the unpaired electrons of the ions. The relative contribution of these interactions depends on the coordinative environment of the ions in the lattice. Spectroscopic studies /12, 13/ have allowed us to conclude that the coordination of Co^{2+} and Ni^{2+} ions supported on Aerosil is either tetrahedral or trigonal. The data on investigation of the complexes in solutions show that a pseudocontact interaction for tetrahedral Co^{2+} ions is small /14, 15/. The latter is coused by the absence of orbital degeneration and relative small values of the g-tensor anisotropy of such ions.

For our systems, the shifts in NMR spectra of adsorbates seems to be caused by contact interaction. This conclusion is confirmed by the different directions of the line shifts for neighboring nuclei in one molecule (for example, propylene) and also for the molecules such as benzene and the fluorobenzenes which have similar geometrical structures but have line shifts going in opposite directions.

Let us consider the trigonal coordination of the ions, which is schematically represented in fig. 6. For such a case the d_{z^2} orbital of Co^{2+} ion contains an unpaired electron whereas the d_{z^2} orbital of Ni^{2+} ion is occupied by a pair of electrons. This fact allows us to explain the much larger value of the contact paramagnetic shift for benzene coordinated with Co^{2+} in comparison with that for Ni^{2+} ion. In the Co^{2+} /Aerosil samples the delocalization of the unpaired electron from the ion's d_{z^2} orbital onto the π-molecular orbital of benzene is predominent because of the large overlaping of the d_{z^2}-orbital with the π-system of benzene. For Ni^{2+} ions this contribution in the contact shift is absent.

Data on the shifts for unsaturated hydrocarbons can be explained by the assumption π-complex formation (see fig. 7). Thus, in the case of propylene delocalization of the unpaired electron onto the π-bond-

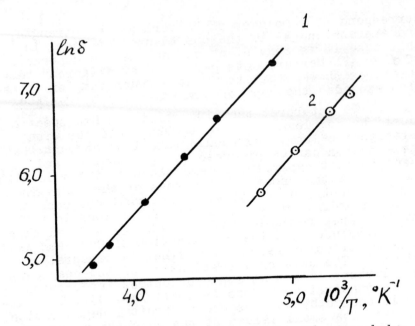

Figure 5. Line shifts to lower fields (Hz) as a function of the inverse absolute temperature in NMR spectra of adsorbed hydrogen on (1) CoNaY zeolite and (2) NiNaY zeolite

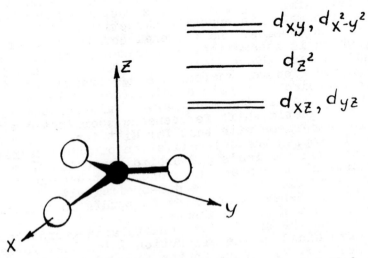

Figure 6.   Energy level splitting and wave functions of a single 3d electron in trigonal coordination

ing orbital of the molecule should lead to the ap-
pearence of a positive spin density on the protons
of the methyl group and a negative one on the protons
at the double bond. Such a distribution of spin densities corresponds to the directions of the experimentally observed shifts.

For the interpretation of the data on shifts
for saturated hydrocarbons we use a σ -structure of
complex, which is schematically represented in fig.8.
Structures of such type were proposed for complexes
of free π -electron N-oxide radicals with organic
ligands /16,17/ . We want to emphasize the possible
anology of such complexes (see fig. 8). This to our
own is confirmed by the direction (towards high
field) of the proton chemical shifts in the complexes
both with radicals and with ions. In our oppinion,
this type of coordination is also supported by the
character of the PMR spectrum is having a single
line with equal shifts for all protons.

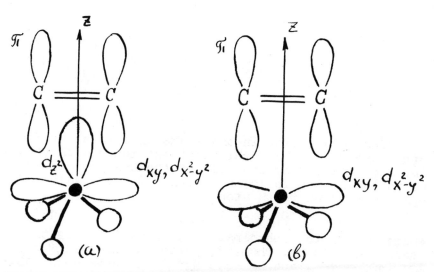

Figure 7.   The bonding in surface π-complexes of unsaturated hydrocarbons with (a) Co^{2+} and (b) Ni^{2+} ions

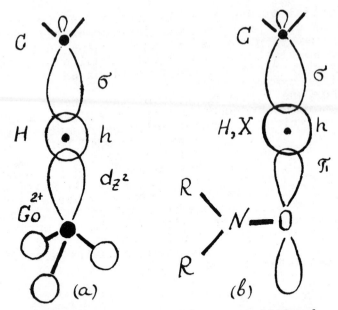

Figure 8. *The bonding in complexes of saturated hydrocarbons with (a) a surface Co^{2+} ion and (b) an N–oxide stable radical*

The calculation in the frame of a three orbital structure /6/ – d_{z^2} orbital of the ion and the bonding and untibonding orbitals of the CH–bond allowed us to obtain the following formula for the value of spin density on the H atom

$$\rho_H \sim - \frac{E_c}{4 E_{CH}} \qquad (12)$$

where E_c – is the coordination energy, E_{CH} – is the energy of CH bond. Thus, for n–hexane assuming $E_{CH} \sim 100 \frac{kcal}{mol}$, and $E_C \sim 2,5 \frac{kcal}{mol}$, we obtain $\rho_H \sim -10^{-2} \div -10^{-3}$. The order of this value corresponds to the experiment.

Literature Cited

1) Borovkov V.Yu., Kazansky V.B., Kinet. Katal. (1972), 13, 1356; 2) Borovkov V.Yu., Kazansky V.B., Kinet. Katal. (1973), 14, 1093; 3) Kazansky V.B., Borovkov V.Yu., Zhidomirov G.M., J. Catal. (1975), 39, 205; 4) Borovkov V.Yu., Kazansky V.B., Kinet. Katal. (1974), 15, 705; 5) Mozharovsky N.N., Borovkov V.Yu., Kazansky V.B., Kinet. Katal., in press; 6) Borovkov V.Yu., Zhidomirov G.M., Kazansky V.B., Zh. Strukt. Khim. (1974), 15, 547; 7) Borovkov V.Yu., Shuklov A. D., Surin C.A., Kazansky V.B., Kinet. Katal. (1973), 14, 1081; 8) Abragam A., Jadernii Magnetizm, IL, 1963; 9) Borovkov V.Yu., Zhidomirov G.M., Kazansky V. B., Zh. Strukt. Khim. (1975), 16, 308; 10) Parmon V. N., Zhidomirov G.M., Teor. Exp. Khim. (1975), 11, 323; 11) McConnel H.M., J. Chem. Phys. (1958), 28, 430; 12) Brotikovsky O.I., Shvets V.A., Kazansky V.B., Kinet. Katal. (1972), 13, 1342; 13) Kaverinsky V.A., Borovkov V.Yu., Shvets V.A., Kazansky V.B., Kinet. Katal. (1974), 15, 819; 14) Wayland B.B., Drago R.S., JACS (1965), 87, 2372; 15) Eaton D.R., Law K., Canad. J. Chem. (1971), 49, 3315; 16) Morishima J., Endo K., Yonezawa T., Chem. Phys. Lett. (1971), 9, 143; 17) Morishima J., Inubushi I., Endo K., Yonezawa T., JACS (1972), 94, 4812.

22

Some Applications of High Resolution NMR to the Study of Gas–Solid Interactions

I. Chemical Shift of Nuclei At Solid Surfaces II. Chemisorption of Hydrogen on Oxide-Supported Platinum

J. L. BONARDET, L. C. MENORVAL, and J. FRAISSARD

Laboratoire de Chimie des Surfaces, Universite P. et M. Curie, 4 Place Jussieu, 75230 Paris Cedex 05

Some years ago, it was pointed out that high-resolution N.M.R. was a promising method for determining the structure of chemically adsorbed phases (1,2,3). Indeed, HR–N.M.R. studies carried out on molecules adsorbed on diamagnetic solid surface indicated that proton resonance frequencies deviate from free molecule values. These variations may represent a disturbance of the electron distribution in the molecule during the adsorption, and consequently allow the determination of the electron distribution in the chemisorbed complex. The intrinsic shift varies with the nature of the functional group. Indeed, the adsorption process can affect the electron environment of a single type of proton when there is preferential orientation of the molecule with respect to the surface of the solid.

However, studies of this sort were for a long time not very numerous. In our opinion, this was due to the fact that it was only possible to follow the evolution of the spectrum of an adsorbed phase in terms of different parameters. But the intrinsic positions of the lines with respect to the standard references were unknown because of the difficulty in calculating the real chemical shift of a heterogeneous sample. Indeed, for a variety of reasons(wide lines, competitive adsorption, etc.) only the method of substitution can be used for measuring N.M.R. chemical shifts of adsorbed phases : the reference and the sample are put into two identical glass tubes and are studied successively. It is necessary, therefore, to correct the observed chemical shift for the bulk magnetic susceptibility of both the solid and the adsorbed gas. Unfortunately, because of the small quantity of powder used, the low sensitivity of the balances and inadequate knowledge of the apparent density of a powder, it is difficult to estimate the small volume susceptibility of powders accurately by conventional techniques (the Faraday balance for example). Now, by means of a new method associating N.M.R. and physical adsorption, the bulk susceptibility of the adsorbent can be measured directly and thus the chemical shift can be corrected accurately (4,5). It become then possible to study heterogeneous systems, in particular equilibria, as for homogeneous systems, even though the number of chemical species present in a heterogeneous system is often greater than that in a homogeneous system. The following two examples demonstrate that N.M.R. can now be used to study chemisorption either on non-conducting oxides or on metals.

I- Determination of the Bronsted acid strength of oxide catalysts

I-1 Theory. The Bronsted acidity of homogeneous samples may be characterized either by the dissociation constant of -OH group or by the NMR chemical shift of the hydrogen atom. The electronic environment of a nucleus is determined by its screening constant which is characterized by the chemical shift δ. Besides, with liquids or gases, the shape and the width of HR-NMR spectra are sensitive to time-dependent processes which occur at rates similar to the difference in spectral frequency measured in Hz. For processes which are slower than the critical rate, spectra appear as superpositions of distinct parts corresponding to individual species. For rapid processes, the spectra are determined by the time-averaged environment of the nuclei under investigation (6).

In solid samples, atom movements are very slow compared to those in liquids and gases. Then the strong dipolar interactions of spins are not reduced to zero by time-averaging and NMR lines are very wide (of the order of one gauss). Consequently the direct measurement of nuclear chemical shifts is impossible. The principle of our work is to make the surface proton take part in heterogeneous equilibrium using the adsorption of a molecule AH able to accept a proton (eq 1)

$$S - OH + AH \rightleftharpoons S-O^- + AH_2^+ \qquad 1$$

When rapid exchange occurs between the proton of the surface S-OH and those of the adsorbed molecule AH, the acid proton must affect the chemical shift of the adsorbed phase. The HR spectrum should contain only one line at frequency ν_e, due to the coalescence of the lines at frequencies ν_{AH}, ν_{AH2}, and ν_{OH}. Then the observed chemical shift δ_{obs} is

$$\delta_{obs} = P_{OH} \, \delta(OH) + P_{AH} \, \delta(AH) + P_{AH_2} \, \delta(AH_2^+) \qquad 2$$

where p_i is the concentration of hydrogen atoms in the group i. Knowing the chemical shift of the two nuclear types in the AH and AH_2^+ species (for example ^1H and ^{15}N or ^{17}O ...), the relative concentrations P_{AH} and $P_{AH_2}^+$, the dissociation coefficient of OH in the equilibrium 1, and the chemical shift $\delta(OH)$ of the surface proton can be calculated.

I-2 Chemical shifts of NH_3 and NH_4^+. The ^1H chemical shift of gaseous NH_3 at 50 torrs is 0.06 ppm relative to gaseous

TMS at a pressure of 1 torr. $\delta(NH_4^+)$ is measured using an aqueous solution of NH_4NH_3 containing dioxane as the internal reference and nitric acid to prevent any proton exchange between NH_4^+ and H_2O. Relative to NH_3 gas : $\delta(NH_4^+) = 6.9$ ppm. The ^{15}N signal of $^{15}NH_3$ has been detected at a pressure of 100 torr. As previously $\delta(^{15}NH_4^+)$ has been measured using $^{15}NH_4$ $^{15}NO_3$ in aqueous solution. Relative to $^{15}NH_3$ gas : $\delta^{15}(NH_4^+) = 43.5$ ppm. This value is in agreement with that calculated from the literature $(\underline{7},\underline{8},\underline{9})$.

I-3 Adsorption of ammonia on zeolithes. First, this hypothesis was checked by means of ammonia adsorption on dehydrated NH_4Y zeolithe. Ammonium cations are found inside the silico-aluminate cages. In a rigid lattice, the half-width of the NH_4^+ signal is about 32 KHz. At room temperature the hydrogen atoms in NH_4Y are able to exchange between different sites ; the nuclear dipolar interactions are time-averaged and the line width is only 8 KHz. However, the maximum sweep field of our apparatus is 4 KHz. Thus the NH_4^+ signal cannot be detected at high resolution. But as soon as the zeolithe adsorbs a small amount of ammonia, a signal is detected whose position δ and width ΔH depend on the relative concentrations of the NH_4^+ and NH_3 species. This signal characterizes the exchange :

$$N_i H_4^+ + N_j H_3 \rightleftharpoons N_i H_3 + N_j H_4^+ \qquad\qquad 3$$

between the protons of ammonium cations and those of the adsorbed molecules. As the NMR experiments are performed at 60 MHZ, the exchange frequency is higher than

$$\Delta\nu = \nu(NH_4^+) - \nu(NH_3) = 420 \text{ Hz}.$$ This illustrates that, from the chemical shift point of view, the influence of the nuclear movement upon the spectral shape is similar, regardless of any overlapping of the lines related to the different sites. Extrapolating to zero concentration of adsorbed ammonia, we find that the real chemical shift of the ammonium zeolithe protons, which cannot be directly measured, is 7.0 ± 0.1 ppm.

Under the same conditions we have studied ammonia adsorption on NaY zeolithe. Compared to the previous results, the width and the chemical shift of the signal are very low, showing that only physisorption occurs in this case (for example, at a pressure of 3 torrs, $\delta = 0.65$ ppm and $\Delta H = 220$ Hz).
The zeolithe HY is obtained by thermal decomposition of the zeolithe NH_4^+Y under vacuum. The HR-NMR spectrum of the OH groups of the zeolithe HY cannot be detected. But after adsorption of a quantity of ammonia greater than that eliminated in reaction 4α a signal whose intensity and position $\underline{\delta}$ depend on the NH_3 concentration, can be detected. Ammonia can be adsorbed on the OH groups (site I) and on other sites II. The possible reactions between NH_3 and the solid surface are the following :— transfer of a surface proton to a molecule $(NH_3)_1$ adsorbed on OH :

$$XOH + (NH_3) \underset{\beta}{\overset{\alpha}{\rightleftharpoons}} XO^- + NH_4^+ \qquad\qquad 4$$

– proton transfer inside the adsorbed phase and exchange of adsorbed molecules on the different sites :

$$(NH_3 - H^+)_I + (NH_3)_{II} \rightleftharpoons (NH_3)_I + (NH_3 - H^+)_{II} \qquad 5$$

$$(NH_3)_I + (NH_3)_{II} \rightleftharpoons (NH_3)_{II} + (NH_3)_I \qquad 6$$

Thus, δ must depend on the O–H dissociation and on the relative concentration in NH_4^+ and NH_3 adsorbed whether on OH(site I) or on other site (site II). The presence of a single line in the spectrum indicates that the lines of these species are in coalescence. During desorption under vacuum at room temperature, the value of δ increases with the amount of ammonia eliminated. But it is observed that : a) there always remains in the solid an amount of ammonia at least equal to that eliminated in reaction 4α,b) the chemical shift, determined for this concentration by extrapolation of the $\delta = f(NH_3)$ plot, is 7 ppm. These results prove that reaction 4β is displaced completely to the right. Consequently, it can be deduced that the acidity of the OH group in zeolithe HY is very high. But the chemical shift δ(OH) of this solid cannot be calculated by means of NH_3 adsorption. A weaker base should be used.

I-4 Adsorption of ammonia on silica gel. Samples of the silica xerogel (10) initially treated for 15 hours under 10^{-4} torr at 150, 400 and 600°C are called Si 150, Si 400 and Si 600. The number of OH groups per 100 $Å^2$ of surface are respectively 5.7, 4.2 and 1.4.

Influence of the pressure. Figure 1 shows that the greater the NH_3 concentration, the smaller the chemical shift δ of the signal detected after NH_3 adsorption on samples Si 400 and Si 600. With Si 150, δ varies little with the NH_3 concentration. Moreover, for the same surface coverage θ, δ varies markedly from one sample to another. For example the values of δ obtained at $\theta = 0.5$ are 3.80, 3.00 and 2.00 ppm for Si 150, Si 400 and Si 600 respectively. That is δ decreases as the pretreatment temperature of the sample is increased ; or, in other words, as the degree of hydration of silica gel is reduced. As in the previous case (zeolithe HY) these results show that δ must depend on the SiO–H dissociation. Equations 4 (X=Si), 5 and 6 express the possible reactions between NH_3 and the surface of the solid. Si 150 has a high superficial OH density ; δ does not depend very much on the amount of adsorbed ammonia, as long as $\theta < 0.8$. On the contrary δ varies with the coverage of Si 400 or Si 600. But for low θ, particularly following the desorption at 30°C under 10^{-4} torr, the chemical shifts tend to a limiting value of δ (δ limit) equal to about 3.9 ppm, whatever the OH concentration. In this case, the number of physically adsorbed molecules must be very small. Moreover, it is obvious that the ratio $(NH_3)_I / (NH_3)_{II}$ cannot be inde-

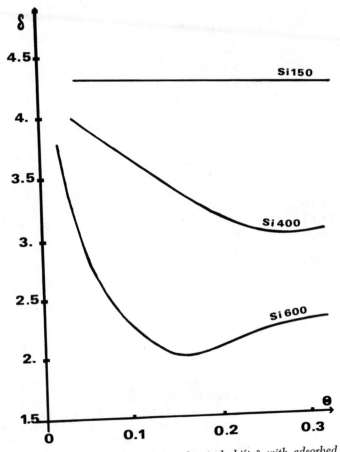

Figure 1. *Dependency of the chemical shift δ with adsorbed ammonia concentration*

pendent of the surface hydration. Thus, $\underline{\delta}$ (limit) must be characteristic of equilibrium 4 and depends on OH bond. Naturally, its value is between $\delta(NH_4^+)= 6.9$ ppm and $\delta(NH_3)_I$ which lies between 0 and 1 ppm because of the SiOH...NH₃ hydrogen bonds. When θ increases, the $(NH_3)_{II}$ concentration increases as well, and consequently $\underline{\delta}$ decreases due to eq.5;eq.6 modifies $\underline{\delta}$ only slightly. At high coverage, $\underline{\delta}$ increases slightly again with the pressure, probably due to a small increase of $\delta(NH_3)_{II}$ due to some H-bonding between the molecules.

   Influence of the temperature. At 60 MHz, the experimental temperature T being 30°C, the line is symmetrical and approximately lorentzian. When T decreases, this line becomes wider and asymmetrical ; it shows a large shoulder downfield while its maximum shifts slightly up field. The lower the temperature the more these changes are important.

ν MHz	Sample	T°C	δ ppm	ΔH Hz	σ
60	Si 150	30	4	300	0.000
		−1	3.95	415	0.010
		−18	3.90	530	0.050
		−30	3.85	660	0.061
	Si 400	30	3.15	265	0.000
		−1	3.15	320	0.026
		−18	3.10	410	0.035
		−34	2.90	480	0.062
250	Si 150	30	4.05	680	0.08
		0	3.80	820	0.00
		−20	3.50	1180	"
		−40	3.10	1740	"
	Si 400	30	3	600	0.074
		0	2.90	670	0.000
		−20	2.70	930	"
		−40	2.40	1480	"

Table 1 . Spectra at various temperature . $(\theta = 0.53)$

For example, considering Si 150 and $\theta=0.53$ (Table 1), when T decreases from 30 to -30°C, δ decreases from 4 to 3.85 ppm, ΔH increases from 300 to 660 Hz while the asymmetry coefficient σ, initially zero, reaches 0,061 (we take the difference of the two half-widths divided by the width as the asymmetry coefficient.) This behavior, similar to that of Si 400, demonstrates that the coalescence of the two lines is lifted : the apparent highest proton concentration line shifts upfield, while the other appears as a shoulder downfield. The two components cannot be separated since the lines become too wide at low temperature. Moreover, when

the lines are no longer in coalescence, at least partially, the
line for superficial NH_4^+ cannot be detected in high-resolution ;
in that case the proton movement is not enough to narrow the line,
originally wide because of dipolar interactions. Neither is the
superficial OH line detectable since it is about 2 KHz wide.

The previous conclusions can be verified at 250 MHz. The theoretical reference frequency $\Delta\nu = \nu(NH4^+) - \nu(NH_3) = 1750\,Hz$ being
definitely higher than $\Delta\nu(60) = 400$ Hz at 60 MHz, the temperature
effect upon the lifting of coalescence must be greater. In fact,
considering the same T variation and the same Si 150 sample, $\underline{\delta}$
decreases more at 250 MHz than at 60 MHz (from 4.00 ppm at 30°C to
3.20 ppm at -30°C) ; but $\underline{J}$ varies in the opposite way (0.08 at 30°C
and 0.0 at 0°C) revealing that coalescence is lifted. Consequently
the spectrum (250 MHz, 30°C) must correspond to the (60 MHz, -50°C)
one.

Coefficient α of OH dissociation and chemical shift δ(OH)

At 30°C, following ammonia desorption at 10^{-4} torr, the chemical
shift tends to a very similar δ (^1H)limit for each sample = 4.2,
3.9 and 3.7 ppm for Si 150, Si 400 and Si 600 respectively. This
value, which characterizes equilibrium 4 , decreases a little with
decreasing OH concentration of the surface, showing that the OH
dissociation and the average acidity of one OH increase with the
OH density. The chemical shift δ(OH) and the constant K of this
equilibrium are unknown. To calculate them, the detection of a
spin other than ^1H is necessary. Because of the quadrupole moment
of ^{14}N, we have chosen the ^{15}N nucleus. With the sample Si 400,
δ(^{15}N) increases from 17.8 to 22.0 ppm when the adsorbed NH_3 decreases from 50 to 5 mg.g^{-1}(correspondig to 3 hours desorption
under 10^{-4} torr). Therefore :

$$\delta(^{15}N)_{limit} = P_{NH_3} \cdot \delta(^{15}NH_3) + P_{NH_4^+} \cdot \delta(^{15}NH_4^+) \quad 8$$

where P_{NH_3} and $P_{NH_4^+}$ are the concentrations of the spins NH_3 and
NH_4^+ :

$$P_{NH_3} + P_{NH_4^+} = 1$$

Ammonia being the reference : $\delta(^{15}NH_3) = 0$;

then $P_{NH_4^+} = \dfrac{\delta(^{15}N)limit}{\delta(^{15}NH_4^+)} = \dfrac{22.0}{43.5} \sim 0.5$ and $\dfrac{NH_3}{NH_4^+} = 1$

Whence the coefficient of OH dissociation is α=0.5 . For comparison, the pH of an equimolar NH_3 and NH_4^+ buffer solution is
about 4.7 . With Si 400, δ(^1H)limit = 3.9 ppm. Using eq 2 and the
coefficient α, the proton shift of the OH group can be calculated:
δ(OH) = 2 ppm
It should be noted that this value is independent of the adsorbed
phase. For comparison, δ(OH) of ethanol diluted in CCL_4 is
0.5 ppm.

II- Chemisorption of hydrogen on silica-supported platinum.

The second application of NMR concerns hydrogen adsorbed by supported platinum ; this has been studied by numerous investigators with the aim of establishing the nature of the active form in the catalysis of hydrogenation. All the experimental results and theoretical considerations suggest that hydrogen exists on platinum in two forms : a strongly irreversibly adsorbed and a weakly reversibly adsorbed species (11-13). But, the irreversibly adsorbed hydrogen could never be detected by spectroscopy and its nature is unknown. Following the first NMR results obtained by TARO Ito et al (14) concerning the adsorption of hydrogen on platinum metal under high pressure, we have undertaken the study of H_2 irreversibly chemisorbed on Pt/SiO_2. The silica-supported platinum was prepared by impregnation of Davison silica with a sufficient amount of chloroplatinic acid solution to give 10 wt % platinum after reduction (15). The particle diameter of platinum is about 200 A. Before hydrogen adsorption, the samples were heated at 350°C under 10^{-4} torr for 12 hours.

II-1 Results. Before any hydrogen adsorption, the spectrum of the sample recorded at room temperature consists of a symmetrical line of Lorentzian shape of width $\Delta H_{ptop} = 0.57$ gauss. Within the limits of experimental error, this line has no chemical shift relative to the usual references. It is due to the proton resonance of the OH groups situated on the silica surface, and it will be used in what follows as the standard. For monolayer coverage (2.10^{19} adsorbed molecules $.g^{-1}$ under 5.10^{-2} torr), the spectrum recorded at room temperature can be broken down into 3 lines (figure 2). One (line 3), similar to that formed before hydrogen adsorption, is due to OH groups. The second (line 1) and the third (line 2) are shifted upfield, slightly (-1 ppm) and very much (-46 ppm) respectively.

After pumping at 10^{-4} torr and room temperature for 12 hours, lines 1 and 2 are slightly weaker and the shift of line 2 is $\delta_1 = -49$ ppm (figure 3C). This spectrum (spectrum 1) corresponds to irreversibly chemisorbed hydrogen at 25°C. After heating the sample under vacuum, the strength of line 1 and 2 decreases and the shift of the latter increases. At the limit of signal detection, corresponding to a desorption temperature of 100°C, the shift of line 2 is $\delta_0 = -57$ ppm (spectrum 0).

When the hydrogen pressure is increased (fig 3), line 3 is unchanged. The intensity of line 1 increases with the pressure until it reaches about 1.5 torr (corresponding to $1.06 . 10^{20}$ adsorbed molecules $.g-1$), after which it remains constant. Line 2 increases continuously with the hydrogen pressure and is shifted downfield ; its δ varies approximately homographically with the number of adsorbed molecules. After hydrogen adsorption at a few torrs and 25°C, the NMR spectrum was determined at 77 K. It can be represented approximately by spectrum 1 on which a relatively nar-

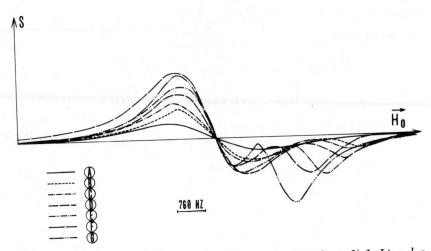

Figure 2. Spectrum of the hydrogen adsorbed on Pt/SiO₂ ($\theta \sim 1$) 1. Line due to the species $H_{(1)}$. 2. Line due to the species $H_{(2)}$. 3. Line due to the OH groups of the support. 4. Recorded spectrum.

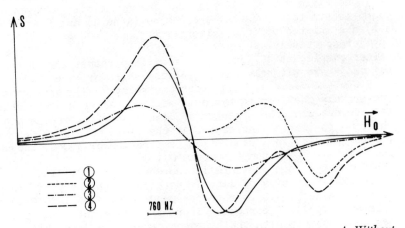

Figure 3. Change of the spectrum with the hydrogen pressure. A. Without hydrogen. B. Hydrogen desorbed at 48°C under vacuum for 2 hr, after G. C. Hydrogen desorbed at 25°C under vacuum for 12 hr. D. Hydrogen desorbed at 25°C under vacuum for 1 hr. E. H_2 adsorbed at 0.32 Torr. F. H_2 adsorbed at 2 torr. G. H_2 adsorbed at 49 Torr.

row non-shifted signal has been superimposed.However, line 2 is less shifted than in spectrum 1.

When the platinum surface is initially covered with strongly chemisorbed CO, neither line 1 nor 2 appears in the spectrum whatever the hydrogen pressure. However, the hydrogen is reversibly adsorbed and is indicated by a narrow signal which is shifted about 7 ppm downfield.

II-2 Discussion. First of all it must be emphasized that, thanks to NMR, for the first time it has been possible to detect spectroscopically hydrogen not only reversibly adsorbed but also irreversibly adsorbed on supported platinum, and to follow the change in the latter when the surface coverage decreases to very low values.

We consider firstly irreversible adsorption. The existence of 2 signals proves that it occurs in two different forms. Let us now examine the different factors which are likely to lead to a chemical shift of the chemisorbed atoms :

- We have determined the bulk magnetic susceptibility X_v of the Pt/SiO$_2$ sample following as previously (5). The effect upon δ of the change of X_v relative to pure silica is a few Hz downfield. It is therefore negligible. Moreover it must be the same for all the components of the spectrum.

- The diamagnetic or orbital effects of the conduction electrons produce shifts to higher field. However they are of the same order as the ordinary chemical shift and must be negligible here.

- The Knight shift in metals arises from the interactions of the conduction electrons near the Fermi surface of the metal with the metal nuclei. In an externally applied magnetic field H, these unpaired electrons have a net spin polarization which contributes to a local magnetic field ΔH at the nucleus. ΔH is proportional to this polarization and to the density $\langle |\Psi_M(0)|^2 \rangle_{av}$ of the conduction electrons at the nucleus M ; therefore only s conduction electrons make a direct contribution. These latter can have a finite density at the H adatoms on platinum which would give rise to a positive shift δ_K opposite to the observed one. It seems likely that the shift of 7 ppm of the line associated with H$_2$ on a surface covered with CO corresponds to the order of magnitude of δ_K.

- A direct overlap between the conduction electrons and electrons localized in s orbitals centred on the protons would pair the spins of the last ones antiparallel to that of the conduction electrons, resulting in an upfield proton shift δ_s. The conduction electrons can also take part in an exchange interactions with the bonding electrons in d orbitals centred on platinum, polarizing these latter parallel to their own directions. The d electrons transmit the spin polarization with a change of sign to the proton via the exchange interactions in the covalent part of the Pt-H bond. Then the proton shift is opposite to the Knight shift.

Therefore the sign and magnitude of δ depend on the relative contribution of the Knight shift and the mechanisms responsible for negative shift. It could therefore be said that for the species corresponding to line 1, the Knight shift neutralizes almost completely the effects of polarization of the bonding electrons, whereas for the species revealed by line 2 the negative spin density is overwhelmingly dominant. Probably because of the insensitivity of their spectrometer and the low surface area of the platinum metal particules, Taro Ito et al studied the adsorption of hydrogen on Pt metal at relatively high pressure (14). They were therefore unable to distinguish reversible and irreversible adsorption. However, since the overall shape of the spectra is maintained when the hydrogen pressure increases, the interpretation of their results is roughly the same as that which we propose for the irreversible phase.

We know that the structure of the bands of a transition metal is characterized by broad s-p bands with low density of states whereas the d-bands are narrow with particularly high density of states. Moreover in platinum, the 5 d-band is close to the Fermi level. On the other hand the shift of line 1 is small and is not changed by increase in the H_2 concentration ; this line then must correspond to the hydrogen atoms $H_{(1)}$ located above the metal surface and bonded with the superficial Pt_s atoms through $\underline{s}$ electrons. On the contrary, δ of line 2 is very large and decreases to a limit value δ_1 when the concentration of irreversible hydrogen increases. Line 2 therefore represents the hydrogen atoms $H_{(2)}$ preferentially adsorbed by means of the $\underline{d}$ electrons of the metal, in the interstices between the Pt_s atoms of the surface, the 1s electron of the $H_{(2)}$ atom being delocalised throughout the neighbouring Pt atoms. For example, on the (100) face the $H_{(1)}$ species will be bonded above the surface to each Pt_s atom, the species $H_{(2)}$ to one Pt atom below the surface and to four Pt atoms. As the interstices of the different planes are being filled, the density of the negative spin at each $H_{(2)}$ atom decreases until the sites are completely occupied. These results show not only the possibility of a localised metal adsorbate bond, as Bond has proposed (16), but also the collective electronic properties of the metal.

This interpretation confirms in more detail the results of Taro ITO and Toru KADOWAKI concerning the participation of $\underline{s}$ and $\underline{d}$ electrons in the chemisorption of hydrogen on metals. For example these authors show that only the $4\underline{s}$ conduction electrons are involved in the chemisorption of hydrogen on copper (17). Moreover they detect in this case a single line with a positive chemical shift (downfield) independant of the adsorbate concentration.

At high pressures, the reversible form is characterized by line 2, which shifts markedly downfield as the hydrogen concentration increases. The evolution of this line is typical of a chemicalexchange (at a frequency higher than 3.10^3 Hz) between the "irreversible" interstitial species $H_{(2)}$ resonating at high field (shift δ_1) and another species $H_{(3)}$, much less shifted. This is confirmed by the separation of line 2 into 2 components (of which one is very slightly shifted) when the spectrum is recorded at 77 K. However we have not enough evidence to state that the $H_{(3)}$ form is that which is adsorbed at room temperature on a CO-covered surface and situated at 7 ppm.

Conclusion

The Bronsted acid strength of an oxide surface of high specific area, and the nature of the hydrogen adsorbed on metals are amongst the most important of classical problems in heterogeneous catalysis. It seems to us that it has not been possible to solve them without the help of NMR. Thus, the Bronsted acidity of oxides has been characterized, as in homogeneous phase, either by calculation of the dissociation constant of the O-H bond in the presence of other molecules, or by determining the chemical shift of the acidic hydrogen, i.e. by "evaluating" its electronic environment, the $\delta(OH)$ value being, of course, independent of the previously adsorbed base.

NMR can also be applied successfully to problems of gas-metal interactions. Thus, we have been able to show spectroscopically the existence of irreversibly adsorbed hydrogen on supported platinum and to describe it accurately : hydrogen atoms located on the surface and interstices. The hydrogen adsorbed subsequently is adsorbed reversibly and exchanges at frequencies above 3.10^3 Hz only with the interstitial H atoms.

ABSTRACT

It is shown for two major chemical problems concerning heterogeneous catalysis that HR-NMR spectroscopy is now suitable for the study of gas-solid interactions. Thus, it is possible to characterize Bronsted acidity of oxides just as in homogeneous phases by calculating the dissociation coefficient of the OH bonds in the presence of other molecules or by determining the chemical shift of the acidic hydrogen. For example, the dissociation coefficient of the OH groups in a silica gel is 0.5 in the presence of NH_3 and the proton chemical shift $\delta(OH)$ is 2 ppm relative to gaseous TMS.

Interactions between hydrogen and supported platinum have also been investigated by NMR. The spectrum of the irreversibly adsorbed hydrogen is composed of two lines which are slightly and greatly shifted upfield, respectively ; they characterize superficial and interstitial hydrogen. Only the latter form exchanges with reversibly adsorbed hydrogen ; the frequency of this exchange is greater than 3.10^3 Hz at room temperature.

Literature Cited

1 - HIROTA K., FUERI K., NAKAI J. and SHINDO K. Bull.Chem.Soc. japan (1959), <u>32</u>, 1261

2 - FRAISSARD J., CAILLAT R., ELSTON J. and IMELIK B. J.Chim.

phys. (1963), 60, 1017

3 - DEROUANE E.G., FRAISSARD J., FRIPIAT J.J. and STONE W.E.E.
Catalysis Reviews (1972), $\underline{7}$ (2), 121-212.

4 - BONARDET J.L., SNOBBERT A. and FRAISSARD J. C.R.Acad.Sci.
$\underline{273}$, Serie C, 1405

5 - BONARDET J.L., FRAISSARD J., J.Mag.Resonance (1976), $\underline{22}$,
001-006

6 - EMSLEY J.W., FEENEY J. and SUTCLIFFE L.H. High-Resolution
Nuclear Magnetic Resonance Spectroscopy (1965) Pergamon Press

7 - WITANOWSKY M. and JARUSZEWSKI H., Canad.J.Chem (1969), $\underline{47}$,
1321

8 - LITCHMAN W., ALEI M. and FLORIN A.E. J.Amer.Chem.Soc. (1969)
$\underline{91}$, 6574

9 - BECKER E.D., BRADLEY R.B. and AXENROD T., J.Mag.Resonance
(1971), $\underline{4}$, 136

10 - PLANCK C.J., Catalysis (1955), 1

11 - MIGNOLET J.C.P., J.Chim.phys. (1957), $\underline{54}$, 19

12 - DIXON L.T., BARTH R., KOKES R.J., and GRYDER J.W. J.Catal.
(1975), $\underline{37}$, 376

13 - PRIMET M., BASSET J.M., MATHIEU M.V. and PRETTRE M. J.Catal.
(1973), $\underline{28}$, 368

14 - ITO T., KADOWAKI T., and TOYA T. Japan J.Appl.Phys. (1974)
suppl. 2 Pt 2, 257

15 - DORLING T.A., LYNCH B.W.J. and MOSS R.L. J.Catal. (1971), $\underline{20}$
190

16 - BOND G.C., Disc.Far.Soc. (1966), 200 and references there
in.

17 - ITO T., KADOWAKI T. Japan J.Appl.phys. (1975), $\underline{14}$, (11).

The NMR Study of Proton Exchange between Adsorbed Species and Oxides and Silicates Surfaces

J. J. FRIPIAT

Centre de Recherche sur les Solides à Organisation Cristalline Imparfaite,
45045 Orleans Cedex, France

Heterogeneous acid catalysts have an important industrial role and in recent years a great deal of research has been devoted to their study and to their improvement. The recognition of the nature of the acid centers specially by the study of the inter action of these acid centers with various molecular species has made important progres. Infrared spectroscopy (IR) has been from this respect a very useful tool in making possible the direct observation of molecules interacting with surface sites. The life-time of a specified molecular configuration revealed by infrared spectroscopy is larger or at least of the order of the inverse of the characteristic frequencies for that configuration, e.g. $\geqslant 10^{-13}$ sec. Any process including intermediates with significantly longer life-times will therefore appear as static for IR.

For example consider an "acid surface site" SH and an activated intermediate AH in the A $\rightarrow$ B transformation:

$$SH + A \underset{\rightarrow}{\leftarrow} S + AH$$
$$AH \rightarrow B + H$$
$$H + S \rightarrow SH$$

If the life-time of AH is, say, 10^{-12} sec, its characteristic vibrational spectrum is observed but infrared will give no information on the rate limiting step though it may well be the proton "turn-over" on A.

The proton jump frequencies measured in homogeneous liquid phases are in general smaller than 10^{+12} sec^{-1} (1) and it may be anticipated that they will be smaller when a solid donor and an adsorbed acceptor are involved. In this domain of frequency nuclear magnetic resonance (N.M.R.) and especially proton pulse N.M.R. are of particular interest. Indeed measurements of the spin-lattice and spin-spin proton relaxation times, T_1 and T_2 respectively, have been used for gaining information on proton transfer reactions. It is the aim of this paper to review some of recent works done in this area and relevant to acid heterogeneous catalysis.

Proton motions generate fluctuations in the local magnetic field and therefore in the spectral density functions J_1 and J_2

at w_0 and $2w_0$ ($w_0 = 2\pi\nu_0$ = proton resonance frequency, radian
sec $^{-1}$). It may be shown that

$$T_1^{-1} \propto \text{(mean square local field)}[J_1(w_0) + J_2(2w_0)] \qquad (1)$$

J_1(or J_2) are the Fourier transforms of the autocorrelation function which for a random walk is

$$G(\tau) = <f(t)\ f^{*}(t + \tau) > = < f(0),\ f^{*}(0) > \exp - \tau/\tau_c \qquad (2)$$

$G(\tau)$ contains the information on the motion with respect to the "coordinates" function $f(0)$ and in that case

$$J(w_0) = \frac{w_0 \tau_c}{1 + w_0^2 \tau_c^2} \qquad (3)$$

The so-called correlation time τ_c measures the persistence of the fluctuation.

If the protonic jumps were the only motions in the system the measurement of the correlation time would provide unambiguous information on the proton transfer reaction. Because the adsorbed species A experiences also other kinds of motions, the main difficulty is to distinguish the various contributions to T_1^{-1} or T_2^{-2}.

T_1^{-1} is usually considered as the sum of two contributions either T_1^{-1} intra or T_1^{-1} inter if the local field modulation (in equation 1) is provoked by rotational or diffusional motion respectively. In some cases isotopic substitutions may greatly help in assigning the diffusional motion to a proton exchange.

For instance in the study of the methanol-silicagel (2,3) or methanol-hydrogen zeolite (4) systems, the diffusional motion was obtained from the proton T_1^{-1} observed for the CH_3 OD-deuterated adsorbent system. Measuring subsequenty the proton T_1^{-1} of the CD_3 OH-OH surface and substracting the contribution of the molecular diffusion, the contribution of the proton exchange was obtained and the correlation time for this exchange was calculated.

When a molecule contains equivalent protons, such as in water and ammonia, this deuteration procedure cannot be used. A favorable situation for calculating proton transfers in case of a preferential orientation of the adsorbate on the surface of the adsorbent results from the following considerations.

Consider a pair of protons, such as in a water for instance, and assume that the interprotonic vector has some privilege orientation with respect to a steady magnetic field H_0. Each proton partner in the pair is characterized by spin number $+$ or $- 1/2$. It may of course exchange with another proton of a opposite spin number. Preferentially oriented protonic pair gives rise to a doublet in the N.M.R. spectra. The proton exchange, if it occurs at a frequency of the order of the doublet splitting, provokes the coalescence of this doublet into a single peak. For that kind of situation, WOESSNER (5) has calculated the transverse nuclear spin magnetization $<\overline{M}_+>$ using the density matrix method where

$$\langle \overline{M}\rangle_+ = T_r(\rho\, M_+) \tag{4}$$

and

$$\rho_{nm} = \overline{C_n\, C_m} \tag{5}$$

$$M_{+nm} = \gamma\, \hbar\, (U_m | I_1^+ + I_2^+ | U_n) \tag{6}$$

(1) and (2) refer to the two protons in the pair.
The total spin function Ψ is

$$\Psi = \sum_{n=1}^{4} C_n U_n = C_1 \alpha_1 \alpha_2 + C_2 \alpha_1 \beta_2 + C_3 \beta_1 \alpha_2 + C_4 \beta_1 \beta_2 \tag{7}$$

where α and β are the individual spin functions.
In an exchange process, two pairs of protons must be considered as one system

$$\psi^{AB} = {}_\psi A \; {}_\psi B \tag{8}$$

The exchange process causes a change in the wave function

$$\psi^{AB} \rightarrow R\, \psi^{AB} \tag{9}$$

Matrix R represent the effect of exchange on ψ^{AB}. The effect of exchange on the density matrix is

$$\rho^{AB} \rightarrow R\, \rho^{AB}\, R \tag{10}$$

the time dependance being expressed as follows

$$\frac{d\,\rho_{nm}}{dt} = \frac{\rho_{nm}\,(\text{after}) - \rho_{nm}\,(\text{before})}{\tau_c} \tag{11}$$

where τ_c is the average time between exchanges. WOESNER has used this method in order to account for the free induction decay (F. I.D.) following a 90° pulse and especially to obtain τ_c from the spacing of the modulation observed in the F.I.D.

Since the F.I.D. is the Fourier transform of the N.M.R. absorption signal, similar information can be directly obtained from the shape of that signal.

Actually (6) the observed spin-spin relaxation rate is

$$T_2^{-1} = T_{2o}^{-1} + \tau_e^{-1}$$

where T_{2o}^{-1} is the dipolar broadning in absence of exchange. Of course it is only for $\tau_e^{-1} \gg T_{2o}^{-1}$ that a direct information on the exchange can be obtained.

For liquids, where dipolar broadning is rather-small (7)

$$T_2^{-1} = T_{2o}^{-1} + A\, \tau_e\, [1 - (\tau_e/t_p)\tanh(t_p/\tau_e)] \tag{13}$$

where t_p is the pulse spacing and $A = \sum_i p_i \delta_i^2$ the product of the probability of occupancy of the ith type of protonic site and δ_i the chemical shift at that site. Equation (13) has been used to determine the temperature dependence of proton exchange in pure water but it has infortunately a limited application for adsorbed phase because of the high T_{2o}^{-1}.

Resing ($\underline{7}$) has extended the model proposed originally by Zimmerman and Brittin ($\underline{8}$) for the exchange between two phases to the proton exchange between water and the OH groups of a partially hydrolyzed zeolite. This means a zeolite in which a fraction of the lattice contains > Al-OH groups. This approach is convenient when the transfer of nuclei occurs between environnements of much different intrinsic relaxation times but approximately the same chemical shift.

Let T_{2A} and T_{2B} be the transverse relaxation times in the two exchanging "phases", C_B^{-1} and C_A^{-1} the life-times in phases A and B and N_A and N_B the number of exchanging "molecules" respectively. For sake of simplicity it is assumed that A and B contain the same number of protons. Then, as shown by Woessner ($\underline{9}$)

$$T_2^{-1} = (N_B/N_A)(T_{2B} + C_B^{-1}) + T_{2A}^{-1} \qquad (14)$$

There may be a temperature domain where C_B^{-1} >>T_{2B} and consequently where

$$T_2^{-1} = (N_B/N_A) C_B^{-1} + T_{2A}^{-1} \qquad (15)$$

If additionaly $T_{2A} >> C_B^{-1}$, in that region $T_2 = C_A^{-1}$ since at equilibrium $N_A C_A = N_B C_B$. At lower temperature where $T_{2A} << C_B^{-1}$, $T_2 = T_{2A}$. It follows that T_2 passes through a maximum and that on the high temperature side of this maximum the life-time in phase A can be approximated by T_2.

These are in short the main N.M.R. techniques that can be used for measuring correlation times of proton exchange processes. Most of them have been used in the systems that are going to be reviewed now.

These examples include surfaces with high (decationated zeolite), medium (hydrated smectites) and weak (silicagel) acid sites and molecules with a very low, low and rather high pKa, namely water, methanol and ammonia.

EXPERIMENTAL RESULTS AND DISCUSSION

Three molecules have been studied : H_2O, CH_3OH and NH_3. The pKa of their protonated forms, namely H_3O^+, $CH_3OH_2^+$ and NH_4^+ are in that order : the smallest is that of H_3O^+, then follows $CH_3OH_2^+$ with a pK$_a$ = - 4.5 and finally NH_4^+ for which pK$_a$ = + 9.25.

1) Proton exchange between acid surfaces and NH_3. Figure 1 shows the correlation times observed for ammonia adsorbed on a silicagel (Aerogel) at degrees of coverage θ between 0.4 and 2 ($\underline{10}$) and for an ammonium Y zeolite ($\underline{11}$) (75% of the initial Na$^+$ content replaced by NH_4^+). These correlation times have been deduced from measurements of the spin-lattice relaxation times.

The pK$_a$ of surface silanols in Aerogel is about 6.5 whereas that of the OH group generated by the deamination of the ammonium

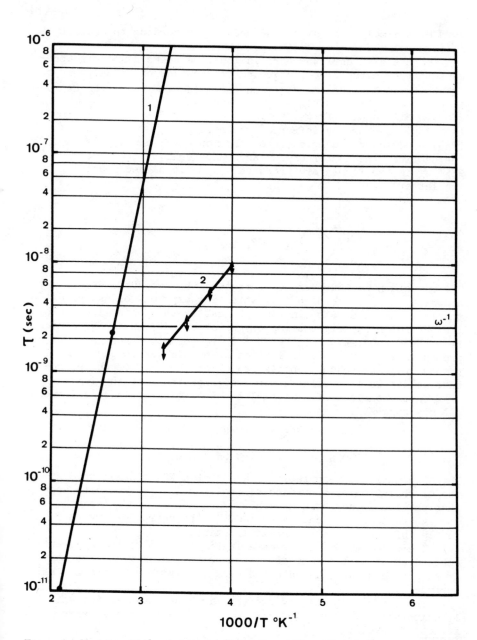

Figure 1. Variation of the proton correlation time with respect to the inverse of temperature: (1) NH₄ Y near faujasite molecular sieve in which 25% of the initial cation Na content has not been replaced by NH₄⁺. There is no adsorbated phase. (2) NH₃ on an Aerogel (silicagel) for degrees of coverage between 0.4 and 2. The inverse of the resonance frequency, wₒ⁻¹, is shown.

form of the molecular sieve is about - 6 according to Rouxhet et
al.(12)

On the silicagel the correlation time seems almost independant of the degree of coverage and the activation enthalpy is between 4.1 and 5.6 kcal mole $^{-1}$. On the molecular sieve the activation enthalpy is much higher ($\sim$ 19 kcal).

On silicagel a fraction of the adsorbed NH_3 is transformed
into NH_4^+ as shown by infrared spectroscopy. At 20°C, $[NH_4^+]/$
$[NH_3] \simeq 0,3$ at the completion of the monolayer content in NH_3. On
the zeolite, the number of surface OH is egal to the number of
NH_4^+ since the former are produced by the local decomposition of
the latter (13).

The situation is thus very different on the two surfaces. In the
ammonium zeolite the high activation enthalpy may be assigned to
the fact that the rate limiting process is the delocalisation of
the OH proton on neighbouring oxygen atoms (14). When NH_3 coming
from the dissociation of the NH_4^+ is temporarily free because the
proton given to the oxygen lattice is not longer in close approach it jumps toward another region of the surface where another acidic OH is available for recombination into NH_4^+.

On the surface of silicagel the situation would be different.
One may suggest for instance

$$Si-OH + NH_3 \rightleftarrows Si-O-NH_4$$

$$Si-ONH_4 + NH_3 \rightleftarrows Si-\bar{O} \overset{+}{NH_4}:NH_3$$

This schematic representation means that a NH_4^+ formed through
the dissociation of a surface silanol could interfere with another physically adsorbed molecule.

Above some critical value (<0.4), the degree of coverage has
little influence on the correlation time.

The observation of the data in Fig.1 leads to the apparent
contradiction that the proton exchange is faster when a relatively strong base (NH_3) is adsorbed on a weakly than a strongly acid
surface.

The two situations reviewed in Fig.1 are however not directly comparable since NH_3 doesn't form an adsorbed phase sensu
stricto in zeolite.

In order to escape this ambiguity let us examine an example
where a weak base (CH_3OH) forms a network of adsorbed molecules
in contact with surface OH.

2) Proton exchange between acid surfaces and methanol. The

correlation times deduced from the spin-lattice relaxation times measurements are shown in Fig.2. ($\underline{4}$). As compared to NH_4^+,$CH_3OH_2^+$ is a much stronger acid. Its strength is comparable with that of OH groups in the decationated zeolite. However the correlation times observed for methanol occluded in the zeolite cage of the decationated sieve are longer that those observed on the silicagel (Xerogel) surface. Thus the proton exchange is faster when a weak base is adsorbed by a weak than by a strong acid surface.

If this observation is compared with that made above for ammonia it must be concluded that the life-time of the acid form of the adsorbed species is not governed by its pK_a. If the pK = $log_{10}K$ of the schematic reaction :

Solid H + adsorbed species → þrot. ads. species + Solid,is calculated for the four systems so far studied, the following values are obtained : pK = - 15.25 for NH_3 adsorbed on the decationated zeolite, pK = - 1.50 for methanol in the same situation; pK = - 2.75 for NH_3 on silicagel and pK = + 11 for methanol on silicagel. By comparing these values with the corresponding correlation times obtained at 298°K, it becomes clear that there is no relationship between the kinetics and equilibrium data.

The life-time of the protonated form of the adsorbate is ruled by the probability of the proton transfer event within the adsorbed phase. e.g. for the following reactions :

$$CH_3OH_2^+ + (CH_3OH)^{\ast} \rightarrow (CH_3OH_2^+)^{\ast} + CH_3OH$$

$$NH_4^+ + (NH_3)^{\ast} \rightarrow (NH_4^+)^{\ast} + NH_3$$

The star stands for purpose of identification. In the homogeneous phase, e.g. pure alcohol ($\underline{15}$) or liquid ammonia ($\underline{16}$) the inverses of the proton jump frequencies are about the same, namely $4 \cdot 10^{-10}$ sec at 295°K.

In the adsorbed phase the rate of the proton transfer is between one and four orders of magnitude lower (Table I).

Beside geometrical reasons relevant to the nature of the bi-dimensional network of hydrogen-bridged molecules formed on surfaces, the reason for such a lowering could be eventually due to the influence of the back donation of protons to the surface.

3) Proton exchange in water adsorbed by layer lattice silicates. The layer-lattice silicates used in the experiments summarized in Fig.3 (vermiculite, montmorillonite and hectorite) share common features :

(a) their specific surface area is about $800m^2/g$ and it is developped to a large extent within the interlamellar space. Most of the adsorbed water is located there and the thickness of the layer is known from the $d_{00\ell}$ X-ray basal reflection;

(b) the interlamellar space accomodates well-characterized "hydrates" e.g., those containing one or two layers of water molecules between the silicate sheets;

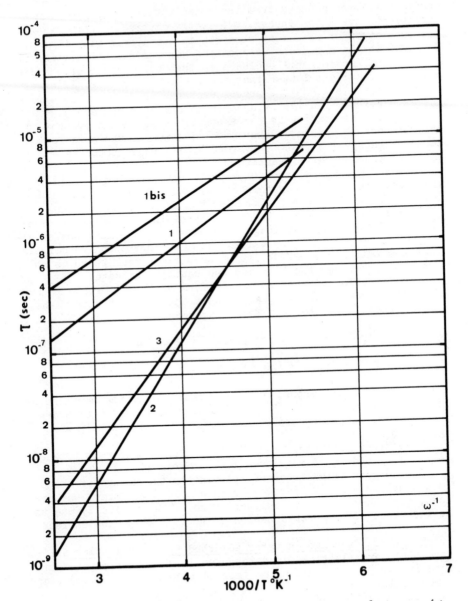

Figure 2. Variation of the proton correlation time with respect to the inverse of temperature: (1) CH_3OH in the HY decationated molecular sieve $\theta = 1$ and 1 bis $\theta = 1/3$; (2) CH_3OH on a Xerogel (silicagel) for degrees of coverage $\theta = 0.8$; (3) $\theta = 1.7$

(c) due to isomorphic substitutions, the lattice is negatively charged. The lattice electrical charge is the highest for vermiculite ($\sim 1.8 \times 10^{-3}$eq/g) and the lowest for hectorite (about 0.9×10^{-3}eq/g). Montmorillonite has a charge of about 1.0×10^{-3}eq/g;

(d) this electrical charge is balanced by cations in the interlamellar space. In the studies to be reviewed, these cations were either Na^+, Li^+ or Ca^{2+};

(e) there are at least two kinds of water molecules in the interlamellar space (17) : those belonging to the cation hydration shell and those forming a network of hydrogen-bridged molecules between these shells;

(f) water in the interlamellar space has a degree of dissociation higher than that in the pure liquid phase (17, 18, 19). The field of the cation is poorly screened by the lattice and consequently, the water molecules in the hydration shell are strongly polarized. It has been proposed that they work as proton donors to the network of surface hydration water as schematically represented below:

$$[M(H_2O)_x]^{z+} + H_2O \rightarrow [M_{(H_2O)_{x-1}}^{OH}]^{(z-1)+} + H_3O^+$$

$$H_3O^+ + H_2O\!:\!: \rightarrow H_3O^+\!:\!: + H_2O$$

Thus two widely different life-times should be measured, namely (a) that of a water proton exchanging in the usual way (as in the liquid) with a neighboring molecule, and (b) that of the transient H_3O^+ species.

Correlation times (1) and (2), in Fig 3, have been assigned to the first mechanism, whereas correlation times (3) and (4) are believed to represent the H_3O^+ life-time.

In a pseudo-monocrystal composed from superimposed films, obtained by slowly sedimenting a clay suspension, the N.M.R. spectrum is, in general, composed from a doublet and from a central line. This was clearly observed for the two-layer hydrate of a Na vermiculite. An octahedral arrangement of water molecules around the Na^+ cations was deduced from the orientation dependence of the doublet splitting. The central line intensity decreased with temperature and, in addition, it was absent in the corresponding D_2O two-layer hydrate. These characteristics were accounted for by assuming that the central line results from the doublet coalescence provoked by an increase in the exchange of one proton in a specified H_2O by another one in the opposite spin state (20). It is in that way that τ_2 in Fig.3 was obtained. τ_1 was obtained by observing that in the one-layer hydrate of a Li hectorite, T_2 goes trough a maximum as the temperature goes from 273°K to 240°K.(21) The life-times described by τ_1 and τ_2 are in general agreement with those obtained under similar conditions by Woessner (22).

As to the life-times τ_3 and τ_4 assigned to H_3O^+, they have been obtained from T_1 measurements in quite a number of very different situations. Actually, at room temperature,($\tau_3 \simeq \tau_4 \simeq 10^{-10}$

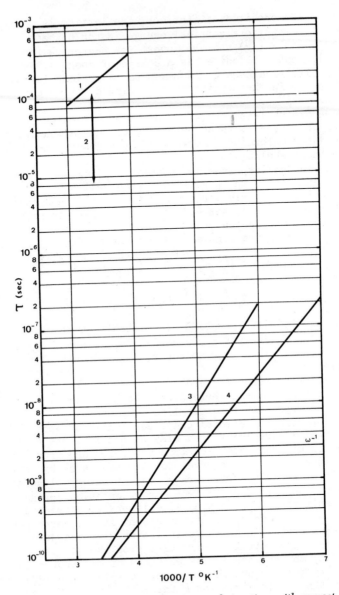

Figure 3. Variation of the proton correlation time with respect
to the inverse of temperature: (1) H_2O in the hydration shell of
$Li^+(H_2O)_3$, in the monolayer hydrate of Li-hectorite; (2) H_2O in
the hydration shell of $Na^+(H_2O)_6$ in the two-layer hydrate of
vermiculite; (3) H_3O^+ in the monolayer hydrate of Na-montmoril-
lonite, in the two-layer hydrate of Na-vermiculite and of Ca-
montmorillonite; (4) H_3O^+ in the monolayer hydrate of Li-hectorite

sec), it does not matter that the proton spin-lattice relaxation occurs through paramagnetic impurities (as in vermiculite and montmorillonite) (20, 22) or through a dipolar mechanism (as in hectorite (21), Fe^{3+} <100 ppm), or that the measurements are obtained for a one-layer or two-layer hydrate in the presence of various cations.

This strongly suggests that τ_3 and τ_4 are not correlation times that could be assigned to a molecular motion. This hypothesis is reinforced by a comparison with relaxation measurements obtained for proton-doped ice. Indeed, in that case, it has been observed (23) that the higher the proton concentration, the higher the T_1. At - 30°C, it could reach 10^2 sec^{-1} for a H^+/H_2O ratio on the order of 10^{-2}, e.g., on the order of that measured by several techniques on the surface of clay (17, 18, 19). Moreover, the activation energy for the proton motion in ice is between those energies obtained for τ_3 and τ_4. Finally, it should also be pointed out that the correlation times deduced from proton diffusion coefficient measurements carried out by neutron inelastic scattering (24) on layer lattice hydrates similar to those studied by N.M.R, are close to τ_3 and τ_4 and room temperature.

At that temperature, it is interesting to point out that the life-time of H_3O^+ is 10^{-12} sec according to Eigen (25), whereas that of a proton in liquid H_2O is about 10^{-3} sec (26).

Therefore, the life-time of H_2O seems to be shortened in the adsorbed state by approximately the same factor as the life-time of H_3O^+ is lengthened.

CONCLUSION

Two different situations have been discussed here : in the first, the surface is a proton donor, as in silica-gel or in a decationated zeolite, while in the other, the adsorbed molecule (H_2O in the cation hydration sphere) acts as proton donor. The two situations share at least one common feature : the proton exchange mechanism obviously requires the presence of a "structured" network of adsorbed species. A good example of the case where this situation is not achieved and where, therefore, the proton exchange frequency is relatively low is that provided by the ammonium Y-zeolite (HY, NH_3 in Table 1).

When the adsorbate has formed a "monolayer" in which molecules are interconnected, by hydrogen bonds for instance, it seems that the rate-limiting process for proton jumps is the back donation to the surface site. For instance, the proton jump frequency is lower in CH_3OH occluded in HY than on silica-gel. The reason may be that a strong acid site favors a strong hydrogen bond with the adsorbed species confining the proton jumps within the two almost symmetrical potential wells along the hydrogen bond direction. Also such a strong bond might prevent, to some extent, the structural rearrangement (occuring for instance by some rotation) that would allow the protonated adsorbate to transmit its proton to a

TABLE I

Average correlation times obtained at 298°K for various adsorbent/ adsorbate pairs. θ : degree of coverage X : xerogel HY : decatio- nated molecular sieve, AG : Aerogel, S : smectite. $\bar{H}$: average activation energy for the exchange process.

Absorbent	Absorbate	θ	τ (sec)	$\bar{H}$ (kcal mole^{-1})	Nature of the Acid site
S	$\begin{cases} H_2O \\ H_2O \end{cases}$	1 - 2	$10^{-5} - 10^{-4}$ $9\ 10^{-11}$	4.4 - 6	Surface hydrated cation
X	CH_3OH	0,8 - 1.7	$\sim 2.5\ 10^{-8}$	5 - 6	silanol
HY	CH_3OH	$1/3 - 1^{(*)}$	$10^{-6} - 4.3\ 10^{-7}$	4	$Al-\overset{H}{O}-Si$
AG	NH_3	0,4 - 2	$2.4\ 10^{-9}$	4.1 - 5.6	Silanol
HY	NH_3	(**)	$1.4\ 10^{-6}$	19	$Al-\overset{H}{O}-Si$

N.B. (*) : full zeolitic cage : $\theta = 1$.
     (**): the NH_3 concentration is equal to that of the $Al-\overset{H}{O}-Si$, acid protons.

neighbour.This would greatly effect the overall jump frequency.

For instance, in a similar vein, Weidemann and Zundel (27) have shown that the interaction between polarisable hydrogen bonds and their interaction with external strong electrical field cause an appreciable decrease in the tunneling frequency. It follows that the tunneling frequency may approach the rearrangement frequency of the solvent molecules. At this stage the reorientation of the solvent molecule in following the fluctuation weakens the field and as a consequence there is a lower limit to the tunneling frequency .

For all these reasons it becomes understandable that there is no direct correlation between the pK_a of either the surface or the protonated surface species and the rate at which protons transfer on the surface.

ABSTRACT

The proton jumps mechanisms contributing to the relaxation of a system of magnetic nuclei may be expressed quantitatively by correlation times measurable by various techniques. The correlation times obtained for various pairs of proton acceptors and proton donors are reviewed critically. These proton donors are the acid sites on the surface of high surface area solids, namely silicagel, decationated zeolite and swelling layer lattice silicates. The proton acceptors are H_2O, CH_3OH and NH_3.

The proton jumps frequency is enhanced by the presence of a mono-molecular network of adsorbed species. The rate limiting step seems to be the proton jump from the protonated onto the non-protonated form of the adsorbed species. However very strong hydrogen bonds such as those formed with the surface OH groups of a decationated zeolite seems to deplete the delocalization of the OH proton.

LITERATURE CITED

(1) Bell R.P. The Proton in Chemistry, Cornell Univ.Press (1959) p.120 et passim
(2) Cruz M.I., Stone W.E. and Fripiat J.J. J.Phys.Chem. (1972) 36, 3078
(3) Seymour S.J., Cruz M.I. and Fripiat J.J. J.Phys.Chem. (1973), 77, 2847
(4) Salvador P. and Fripiat J.J. J.Phys.Chem. (1975) 79, 1842
(5) Woessner D.E. J.Magn.Res. (1974) 16, 483
(6) Knispel R.R. and Pintar M.M. Chem.Phys.Letters (1975) 32, 238
(7) Resing H.A. J.Physic.Chem. (1974), 78, 1279.
(8) Zimmerman J.R. and Brittin W.E. J.Phys.Chem. (1957) 61, 1328.
(9) Woessner D.E. J.Chem.Phys. (1961) 35, 41.
(10) Fripiat J.J., Van der Meersche C., Touillaux R. and Jelli A. J.Phys.Chem. (1970), 74, 382.
(11) Mestdagh M.M., Stone W.E. and Fripiat J.J. J.of Catalysis

(1975), $\underline{38}$, 358.

(12) Rouxhet P.G. and Sempels R.E. J.Chem.Soc.Farad.Trans. I (1974) $\underline{70}$, 2021.

(13) Uytterhoeven J.B., Christner L.G. and Hall W.K. J.Phys. Chem. (1965) $\underline{69}$, 2117.

(14) Mestdagh M.M., Stone W.E. and Fripiat J.J. J.Chem.Soc.Faraday Trans. I (1976) $\underline{72}$, 154.

(15) Lutz Z., Gill D. and Meiboom. J.Chem.Phys. (1959) $\underline{30}$, 1540.

(16) Clutter D.R. and Swift T.J. J.Amer.Chem.Soc. (1968), $\underline{90}$, 601.

(17) Fripiat J.J., Jelli A.N., Poncelet G and André J. J.Phys. Chem. (1965) $\underline{69}$, 2185.

(18) Salvador P., Touillaux R., Van der Meersche C and Fripiat J.J Israel J.Chem. (1968) $\underline{6}$, 337.

(19) Mortland M.M. and Raman K.V. Clays and Clay Min. (1968) $\underline{16}$, 393.

(20) Hougardy J., Stone W.E. and Fripiat J.J. J.Chem.Phys. (1976) to be published

(21) Kadi Hanifi, personal communication.

(22) Woessner D.E. and Snowden B.S.J.,Coll.Interf.Sci. (1969) $\underline{30}$, 54.

(23) Walley E., Jones J.J. and Gold L.W. Physics and Chemistry of ice, Royal Society of Canada, Ottawa (1973), p.222.

(24) Olejnik S., Stirling G.S. and White J.H. Spec.disc.Farad.Soc. (1970) $\underline{1}$, 188.

(25) Eigen M. Angew Chem. (Int.Edn) (1964) $\underline{3}$, 1.

(26) Meiboom S. J.Chem.Phys. (1961), $\underline{34}$, 375.

(27) Weidemann E.G. and Zundel G. Zeit.für Naturforsch. (1973) $\underline{28a}$, 236.

Use of High Resolution Solid State NMR Techniques for the Study of Adsorbed Species

R. W. VAUGHAN and L. B. SCHREIBER

Division of Chemistry and Chemical Engineering, California Institute of Technology, Pasadena, Calif. 91125

J. A. SCHWARZ*

Chevron Research Company, Richmond, Calif. 94802

Recent years have seen the development of a variety of high resolution solid state nuclear magnetic resonance (NMR) techniques (1-7) that are allowing detailed characterization of the physical and chemical environment of adsorbed molecules or ions. Many of these techniques are still in the developing stages, and only preliminary results are available at this time. It will be the purpose of this paper to discuss recent results which we have obtained in the application of multiple pulse techniques to the characterization of protons, hydroxyl groups, on oxide-type surfaces (8), and to describe a recently introduced (9) multiple pulse double resonance (^{13}C - ^{1}H) technique for characterization of both chemical and local geometrical changes that occur upon adsorption of hydrocarbons on an oxide-type surface.

Nuclear Magnetic Resonance Techniques

The use of conventional pulsed NMR techniques (10), multiple pulse NMR techniques (1-3), and double resonance high resolution techniques (4-7,9) has been widely discussed, and the reader is referred to these references for more detailed discussions than can be presented here. In the normal situation, a number of interactions contribute simultaneously to both the lineshape and associated relaxation effects of the NMR resonance of an adsorbed species. Although each of these interactions contains useful information, usually the NMR spectrum can be interpreted to give only the largest one or some complex combination of several. To allow further separation, simplification, and characterization of the interactions present, a variety of recently developed multiple pulse NMR techniques are used here. These are, in general, characterized by the application of periodic and cyclic chains of intense rf pulses and have the effect of altering the effective size and character of specific interactions. Thus, use of these techniques can allow measurement of small interactions which are normally not measurable due to the presence of much larger interactions and can allow separation of interactions normally only measurable as a sum. Three such sequences were used in the study of hydroxyl protons described below. An eight-pulse cycle, illustrated in Figure 1 and discussed in detail previously (3), was used to suppress the effects of static homonuclear dipolar broadening and allow measurement of the proton chemical shift tensor. Two phase-altered versions of the same eight-pulse cycle (8,11) have been used to simultaneously suppress the effects of static magnetic field inhomogeneities, in addition to homonuclear dipolar interactions from the NMR spectra, and thus allow a separation of contributions to the eight-pulse linewidth from chemical shift anisotropy, motional effects, and any residual homonuclear dipolar or experimental problems. For the double resonance experiments to be discussed, heteronuclear dipolar modulated chemical shift spectra are obtained (9) by using the eight-pulse cycle to remove homonuclear broadening between the abundant spin protons and allowing heteronuclear contact between the carbon and proton spin systems. These techniques are in the developing stage and will be discussed in more detail in a later portion of the paper.

Experimental Techniques

These techniques all involve the application of rather large rf fields (20 - 50 gauss) to the sample and involve irradiation with complex cycles of rf pulses. The spectrometer used for this work is, however, relatively simple and has been described in detail (12) elsewhere. The large rf fields needed are obtained under low power conditions (100 watts rf) by using small, 5 mm, coils and high Q circuitry. In order to allow performance of the double resonance multiple pulse experiments, the spectrometer was augmented with the capability of handling a second rf frequency, and a special single-coil double resonance probe was designed (13) which allows rotating H_1 fields in the 40 - 50 gauss range to be obtained over a 5 mm sample for both frequencies while still keeping the rf amplifier requirements in the 150 watt range.

The spectrometer used for these studies differs from a conventional pulsed apparatus primarily in that: (i) care is taken with the rf power supply to assure a high degree of stability in the rf pulses over the time scale of the experiment, (ii) careful matching and probe design has allowed optimum use of the rf power available, (iii) the preamplifier and amplifier in the receiver are designed to allow rapid recovery after pulse overload, and (iv) the digital electronics, while simple, furnish the capability of a variety of pulsing sequences.

Hydroxyl Groups on High Surface Area Silica-Aluminas

The results of a study using a combination of conventional and multiple pulse NMR techniques to examine the hydroxy protons on a series of high surface area silica-aluminas have been recently reported (8), and these results will be discussed here as a demonstration of the use of these NMR techniques in complex surface environments. Seven samples ranging in composition from 0 - 100% silica were prepared by Dr. D. A. Hickson at the Chevron Laboratories in Richmond, California. The silica-aluminas were prepared by mixing desired amounts of $AlCl_3$ and tetraethyl orthosilicate in methanolic solution. The resulting solution was homogeneously gelled by the addition of propylene oxide, thus raising the pH. The methanol was then displaced by rinsing in diethyl ether. The material was then placed in an autoclave, and the temperature was brought to the critical point for diethyl ether. The vapor was flushed away with dry nitrogen. This resulted in an extremely high surface area material. Infrared techniques were used to demonstrate that hydroxyl groups in these materials rapidly exchanged with D_2O vapor and thus represent surface groups. Two samples of each composition were prepared. The first sample was calcined in 152 torr of oxygen for one hour at 500° C and then sealed. The second sample, after being calcined for one hour at 500° C, was cooled to 150° C and exposed to water vapor at 4.6 torr for one hour, evacuated, and then sealed.

Results and Discussion

100% Silica. The semilogarithmic plot of the free induction decay for the calcined sample of pure silica is shown in Figure 2. Free induction decay spectra were taken well off resonance (50 KHz) and the envelope of the observed oscillation used to determine the decay. This technique involves giving up a substantial portion of the available signal-to-noise and accounts for much of the experimental scatter observed in Figures 2 to 5. However, it assures that the reported decays are not contaminated by artifacts. As is shown, the decay is exponential within the precision of the data. By extrapolating the line in Figure 2 back to zero time, an initial intensity is obtained which indicated 1.6 protons/ 100 $\AA^2 \pm$ 10% and a ratio of silicon atoms to protons of 11. The hydroxyl concentrations for silica samples treated at 500° were found in the range 1.6 to 2.3 protons/100 $\AA^2$ by Davydov and coworkers (14), using chemical methods of analysis and a variety of preparation techniques. O'Reilly (15) found 2.6 protons/100 $\AA^2$ and more recently Freude (16) measured 0.87 protons/100 $\AA^2$, both using NMR techniques.

The time constant, T_2, for the free induction decay is 200 microseconds, and as an exponential decay in time space is equivalent to a Lorentzian line in frequency space; this is equivalent to a Lorentzian absorption spectrum with a full width at half height of 1600 Hz. A Lorentzian lineshape has been previously observed by O'Reilly and coworkers (15,17), who reported a T_2 of 180 microseconds which did not vary as a function of temperature between -210° and 280° C. Both the lack of a temperature dependence and results of the multiple pulse NMR measurements to be discussed below indicate that the Lorentzian lineshape is not the

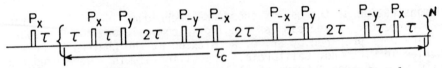

Figure 1.   Eight-pulse cycle; all pulses are π/2 and phases are as indicated

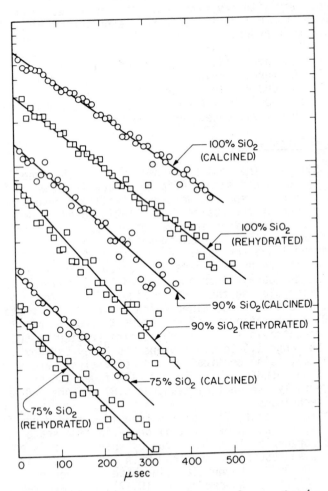

Figure 2.   Plot of free induction decay envelope vs. time for samples of varying SiO₂ composition. Data from the various samples have been displaced vertically in order to clarify presentation. The values of T₂ reported in Table 1 were obtained from the slopes of the lines placed through the data.

result of motional averaging. Lorentzian lineshapes have been predicted for a number of geometries where it is assumed that spins are distributed in a random fashion over only a small fraction of possible sites (18). It does not appear possible to use the results of such analytic treatments to extract any quantitative information on the distribution of hydroxyls on the silica. However, it is possible to draw the following quantitative conclusions: the hydroxyl groups are isolated, that is, the protons do not exist in pairs (i.e., not as adsorbed water), and the observed lineshapes are qualitatively consistent with the dilute spin model (16). These conclusions are in agreement with a number of infrared investigations which conclude that the remaining protons on silica surfaces after treatment at $500°$ C are in isolated hydroxyl groups (14,19-21).

The eight-pulse cycle, with a cycle time of 96 microseconds, was applied to the calcined 100% SiO_2 sample. As can be seen in Figure 3, the linewidth collapsed to near 250 Hz. One can conclude that the width and Lorentzian shape of the free induction decay were determined almost completely by dipolar interactions and furthermore that the dipolar interactions were essentially static on the timescale of 10^{-4} to 10^{-6} seconds. This furnishes evidence for the interpretation given above of the free induction decay lineshape.

The lineshape produced by the multiple pulse experiment has several potential sources: (i) inhomogeneous broadening, primarily chemical shift coupling, (ii) time dependence, particularly in the dipolar interaction, and (iii) residual instrumental broadening. The phase altered eight-pulse cycle was applied to further separate these effects since it will remove inhomogeneous broadening as well as static dipolar broadening. The line collapsed further to 32 Hz under the influence of the phase altered eight-pulse sequence, thus indicating that the lineshape in Figure 3 can be attributed to a chemical shift powder pattern broadened by bulk susceptibility effects. The smooth line through the data points in Figure 3 is a computer fit for an axially symmetric chemical shift tensor with a Lorentzian broadening function. The principal values furnished by the fit are $\sigma_\perp = -5.1$ ppm and $\sigma_\parallel = +1.8$ ppm relative to a spherical sample of tetramethylsilane ($\pm$ 1 ppm). In order to correct values for the effects of bulk susceptibility, the sample was approximated by an ellipsoid (a = b = 1/2 c, where a, b, and c are the semiaxes and with the magnetic field being along a), and the tabulated values of Osborn (22) were used with a value of -1.13×10^{-6} for the volume susceptibility to calculate a susceptibility correction of -1.4 ppm. Thus, susceptibility independent principal values for the hydroxyl protons on silica are $\sigma_\perp = -6.5$ ppm and $\sigma_\parallel = +0.4$ ppm, and a susceptibility independent values for the trace, or isotropic portion, is -4.2 ppm (relative to TMS). A Lorentzian broadening function with a width of near 6 ppm was required to fit the spectrum in Figure 3, and this width is attributed

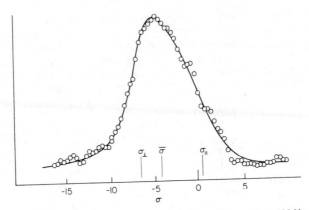

Figure 3. *Powder pattern using eight-pulse cycle of 100% SiO₂ calcined sample. The signal amplitude is plotted vs. chemical shift (ppm) relative to TMS.*

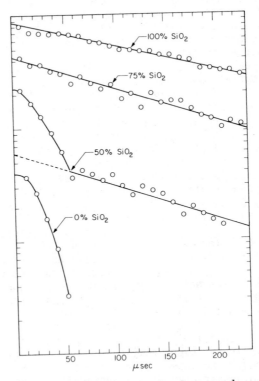

Figure 4. *Plot of free induction decay envelope vs. time for calcined samples of varying SiO₂ composition. Data from the various samples have been displaced vertically in order to clarify presentation.*

primarily to smearing of the magnetic field at the surface of a particle whose magnetic susceptibility is different from the medium it is in.

A number of studies have recently been reported of the chemical shift tensor of both hydrogen bonded and non-hydrogen bonded hydroxyl protons (23), and all have reported tensors which exhibit an anisotropy ($\sigma_{\parallel} - \sigma_{\perp}$) from two to four times the size of the 6.9 ppm found in this study. One can conclude either that the surface hydroxyl groups studied here have substantially altered chemical bonding from these solid state systems or that the value measured here has been partially averaged by an angular motion of the O--H vector (i.e., a wagging or rotation of the hydroxyl bond around the oxygen). Thus, while NMR data indicate that major motion, diffusion from oxygen site to oxygen site, does not occur, there is some indication that a local reorientation of the hydrogen around its oxygen may occur in this system. Hopefully, this point can be clarified by performing the multiple pulse measurements at low temperatures.

The same series of NMR measurements has been made on the rehydrated sample of 100% silica with almost identical results. A surface concentration of 1.9 protons/100 Å^2 was determined from the free induction decay. This small increase of 0.3 protons/100 Å^2 suggests that the rehydration by water vapor is not reversible. Such irreversible rehydration has been reported earlier for silicas heat treated at temperatures over 400° C both by Young (24) and by Hockey and Pethica (25).

The lineshape for the rehydrated sample was Lorentzian within the precision of the data and indicates that the water molecules which were adsorbed went on the surface as isolated hydroxyl groups. The principal values for the chemical shift tensor were determined as above and found to be $\sigma_{\perp}$ = -6.2 ppm and $\sigma_{\parallel}$ = +0.6 ppm, which are identical within experimental accuracy with the non-hydrated sample. Thus, essentially no difference between the two samples could be determined except for the small concentration difference noted above.

   0% Silica (100% Alumina). The logarithm of the amplitude of the free induction decay for the calcined alumina sample is shown versus time in Figure 4 and versus time squared in Figure 5. The proton concentration was 3.1 protons/100 Å^2, and the ratio of aluminum atoms to protons was 6.7. This higher concentration of protons on alumina compared to silica has been pointed out before (26). The linear plot in Figure 5 is the time response expected of a decay associated with a Gaussian lineshape, and from the slope of the line through the data points, one estimates a second moment of 2.8 gauss^2 for the proton line. This second moment has contributions from proton-proton interactions (homonuclear) and from proton-aluminum interactions (heteronuclear), and in order to separate these, a Carr-Purcell-Meiboom-Gill cycle was applied with a pulse spacing of 50 μseconds. Assuming

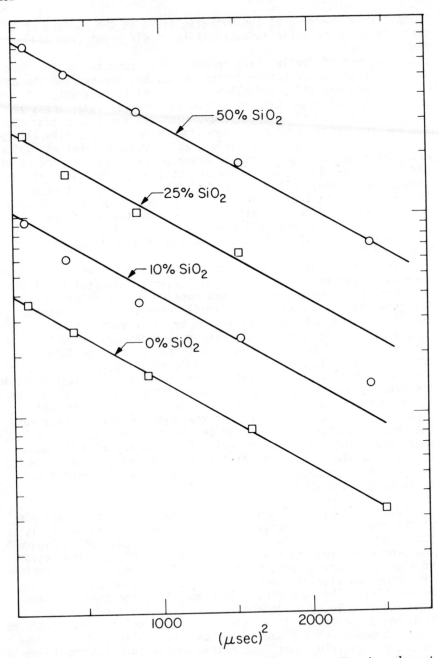

Figure 5. Plot of decay envelope of varying SiO_2 composition. Data from the various samples have been displaced vertically in order to clarify presentation. The values of second moment reported in Table 1 were obtained from the slopes of the lines placed through the data.

a Gaussian lineshape for the envelope of the echo maxima, an estimate of near 0.2 gauss^2 was obtained for the second moment. As the
Carr-Purcell-Meiboom-Gill sequence removes static heteronuclear
dipolar broadening as well as magnetic field inhomogeneity effects,
the 0.2 gauss^2 can be attributed to homonuclear dipolar broadening
(i.e., proton-proton broadening). This would imply that 2.6 gauss^2
of the proton second moment (i.e., essentially all of the 2.8
gauss^2 second moment measured) can be attributed to the heteronuclear (proton-aluminum) dipolar interaction. Also note that the
small size of the proton-proton second moment implies widely
separated protons on the alumina surface.

The rehydrated sample of alumina has a still higher concentration of hydroxyl groups, 4.4/100 Å^2, and was Gaussian within
the experimental error. The second moment estimated from the
slope of the line through the data points was 3.7 gauss^2.

Silica-Aluminas. Compositions of 10, 25, 50, 75, and 90%
silica were prepared, treated, and investigated as described
above for the pure SiO_2 and Al_2O_3. Experimental results are
presented in Figures 2, 4, and 5 and in Table I. A number of
conclusions can be drawn from these results.

The 90 and 75% SiO_2 samples exhibited a single exponential
decay for both calcined and rehydrated samples as can be seen in
Figure 2. The concentrations of protons on these samples and
estimates of the decay constants are given in Table 1. Using
results and arguments from the discussion of the pure silica
sample, the measured decay constants indicate that protons on
these samples do not exist as closely spaced pairs (i.e,, not as
adsorbed water) and do not exist in close proximity to an
aluminum ion (i.e., not as AlOH groups). Again one notes that
exponential decays, or Lorentzian lineshapes, have been predicted
for a number of geometries where it is assumed spins are
distributed in a random fashion over only a small fraction of possible sites (18). As in the case of pure silica, the eight-pulse
sequence was applied to both the calcined and rehydrated 90%
silica samples with results similar to those reported above for
the pure silica, confirming that the observed decays were
produced by static dipolar interactions.

The 50, 25, and 10% SiO_2 samples are quite different from
the 75 and 90% samples. Figure 3 illustrates this point for
the 50% SiO_2 calcined sample where a logarithmic plot is shown
for the 100, 75, 50, and 0% calcined samples. The 50% SiO_2 decay
consists of two components. At long times (>70 microseconds) it
is qualitatively and quantitatively similar to the 75 and 100%
SiO_2 decays, while at short times it is similar to the 0% (pure
alumina) decay. In the analysis of the data of the pure alumina,
the short decay was found to be due to the aluminum-proton dipolar
interaction and was associated with hydroxyl groups attached to
aluminum ions, while in the analysis of the 100, 90, and 75% SiO_2,
the long decays were associated with isolated hydroxyls attached

to silicon ions, i.e., SiOH groups. It is thus possible to separate quantitatively the fraction of protons bound to silicon and that fraction bound to aluminum. The intercept of the extrapolated line which fits the long-time behavior, as shown in Figure 4, is a measure of the SiOH density. The difference between this extrapolated intercept and the total intercept is a measure of the AlOH density on the sample. In order to compute this difference accurately, the 50% SiO_2 curve in Figure 5 was constructed by subtracting the amplitudes of the line extrapolated from the long-time portion shown in Figure 4 from the amplitudes of the data points in the short-time region. Thus, the 50% SiO_2 curve shown in Figure 5 represents the signal from the AlOH groups, and the intercept is directly proportional to the concentration of hydroxyl groups bound to aluminum ions

The decays observed for the 50% SiO_2 rehydrated sample and both of the 25% SiO_2 samples also consisted of two components. Like the 50% SiO_2 calcined sample, the long-time portion was similar to the 100% Al_2O_3 decay. The number of hydroxyl groups bound to silicon ions and the number to aluminum ions were calculated in the same manner as above for the 50% SiO_2 calcined sample. For both 10% SiO_2 samples, the decays were similar to the 0% SiO_2 samples.

In order to compute second moments for the AlOH groups, a Gaussian decay was assumed. As can be seen from Figure 5 the straight lines, which are indicative of Gaussian behavior, fit the data, and from the slopes of these lines, second moments were computed and are presented in Table 1. The second moments for the AlOH groups were all between 2.7 and 3 $gauss^2$ except for the rehydrated 100% Al_2O_3 sample.

The entire results for the surface concentrations of AlOH groups and SiOH groups are presented in Table 1. It appears to be of substantial significance that AlOH groups were not detected on the 75% silica sample while nearly 75% of the protons on the 50% silica sample appear to be AlOH groups. Previous infrared work by Basila (27) has concluded that the existence of a single hydroxyl stretching frequency on a 75% silica implied only SiOH groups. Prior NMR work by O'Reilly, et al. (15) and Hall, et al. (28) on an 89% SiO_2 sample had limited the percent of surface protons in AlOH groups to less than 20%. The 75% silica sample contains one aluminum ion for every 2-1/2 silicon ions, but no measurable AlOH groups; the 50% silica contains one aluminum ion for each 0.85 silicon ion (54% aluminum ions), yet 75% of the hydroxyls are bound to the aluminum. These samples are all amorphous in the sense that no X-ray spectra are obtainable and infrared data on Al-O and Si-O are broadened, yet the results presented here imply substantial order, specifically in the attachment of the hydroxyls.

There appears to be a clear qualitative change occurring in the nature of the hydroxyl groups between the 75 and 50% silica samples, with the 100, 90, and 75% being similar and exhibiting no detectable AlOH groups while the 50, 25, 10, and 0% SiO_2 have the bulk of their protons attached in close proximity to an aluminum ion.

The change in the surface environment of the surface hydroxyl groups in going from 75% SiO_2 to 50% SiO_2 appears to correlate with other properties of the silica-aluminas including their catalytic activity. This is illustrated in Figure 6 which compares several measured characteristics of these seven samples: cation exchange capacity, Bronsted acidity (29), relative activity for butene isomerization, and the local environment of the hydroxyl groups. Thus, it appears that a surface structure in which the hydroxyl groups are attached as AlOH groups is not a desirable configuration for the presence of catalytic sites for butene isomerization.

C-13 Double Resonance Multiple Pulse Techniques

Nuclear magnetic resonance techniques are potentially among the most informative means to obtain information on the chemical bonding, or electronic structure, molecular geometry, and motional properties of adsorbed species; however, a number of factors have worked to make extraction of such information difficult or impossible.

Attempting to characterize adsorbed hydrocarbon molecules by conventional ^{13}C NMR spectroscopy is difficult since the small number of carbon-13 spins results in a very weak signal, and the spin-spin interactions between the protons and carbon-13 nuclei broaden what spectrum one can obtain to the point that all electronic structure, or chemical bonding, information is washed out. This latter situation is particularly distressing since the spin-spin interaction contains a wealth of detailed geometrical information. Not only is this information folded together in such a fashion that it is not extractable, but it effectively smears the carbon-13 spectra to the point that little additional information can be obtained. Molecular motion of the adsorbed species will alter the spin-spin broadening, and rapid isotropic motion will eliminate it altogether. Consequently, uses of conventional NMR technique in studies of adsorbed species have, in general, been limited to the detection of such molecular motion and characterization of species which are in rapid enough motion to eliminate the line-broadening effects of spin-spin inter-actions.

In order to be able to obtain information inherently available with the NMR techniques, one needs: (i) means to enhance the sensitivity, particularly for the detection of isotopically dilute species such as carbon-13, (ii) means to eliminate the spin-spin broadening from the NMR spectra to allow characterization of other interactions, and (iii) means to unfold the geometrical information contained in the spin-spin interaction because this is capable of furnishing highly precise geometrical parameters, bond angles, and bond distances. Efficient means of accomplishing the first two of these goals have been developed over the past five years (4), and have been applied to surface systems in at least two cases (30, 31). Means of using the heteronuclear dipolar interaction between two spin 1/2 nuclei together with chemical shift information to furnish geometrical information have been discussed in the literature in the past year or two (5,6,32), and a recent paper by Stoll, Vega, and Vaughan (9) describes in detail a scheme which accomplishes the third of these desired goals. This double resonance scheme is illustrated in Figure 7 and furnishes a scheme for determining proton-carbon-proton bond angles and bond distances as well as orienting the carbon chemical shift tensor in the molecular frame. The scheme proposed was designed for use in polycrystalline samples and was designed for optimization of the signal-to-noise in the final spectra, thus making it ideally suitable for surface systems where only poly-crystalline spectra are available and where the low concentration of adsorbed species must be dealt with.

The scheme (9) illustrated in Figure 7 can be discussed in three sections. In the first, preparatory, period a transverse magnetization for the S (^{13}C-dilute) spin species is produced. The S spin transverse magnetization is created here by a Hartmann-Hahn (33) transfer of polarization from the I (^{1}H-abundant) spin system to make use of the signal-to-noise enhancement gained with this procedure although a number of other means could be used (4).

In the second period an evolution of the I-S heteronuclear dipolar Hamiltonian is allowed to take place for a period, τ, while simultaneously suppressing the abundant spin dipole-dipole interaction with the eight-pulse sequence.

After the desired amount of dipolar evolution, τ, has occurred, the eight-pulse sequence is replaced by a simple decoupling field on the I spin system, and this causes the heteronuclear dipole-dipole interaction to go to zero. The S spin system is still evolving, however, under its off-resonance and chemical shift Hamiltonians. The next step within this evolutionary period is to refocus the time development of the off-resonance and chemical shift Hamiltonians in the S spin system by the application of a 180° pulse to the S spin system, causing the production of a spin echo at a later time.

The third period involves digitally recording the second half of the S system echo, which provides the signal for further processing.

Thus, the total net time evolution of the S spin system that has occurred at the point where data collection is started, t = 0, is only the heteronuclear dipole-dipole interaction, Equation 3, for a time period τ. During the data collection period, t, however, only the off-resonance and chemical shift Hamiltonians are producing the time evolution of the S spin system. Thus, the data collected are easily interpreted since the heteronuclear dipolar information is in the initial amplitudes, and the chemical

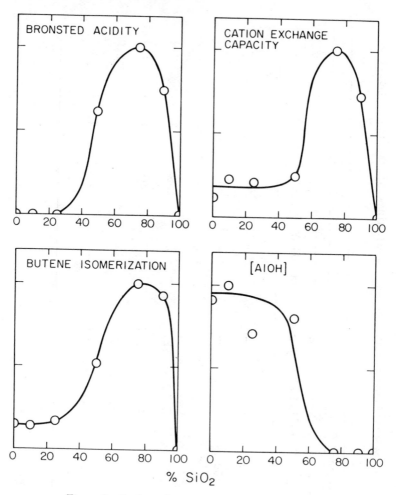

Figure 6. Surface characteristics of silica-aluminas

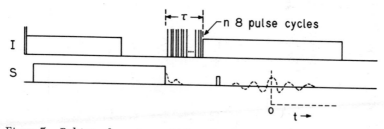

Figure 7. Pulsing scheme for multiple pulse double resonance experiments

Table I

Summary of proton NMR results for calcined and rehydrated silica-aluminas of varying composition.

Weight % SiO$_2$	Surface Area m^2/gm	Calcined Samples				Rehydrated Samples			
		SiOH		AlOH		SiOH		AlOH	
		#H/100 Å^2	T$_2$ μsec	#H/100 Å^2	M$_2$ gauss^2	#H/100 Å^2	T$_2$ μsec	#H/100 Å^2	M$_2$ gauss^2
100	576	1.6	200	---	---	1.9	170	---	---
90	693	1.5	160	---	---	3.2	120	---	---
75	615	1.2	150	---	---	1.8	140	---	---
50	727	1.0	140	2.7	2.7	1.0	105	3.5	2.9
25	839	0.2	160	2.4	2.7	.5	80	3.7	2.9
10	768	---	---	3.4	2.7	---	---	4.4	3.0
0	571	---	---	3.1	2.8	---	---	4.4	3.7

shift information is in the time evolution during the data collection time period, t. The chemical shift powder patterns obtained from this technique are modulated by the heteronuclear dipolar interaction, and one can use the modulation patterns produced to determine the length and orientation of the I-S (proton-carbon) vectors in the chemical shift principal axis frame to the molecular frame. This double resonance, multiple pulse scheme is described in detail elsewhere (9) and its usefulness demonstrated there by experiments performed on polycrystalline benzene and calcium formate samples.

Portions of the work discussed in this paper were supported by the National Science Foundation and the Office of Naval Research.

*Now at Exxon Research Laboratories, Linden, New Jersey 07036.

Literature Cited

1. Vaughan, R. W., "Annual Reviews in Materials Science," Vol. 4, p. 21 (ed. by R. A. Huggins, R. H. Bube, and R. W. Roberts). Annual Reviews, Palo Alto, 1974, and references therein.
2. Waugh, J. S., Huber, L. M., and Haeberlen, U., Phys. Rev. Lett. (1968), 20, 180.
3. Rhim, W-K., Elleman, D. D., Schreiber, L. B., and Vaughan, R. W., J. Chem. Phys. (1974), 60, 4595, and references therein.
4. Pines, A., Gibby, M. G., and Waugh, J. S., J. Chem. Phys. (1973), 59, 569, and many references therein.
5. Hester, R. K., Ackerman, J. L., Cross, V. R., and Waugh, J.S., Phys. Rev. Lett. (1975), 34, 993.
6. Hester, R. K., Cross, V. R., Ackerman, J. L., and Waugh, J.S., J. Chem. Phys. (1975), 63, 3606.
7. Stoll, M. E., Rhim, W-K., and Vaughan, R. W., J. Chem. Phys., scheduled for June 1, 1976, issue.
8. Schreiber, L. B., and Vaughan, R. W., J. Catalysis (1975), 40, 226.
9. Stoll, M. E., Vega, A. J., and Vaughan, R. W., submitted to J. Chem. Phys.
10. Farrar, T. C., and Becker, E. D., "Pulse and Fourier Transform NMR," Academic Press, New York, 1971.
11. Dybowski, C. R., and Vaughan, R. W., Macromolecules (1975), 8, 50.
12. Vaughan, R. W., Elleman, D. D., Stacey, L. M., Rhim, W-K., and Lee, J. W., Rev. Sci. Instr. (1972), 43, 1356.
13. Stoll, M. E., Vega, A. J., and Vaughan, R. W., in preparation.
14. Davydov, V. Ya., Zhuravlev, L. T., and Kiselev, A. V., Trans. Faraday Soc. (1964), 60, 2254.
15. O'Reilly, D. E., Leftin, H. P., and Hall, W. K., J. Chem. Phys. (1958), 29, 970.
16. Freude, D., Müller, D., and Schmiedel, H., Surface Sci. (1971), 25, 289.

17. O'Reilly, D. E., in "Advances in Catalysis," Vol. 12, p. 31 (ed. by D. D. Eley, P. W. Selwood, and P. B. Weisz). Academic Press, New York, 1960.
18. Rakvin, B., and Herak, J. N., J. Magn. Resonance (1974), 13, 94, and references therein.
19. McDonald, R. S., J. Phys. Chem. (1958), 62, 1168.
20. Hair, M. H., "Infrared Spectroscopy in Surface Chemistry," Marcel Dekker, New York, 1967.
21. Little, L. H., "Infrared Spectroscopy of Adsorbed Species," Academic Press, New York, 1966.
22. Osborn, J. A., Phys. Rev. (1945), 67, 351.
23. Lau, K. F., and Vaughan, R. W., Chem. Phys. Letters (1975), 33, 550. References to numerous proton chemical shift studies can be found here.
24. Young, G. J., J. Colloid Sci. (1958), 13, 67.
25. Hockey, J. A., and Pethica, B. A., Trans. Faraday Soc. (1961), 57, 2247.
26. Haldeman, R. G., and Emmett, P. H., J. Amer. Chem. Soc. (1956), 78, 2917.
27. Basila, M. R., J. Phys. Chem. (1962), 66, 2223.
28. Hall, W. K., Leftin, H. P., Cheselske, F. J., and O'Reilly, D. E., J. Catal. (1963), 2, 506.
29. Schwarz, J. A., J. Vac. Sci. Technol. (1975), 12, 321.
30. Kaplan, S., Resing, H. A., and Waugh, J. S., J. Chem. Phys. (1973), 59, 5681.
31. Stejskal, E. O., Schaefer, J., Henis, J.M. S., and Tripodi, M. K., J. Chem. Phys. (1974), 61, 2351.
32. Müller, J., Kumar, A., Baumann, T., and Ernst, R. R., Phys. Rev. Lett. (1974), 32, 1402.
33. Hartmann, S. R., and Hahn, E. L., Phys. Rev. (1962), 128, 2042.

NMR Studies of Water and Ammonia inside the Cubooctahedra of Different Zeolite Structures

W. D. BASLER

Institute of Physical Chemistry, University of Hamburg,
Laufgraben 24, 2000 Hamburg 13, W. Germany

Zeolites of faujasite-type (Figure 1) are alumo-silicates, composed of cubooctahedral units, which are joined by double six-rings, forming a diamond-like structure. Thus two systems of intracristalline voids and channels are formed: The first system consists of large cavities of 13 Å diameter, which are joined by windows of 8 Å diameter. The second system consists of the interior of the cubooctahedra, joined by the double six-rings. On the other hand, a passage from the large cavities into the interior of the cubo-octahedra is possible through oxygen six-rings. The cations, which are necessary to compensate the smaller charge of the Al-ions, are partly localized before or in these six-rings. At room temperature a large cavity can take up 28, a cubooctahedra 4 water molecules(1-5).
    In previous NMR-studies(6-12) the water molecules in the large cavities were studied, the water inside the cubooctahedra was omitted in the discussions. Perhaps, this was caused by the observation of only one single uniform NMR-signal, at least at room temperature and in zeolites, where the amount of OH-groups could be neglected. However, recently a second NMR-signal was discovered and attributed to the water inside the cubooctahedra, independently by two other groups and us(13-15). Here we want to report our NMR-studies dealing with a) proving that the observed second NMR-signal originates from water inside the cubooctahedra, b) the equilibrium properties of this water, c) the kinetics of the water molecules passing from the large cavities into the interior of the cubo-octahedra and d) first results using ammonia as sorptiv.
    Zeolites of faujasite-type and related structures containing cubooctahedral units, with different Si/Al-ratios and Fe^{3+}-content were prepared by Kacirek,

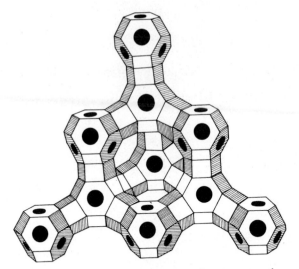

Figure 1. Model of the crystal structure of faujasite

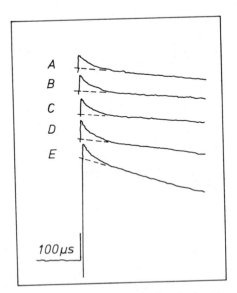

Figure 2. Free induction decay of water protons in various faujasites, denoted as in Table I. Full loading, T = 295°K.

using newly developed methods of zeolite cristalli-
sation(16,17). Exchange of cations was performed as
usually. The zeolite powder was outgassed at 400 °C
until the pressure was less 10^{-5} Torr. The sorption
was performed using the corresponding vapor, the
amount was determined by weight. The measurements by
pulsed proton-NMR were done at 60 MHz, using a Bruker
B-KR 322 spectrometer.

Whereas the outgassed zeolites showed no signal
after the deadtime of 9 µs at a detection limit of
1-2 mg H20/g zeolite, the observed free induction decay after a 90°-pulse for water in different zeolites
of faujasite-type at room temperature and full loading (conditioned over saturated NH_4Cl-solution) is
shown in Figure 2. Beside the well known relaxation
of the water in the large cavities, here named phase
A, with relaxation times of several ms depending
strongly on Fe^{3+}-content, there is a fast decaying
component, named phase B, with T2B=25 µs and T1B=100
ms. Table I shows that the intensity IB (13%, extrapolated 17%) and transverse relaxation time T2B of
this component are independent of Si/Al-ratio and of
Fe^{3+}-concentration. This is a first hint that phase

Table I. Transverse(T2B) and longitudinal(T1B) relaxation times of phase B in different zeolites of faujasite-type. T=295 K, full loading. Intensity IB1
(after the dead-time) and IB2(extrapolated) of phase
B.

Zeolite	Si/Al	Fe3+ ppm	T2B µs	T1B ms	IB1 %	IB2 %
A Linde 13X	1.18	150	25	70	12	16
E Linde 13Y	2.5	400	23	110	12	17
B self made	1.22	3	27	100	13	17
C " "	2.34	7	23	210	13	17
D " "	2.74	90	25	210	14	18

A and B correspond to the water molecules in the
large cavities and in the interior of the cubooctahedra, for these structural characteristics are the
only quantities common to all zeolites of Table I.

The second hint is the fact that the intensity
IB gives within the limits of error(15%) a number of
4 water molecules per cubooctahedra, just what is
known from X-ray-data(18). Whereas the quotient T1A/
T2A=2-3 shows that the molecules of phase A are well
mobile and in a liquid-like state, the corresponding
T1B/T2B=4000 indicates molecules with strongly restricted motion. This also favours the identification
of phase B with the interior of the cubooctahedra,
where only rotational tumbling motions are possible.

Thus for a water molecule rotating about one axis one obtains T2=16 μs, which can be compared with the experimental T2B=25 μs.

By D2O-dilution it could be proved that phase B does not consist of isolated OH-groups, but that each proton has its neighbour-proton: Replacing 80% of the protons by deuterons increased the transverse relaxation time T2B about 25 and T1B about 2.5 times. Therefore, phase B relaxes predominantly by H-H-interaction. This H-H-interaction is intramolecular, as T2B was found to be independent of the filling factor, too: Table II. Transverse(T2B) and longitudinal(T1B) relaxation times of phase B at different loading and D2O-dilution. Faujasite of Si/Al=2.74, self made, T=295 K.

Loading mg/g	T2B μs	T1B ms
60	25	150
120	25	140
250	25	110
355	25	210
60 H2O/240 D2O	600	500

Summarizing, all experimental results on water in faujasites support the assumption that phase B are the water molecules in the interior of the cubooctahedra. Consequently, this phase B should be found in other zeolites build up of cubooctahedral units, too.

Therefore, we studied zeolite A and ZK 4, where the cubooctahedra are arranged in a primitive cubic lattice, and zeolite sodalite, which consists only of cubooctahedra without large cavities. The results in Table III show, as expected, in the case of zeolite A and ZK 4 two-phase-behaviour with relaxation times and intensities close to the values found in faujasites. Sodalite, on the other hand, showed only one relaxation with values close to those of phase B:
Table III. Intensity IB and relaxation times T1 and T2 of phases A and B in zeolites containing cubooctahedral units. Full loading, T=295 K.

Zeolite	Si/Al	T2B μs	T1B ms	IB %	T2A ms	T1A ms
Linde 13X	1.18	25	70	16	5.6	8.7
Linde 13Y	2.5	23	110	17	1.7	2.5
NaA	1	100	70	16	2.5	20
ZK 4	1.61	70	60	18	1.2	4.0
Sodalite	1	50	65	100	-	-

As it was proved that water inside the cubeocta-
hedra could be studied separately by the NMR-signal
named phase B, we studied the distribution of the water molecules between the large cavities and the interior of the cubooctahedra of a faujasite of Si/Al=
2.74 for different loading and its temperature dependence. Figure 3 shows that the interior of the cubooctahedra is filled up predominantly: At low filling
nearly half of the water is found inside the cubeocta-
hedra. This fraction decreases with increasing coverage, in agreement with the results of Pfeifer(13).
This equilibrium of sorption is shifted by temperature
in favour of the large cavities(Figure 4). This indicates that the heat of sorption inside the cubeocta-
hedra is slightly greater(1 kcal/mol) than in the
large cavities.

As we found two-phase-behaviour even in the longitudinal relaxation, it follows that the exchange
rate between phase A and B must be at least in the
range of seconds(19). To study this rate, the sorption was allowed to take place slowly(appox. 1 h)
while the zeolite was held at $0^{\circ}C$ by a water-ice bath
to avoid any warming. Loaded by this technique, the
zeolite showed only phase A and no phase B-signal
immediately after sorption. Thus it was possible to
study the appearing of phase B. The approach to the
saturation value is within the limits of error expo-
nential(Figure 5).

We studied the growing of phase B for faujasites
NaX(Si/Al=1.18) and NaY(Si/Al=2.36) and a coverage of
100 mg H2O/g and at different temperatures. The time
needed for growing of half of phase B varied between
1 month at room temperature and 1 hour at $80^{\circ}C$. No
difference was observed between NaX and NaY. The Ar-
rhenius-plot(Figure 6) gives an activation energy of
24 kcal/mol for this process.

To pass into the interior of the cubooctahedra,
the water molecules have to get through the oxygen
six-rings, where some of the cations are localized
(S2-sites). Removing these S2-cations should increase
the rate of growing of phase B. This can be done by
three ways: 1) High water loading increases the mobility of the cations. 2) High Si/Al-ratios reduce the
number of cations. When Si/Al>2.43 there are not
enough cations to occupy all sites. 3) Exchange of Na
by cations of higher charge has the same result.

We have studied all these three influences and
found an acceleration of the rate in every case: 1)At
a loading of 200 mg/g phase B is build up 5 times
faster at room temperature compared with the loading

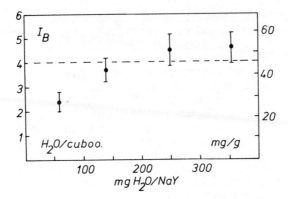

Figure 3. Intensity I_B of phase B for different loading. Faujasite NaY, (Si/Al = 2.74) T = 295°K.

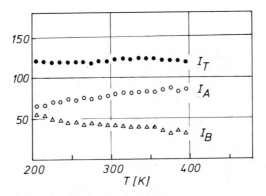

Figure 4.   Intensities I_A and I_B of phases A and
B and total intensity I_T for different temperatures.
Zeolite NaY (Si/Al = 2.74) 120 mg H_2O/g.

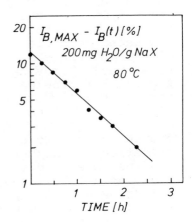

Figure 5.   Evolution of phase B, showing exponential approach to equilibrium value

of 100 mg/g, and the activation energy seems to be smaller(20 kcal/mol, Figure 7). At full loading, phase B showed its equilibrium value immediately after the sorption, which lasted one hour. This is in good agreement with the observation of mobile Na-ions for higher loading(20).

2) Whereas in NaX and NaY all S2-sites are occupied, and we consequently find the same rate, a faster rate is expected for Si/Al>2.5. We studied a faujasite with Si/Al=3.40 and found a half time of 200 hours (Table 4).

3) When Na1+ is exchanged against La3+, the number of cations is reduced by a factor 3. We prepared a faujasite with Si/Al=2.93, where 80% of the Na was replaced by La. As it is known that the La-ions are in the double six-rings, almost every passage from the large cavity into the cubooctahedra is open now. Phase B was fully observed immediately after sorption, indicating that the equilibrium was reached in less one hour. Further, two-phase-behaviour begins to vanish at 150-200 °C and rapid exchange on the time-scale of NMR begins. The removing of the Na-ions at the S2-sites increased the rate of passage through the oxygen six-rings by more than 1000 times. The same was found using Ca2+.

Table IV. Time t1/2 for forming half of phase B. T=295 K, loading 100 mg H2O/g.

Zeolite	Si/Al	t1/2(h)
NaX	1.18	800
NaY	2.36	800
NaY	3.40	200
LaY	2.93	<1
CaY	2.93	<1

After water, we studied ammonia, which is known to enter the cubooctahedral units(21). Under the same conditions as in the case of water, NaX and LaY were loaded with 65 mg NH3/g. 1 hour after sorption, LaY showed two phases with 20% phase B(T1B=85 ms, T2B= 50 μs) and 80% phase A(T1A=34 ms, T2A=0.75 ms). A temperature treatment(80 °C for 24 h) did not change anything. This means that ammonia, like water, passes the open oxygen six-rings of LaY readily into the cubooctahedra. From IB=20% it follows that approx. 1 molecule of ammonia is inside the cubooctahedra and 5 in the large cavity. NaX showed only phase A(Figure 8) with T1A=7.7 ms and T2A=3.8 ms.

The results of heat treatment are shown in Figure 9: 7 days at room temperature have no significant

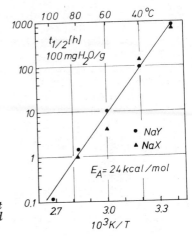

Figure 6. *Arrhenius plot of the time $t_{1/2}$ to get half of phase B. Faujasites NaX(Si/Al = 1.18) and NaY(Si/Al = 2.36), 100 Mg H_2O/g.*

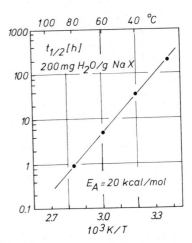

Figure 7. *Arrhenius plot of the time $t_{1/2}$ to get half of phase B. Faujasite NaX(Si/Al = 1.18), 200 mg H_2O/g.*

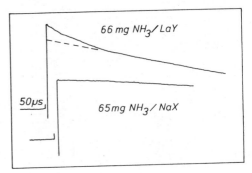

Figure 8. *Free induction decay of ammonia protons in NaX and LaY 1 hr after sorption, T = 295°K*

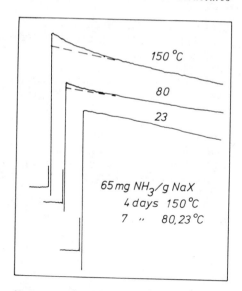

Figure 9. Free induction decay of ammonia protons in NaX, 65 mg NH₃/g NaX, after various heat treatments

effect, but after 7 days at 80 and 4 days at 150 °C we find 5 and 9% phase B with T2B=70 µs and T1B=85 ms. Compared with water, the rate is very slowly. Further work on ammonia is in progress and will be reported orally.

Literature cited

1) Barrer, R M.,Endeavour(1964)23,122
2) Fischer, K.F. and Meier, W.M.,Fortschr.Mineral. (1965)42,50
3) Olson,D.H.,J.phys.chem.(1970)74,2758
4) Mortier,W.J. and Bosmans,H.J.,J.phys.chem.(1971) 75,3327
5) Barrer,R.M. and Bratt,C.C.,J.phys.chem.sol.(1959) 12,130
6) Lechert,H. and Henneke,H.W.,Surface Sci.(1975)51, 189
7) Resing,H.A., 8.Coll.NMR-spectrosc.(1971)Aachen
8) Resing,H.A. and Thompson,J.K.,2.Int.Conf.Molecular Sieve a. Zeolites(1971)Worcester
9) Pfeifer,H., Przyborowski,F., Schirmer,W. and Stach,H.,Z.phys.chem.(1969)236,345
10)Pfeifer,H., Gutze,A. and Shdanov,S.P.,Z.phys.chem. in press

11)Kärger,J.Z.phys.chem. in press
12)Pfeifer,H.,in NMR Basic Principles and Progress $\underline{7}$,
   53,Springer,Heidelberg,1972
13)Pfeifer,H.,Surface Sci.(1975)$\underline{52}$,434
14)Murday,J.S.,Patterson,R.L.,Resing,H.A.,Thompson,
   J.K. and Turner,N.H.,J.phys.chem.(1975)$\underline{79}$,2674
15)Basler,W.D.,Lechert,H. and Kacirek,H.,Ber.Bunsen-
   ges.phys.chem.(1976) in press
16)Kacirek,H.,Thesis(1974)Hamburg
17)Kacirek,H. and Lechert,H.,J.phys.chem.(1975)$\underline{79}$,1589
18)Baur,W.H.,Amer.Mineralogist(1964)$\underline{49}$,697
19)Zimmermann,J.R. and Brittin,W.E.,J.phys.chem.
   (1957)$\underline{61}$,1328
20)Lechert,H., Habilitation Thesis(1973)Hamburg
21)Yanagida,R.Y. and Seff,K.,J.phys.chem.(1972)$\underline{76}$,2597

Mössbauer Studies of Ferrous-A Zeolite

LOVAT V. C. REES

Physical Chemistry Laboratories, Imperial College of Science and Technology, London SW7 2AY, England

Mössbauer spectroscopy is a versatile technique for determining cation locations and coordination geometries in materials where the Mössbauer ions may be found in several sites. Such is the case in zeolites where the cations required to balance the negative charge of the aluminosilicate lattice can take up a number of different positions to produce a distribution with the lowest possible free energy. Mössbauer studies have been made on a number of zeolites, most interest being shown in the catalytically active Y-zeolites. (1)(2) Because of these catalytic properties the interaction between the cations in the zeolite and adsorbed gases is also of interest and one that is particularly amenable to study by Mössbauer spectroscopy.

Unfortunately, in ferrous-Y (in these samples only partial exchange of sodium-Y with ferrous ions has taken place) most of the ferrous ions occupy sites which are not accessible to organic sorbates. Essentially, the spectra of dehydrated ferrous-Y consist of two doublets; one with an isomer shift (i.s.)of 1.54 mm s^{-1} and a quadrupole splitting (q.s)of 2.35 mm s^{-1} and a smaller doublet with an i.s. of 1.16 mm s^{-1} and a q.s. of 0.62 mm s^{-1}. (1-4) It is the ferrous ions giving the inner doublet which are accessible to sorbates but they produce less than 15% of the total Mössbauer absorbance. Thus in the spectra of ferrous-Y after adsorption of various gases, the peaks due to the ferrous-sorbate complexes are not easily resolved, though changes to the parameters of the inner doublet indicated complexes were being formed. (1)(2)

Type A zeolite was studied in an attempt to overcome this problem. This zeolite is similar to Y zeolite: both contain sodalite cages as their basic structural unit. (5) These sodalite cages consist of $24(Si/AlO_4)$ tetrahedra stacked to give a hollow truncated octahedron. This has eight faces made up of six oxygen atoms forming a puckered ring with a free diameter of 0.22 nm and six faces made up of four oxygen atoms. In Y zeolite four of these six-membered rings are joined to neighbouring sodalite cages by oxygen bridges to give a tetrahedral array whilst in

A zeolite the six four-membered rings are used to give an octahedral array. Thus, the hexagonal prism between the sodalite
cages in Y zeolite, which is the major internal site for cations,
does not occur in A zeolite. Both zeolites contain the sodalite
window site (the six-membered ring)and it is in this site that
cations accessible to sorbed gases are found.

Experimental

The ferrous exchanged form was prepared by treating 0.5g of
the zeolite with approximately 45 cm^3 of 0.15 mol dm^{-3} ferrous
sulphate solution under a nitrogen atmosphere. The solution was
initially adjusted to pH 5 with dilute sulphuric acid. A few
grains of ascorbic acid were added to reduce any ferric ions present. The exchange was allowed to proceed for 2 h before the
zeolite was washed with oxygen free distilled water. Chemical
analysis for iron, aluminium and silicon showed that 40-50% of
the exchangeable sodium ions had been replaced by the treatment.
A typical sample of the dehydrated zeolite was analysed and found
to have the unit cell composition:

$$Fe_{2.66}Na_{6.68}Al_{12}Si_{12}O_{48}0.9NaAlO_2$$

In accordance with previous work (6)(7)ordering of the Si and Al
atoms in the zeolite is assumed and the excess aluminium is taken
to be present in the form of occluded sodium aluminate molecules
in the sodalite cages.

The crystallinity of the ferrous-A after dehydration under
vacuum at 360°C for 18 h was examined by X-ray powder diffraction
and by oxygen adsorption measurements. Sharp lines in the X-ray
powder patterns indicated no structural breakdown had occurred.
The unit cell of the dehydrated ferrous-A was found to be 1.219 $\pm$
0.004 nm; a value between that of dehydrated sodium-A (1.228 nm
(5))and fully exchanged cobalt-A (1.208 nm (8)), i.e. consistent
with a 50% exchanged ferrous-A.

The amount of oxygen adsorbed by dehydrated ferrous-A at
78 K under an equilibrium pressure of 14.49 kN m^{-2} was 206 $\pm$
2 cm^3 (s.t.p.)g^{-1}. Under similar conditions dehydrated sodium-A
adsorbed 195 $\pm$ 2 cm^3 (s.t.p.)g^{-1}. The slightly higher value for
ferrous-A results from the increase in free volume in the zeolite
when two sodium ions are replaced by one ferrous ion. These results show clearly that the crystallinity of the zeolite is retained.

The freshly prepared zeolite was contained in a glass cell
(4 cm x 1.5 cm diameter) which had two thin (ca. 0.015 cm) windows
that allowed about 40% transmission of the 14.4 keV γ-rays. These
cells could be attached to a vacuum line to enable either dehydration of, or adsorption of gases into, the zeolite. A standard
volumetric apparatus was used to titrate doses of gases into the
zeolite.

The spectra taken at and above room temperature were made using an Elron MD-1 constant acceleration spectrometer in conjunction with an Intertechnique SA-40B multichannel analyser. Spectra taken at below room temperature were made with the glass cells mounted in a Ricor MCH-5 cryostat. These measurements were made with an Elron MD-3 spectrometer and a Nuclear Data ND-2200 multichannel analyser. Further details of the apparatus are given in ref.(4). The data from each spectrum were analysed by a constrained non-linear least squares fitting program. Area measurements were corrected by the use of a 0.1 mm brass foil filter to measure the high energy contribution to the background (9). The resolution of the Reuter-Stokes Kr/CO_2 proportional counters and Nuclear Data 522 single channel analysers was sufficient to prevent any significant low energy background. Isomer shifts reported here are with respect to a sodium nitroprusside [trisodium pentacyanonitrosylferrate(II)] absorber.

Results

The Mössbauer spectra taken of the freshly prepared ferrous-A zeolite and after various stages of dehydration are shown in Figure 1. Spectrum 1A of the hydrated zeolite under an equilibrium pressure of 2.1 kN m^{-2} of water contains a doublet with broadened lines. The i.s. of 1.50 mm s^{-1} and q.s. of 2.10 mm s^{-1} of this doublet are similar to the values found for hydrated ferrous-X (4) and Y(1)(2)(4) zeolites indicating that in all three zeolites the hydrated ferrous ions have a similar coordination.

Pumping the zeolite at room temperature caused the i.s. of ferrous-A to decrease to 1.40-1.43 mm s^{-1} and the q.s. to 1.90-2.02 mm s^{-1}. Some variation between different samples was seen and the linewidths of the peaks in these spectra also varied between 0.60 and 0.85 mm s^{-1}. The spectra 1B, 1C and 1D were taken with the zeolite at intermediate stages of dehydration while in 1E the zeolite was fully dehydrated. Full dehydration of the zeolite appears to occur near 100°C. Some 22 spectra were taken of 12 different samples of ferrous-A dehydrated in vacuo between 100 and 360°C. These showed no greater variance from the average i.s. value of 1.08 mm s^{-1} and q.s. value of 0.46 mm s^{-1} than ± 0.02 mm s^{-1}. Linewidths showed some variation between 0.32 and 0.40 mm s^{-1}. Further heating of the zeolite to 525°C for periods of up to 24 h gave small (< 0.02 mm s^{-1}) changes in the i.s. and q.s. indicating that complete dehydration has taken place at the lower temperatures. The dehydration was found to give a 20% decrease in the absorption area between spectra 1A and 1D in marked contrast to ferrous-Y zeolites where dehydration gave a large increase in the absorption area.(1)(2)(4)

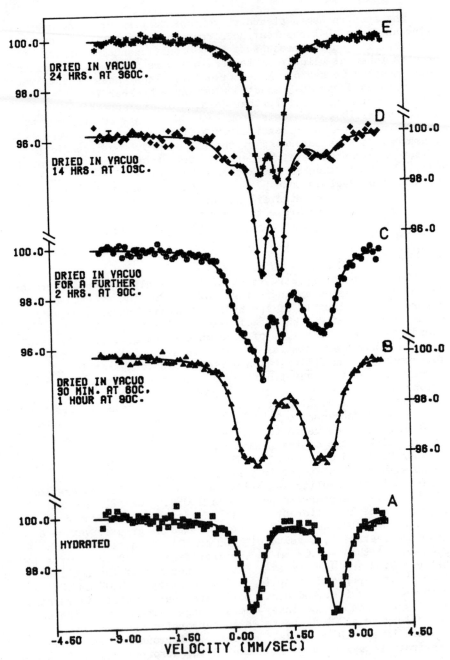

Figure 1.   Spectra of ferrous-A at different states of dehydration

The temperature dependence of the spectrum of dehydrated ferrous-A was studied over the range 80 to 625 K. Three samples were used; one to cover the temperatures between 80 and 375 K and two for the high temperatures. The q.s. was found to be almost temperature independent. The first sample gave a q.s. of 0.443 ± 0.008 mm s^{-1} at 375 K and 0.461 ± 0.008 at 80 K. The second sample had a q.s. of 0.470 mm s^{-1} at 296 K decreasing to 0.430 at 626 K. Though the samples give slightly different q.s. values at 296 K the q.s. variation with temperature of each sample is very small. Adjusting the three sets of results to 0.460 mm s^{-1} at 296 K gives a total q.s. decrease of 0.03 mm s^{-1} for a temperature change from 80 to 625 K.

The i.s. values show a non-linear temperature dependence. The results, along with the q.s. results are shown in Figure 2. The i.s. results were also adjusted to the same room temperature value of 1.08 mm s^{-1}.

The spectra of the partially dehydrated ferrous-A (1B, 1C and 1D) show an unexpected splitting of the outer peaks which could be fitted as pairs of doublets in several ways. A study of the temperature dependence of the spectrum from a sample of ferrous-A dehydrated at 90°C for one hour was made to resolve this problem. The spectra obtained at 300, 247, 177 and 80 K show the two inner peaks decrease in intensity whilst the outer peaks increase as the temperature decreased. The room temperature spectra thus consist of two nested doublets with similar i.s values. The parameters fitted to the spectra of Figure 1 on this basis are given in Table I.

The readsorption of water into dehydrated ferrous-A was investigated to determine the number of water molecules coordinated to each ferrous ion. The sample used had 44.3% of the sodium ions replaced and thus contained 1.46 mmol iron per g dehydrated zeolite. The sample weighed 0.265 g dry. Water, purified by several freeze, pump, thaw cycles, was titrated onto the zeolite. The spectra are shown in Figure 3.

The spectra were very similar to those obtained with the partially dehydrated ferrous-A. The fully rehydrated ferrous-A gives the same spectrum as freshly prepared ferrous-A confirming the stability of ferrous-A to dehydration. The parameters fitted to the spectra are given in Table II. In the case of the spectra 3A and 3B the outer peaks were insufficiently resolved to be fitted as two doublets. Spectrum 3E could be equally fitted as one or two doublets. The parameters for the one doublet fit; i.s. = 1.43 mm s^{-1} and q.s. = 1.99 mm s^{-1} are similar to those found with the zeolite that had been pumped at room temperature. This indicates that pumping removes approximately 67% of the adsorbed water from a fully hydrated zeolite.

Figure 4 shows the area under the inner doublet and the outer doublets and the total absorption area plotted as a function of the amount of water adsorbed. The point at which all the ferrous ions are complexed corresponds to 2.94 mmol water per g zeolite or to two water molecules attached to each ferrous ion.

Table I. Parameters Fitted To Spectra Of Dehydrated Ferrous-A In Figure 1.

spectrum	sample treatment	isomer shift[a]/ mm s^{-1} ($\pm$0.02)	quadrupole splitting/ mm s^{-1} ($\pm$0.02)	f.w.h.m./ mm s^{-1} ($\pm$0.02)	area%
A	fully hydrated P(H_2O) = 2.1 kN m^{-2}	1.50	2.10	0.55	100
B	heat in vacuo 30 min at 60°C/ 1 h at 90°C	1.40 1.40	1.41 2.27	0.77 0.48	73 27
C	further 2 h at 90°C	1.06 1.38 1.37	0.48 2.22 1.44+0.6	0.28 0.52 0.74	19 29 52
D	heat in vacuo 14 h at 103°C	1.07 1.30	0.48 2.39+0.05	0.35 0.67+0.05	75 25
E	heat in vacuo 24 h at 360°C	1.08	0.46	0.37	100

[a] w.r.t. sodium nitroprusside

Table II. Parameters Fitted To Spectra Of Figure 3.

spectrum	number of water molecules per unit cell	isomer shift[a]/ mm s^{-1} ($\pm$0.02)	quadrupole splitting/ mm s^{-1} ($\pm$0.02)	f.w.h.m./ mm s^{-1} ($\pm$0.03)
A	0.80	1.11 1.51+0.05	0.47 1.74+0.05	0.38 1.40+0.15
B	2.84	1.10 1.42+0.03	0.49 1.66+0.04	0.38 1.10+0.05
C	4.75	1.03 1.41 1.43	0.52 1.35+0.08 2.20+0.06	0.24+0.06 0.82+0.06 0.48+0.10
D	6.50	1.41 1.42	1.44 2.32	0.74 0.50
E	10.5	1.44 1.41	1.83 2.36+0.07	0.60 0.52+0.05
F	≈ 28	1.50	2.11	0.63

[a] w.r.t. sodium nitroprusside

The total area in the spectrum of the hydrated ferrous-A is also found to be 20% greater than in the spectrum of the dehydrated ferrous-A. This is in agreement with the decease in area seen after dehydration of the zeolite.

The initial rapid increase in the total area is considered to be an artifact of the use of the parameters fitted to the spectra to measure the areas. It is difficult to fit small peaks in a spectrum without distorting the baseline parabola. The

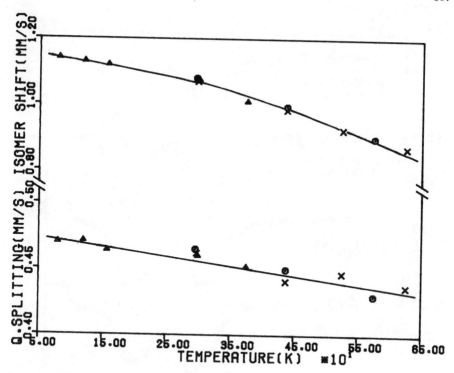

Figure 2. Effect of temperature on the isomer shift and quadrupole splitting of dehydrated ferrous-A

linewidths of the small peaks tend to broaden, as can be seen in the data of Table II and this leads to unrealistically large areas.

The adsorption of gases other than water into dehydrated ferrous-A can be used to help locate the position of the ferrous ions. Molecules such as methanol, acetonitrile and ethylene are restricted to the large cavities in the zeolite. At saturation of the ferrous-A with these three gases the spectra obtained all single doublets with parameters as given below.

Gas	isomer shift/ mm s^{-1}	quadrupole splitting/ mm s^{-1}	linewidth/ mm s^{-1}
methanol	1.38	2.00	0.90
acetonitrile	1.29	1.18	0.50
ethylene	1.19	1.00	0.38

All three gases are able to interact with all the ferrous ions. This shows the ferrous ions to be located on or near the walls of the large cavities. Further details on the adsorption of ethylene into dehydrated ferrous-A ,are given below.

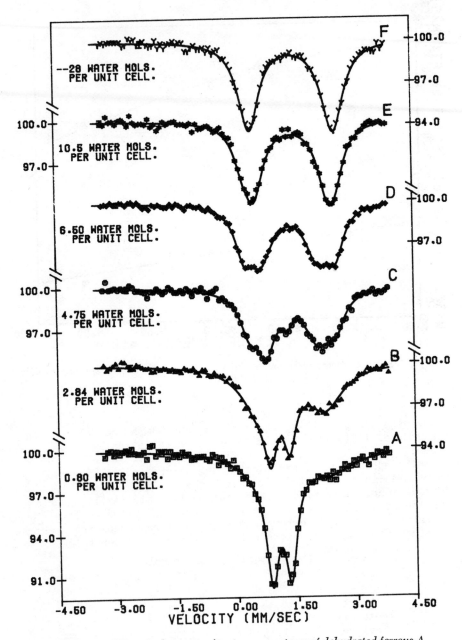

Figure 3.   *Effect of adsorption of water on spectrum of dehydrated ferrous-A*

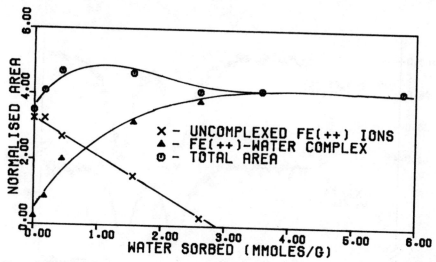

Figure 4. Area under the inner and outer doublets and total absorption area of the spectra in Figure 3

Representative spectra obtained from adsorbing ethylene into dehydrated ferrous-A zeolite are shown in Figure 5. The spectra show no sign of a product doublet and attempts to fit two doublets to the spectra were unsuccessful. Single doublets were fitted to the spectra and the isomer shift (i.s.), quadrupole splitting (q.s.) and the absorption area values obtained are plotted in Figure 6. This area shows a remarkable decrease and then increase as ethylene is added. The linewidth in the starting spectrum of ferrous-A zeolite was 0.34 ± 0.02 mm s^{-1} and increased slightly with the addition of ethylene to 0.38 ± 0.02 mm s^{-1}.

The ethylene is only weakly bonded in the zeolite. The last two points in Figure 6 were taken at pressures of 19 and 38 kN m^2 respectively. The coverage at the higher pressure corresponds to ~ 5 molecules of ethylene in each large cavity. All the ethylene may be desorbed at room temperature by pumping for short periods. The results obtained from several samples after pumping are shown in Figure 6. No areas are given for two of these spectra as a large change in counting rate occurred whilst these were being taken.

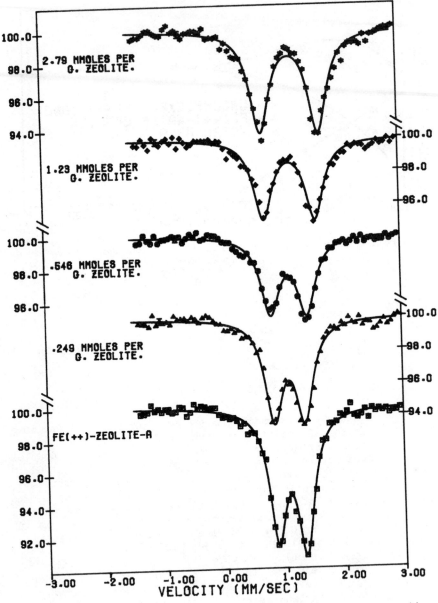

Figure 5. Room temperature spectra of ferrous-A containing increasing quantities of adsorbed ethylene

The temperature dependence of the spectrum of a sample which had 0.8 mmol ethylene sorbed per g zeolite is shown in Figure 7. The spectra taken at temperatures below 200 K show two doublets. At the lowest temperatures the inner of the two doublets has the parameters of uncomplexed ferrous-A zeolite. The other doublet must be that of the ferrous-ethylene complex.

Discussion

Dehydrated Ferrous-A. The accessibility of the ferrous ions in the dehydrated zeolite to the organic gases show them to be located on the walls of the large cavity. There are three possible sites fulfilling this condition. One site, as found in sodium-A (5) is in the plane of the 0.4 nm free diameter window connecting large cavities. As 50% exchanged ferrous-A can absorb n-pentane this site can be ruled out. If this window contains a cation n-pentane will not be adsorbed as is found with sodium-A (5) (The effect of n-pentane on the ferrous-A spectrum is to increase the q.s. to 0.51 mm s^{-1}.) Another site occupied by one of the twelve ions in sodium-A is on the side of the cubic cage joining sodalite cages and requires the cation to sit above the plane of the four oxygen ions.(10) This is known to be an unfavourable site. The major cation site in A zeolite is in the window between the large cavity and the sodalite cage. A small cation can sit in the plane of the window in close contact with three oxygen ions at ~0.22 nm and with the other three ions at ~0.30 nm.(10)

A cation in this site should show a high vibrational anisotropy. No distinct asymmetry in the peak amplitudes of the ferrous-A spectrum was seen as would be expected if the Goldanskii-Karyagin effect was present. However, the low counting statistics could hide a small effect. The decrease in the absorption area upon dehydration of ferrous-A indicates that the ferrous-A ions in the dehydrated zeolite have a low recoilless fraction. When water molecules coordinate to the ferrous ions they must also coordinate to the lattice through hydrogen bonding. This will hold the ferrous ions more firmly in place and so lead to an increase in area. A similar behaviour has been found for manganese ions in A zeolite. (11) The thermal ellipsoids of the manganese ions in the sodalite window site as determined in a single crystal X-ray study are larger in the dehydrated zeolite than in the hydrated form.

The position of the ferrous ions has been confirmed by a Mössbauer study (12) of sodium-A doped with ^{57}Co and a recent X-ray study (13) of cobalt-A. The Mössbauer spectrum of the ^{57}Co-A source contained two doublets; an inner one which had a temperature independent q.s. of 0.50±0.02 mm s^{-1} and a room temperature i.s. of 1.04±0.02 mm s^{-1}. The outer doublet was considered to arise from recoil effects in the decay of ^{57}Co.

The X-ray study of cobalt-A showed all the cobalt ions to be sited in the sodalite window-site.(13) The similarity of the spectra of ferrous-A and the inner doublet of the ^{57}Co-A source suggest that the position of the ferrous and cobalt ions is the same, i.e. in the sodalite window site.

The i.s. of 1.08 mm s^{-1} is low for high spin ferrous iron but is in agreement with the ferrous ion being in a site of low coordination number. For example ferrous ions in the mineral gillespite (14) are in a square planar site and show an i.s. of 1.04 mm s^{-1}. They also give a small temperature independent q.s. as seen with ferrous-A. Simple crystal field theory shows this to result from near cancellation of the valence and lattice contributions to the electric field gradient at the ferrous nucleus. A value of 2.6 mm s^{-1} is obtained for $(q.s.)_{lat}$, and between -2 and -3 mm s^{-1} for $(q.s.)_{val}$. As the ferrous ion site has three fold symmetry the total q.s. is determined as

$$q.s. = (q.s.)_{val} + (q.s.)_{lat}.$$

Thus the use of crystal field theory shows that the planar geometry of a ferrous ion in the window site will lead to a small q.s. value. Refinements could be made to the calculation to obtain a better value of the q.s. but a more accurate knowledge of the position of the ion in the window would be needed. Other problems such as estimating the effect of covalency and the actual charge on the oxygen ions make more refined calculations pointless.

Hydrated Ferrous-A. In Figure 4 the linear decrease in the absorption area of the uncomplexed ferrous ions upon adsorption of water indicates that two molecules of water are coordinating to each ferrous ion. The complexed ions then have the coordination $(Z-O)_3Fe^{2+}(H_2O)_2$ where $(Z-O)$ represents a zeolite lattice oxygen ion. However, the spectra of these complexed ions in Figure 3 indicate that there are two types of iron present and these two types have been shown to be related by a temperature dependent process. Initially addition of more water beyond the amount needed to cover all the ferrous ions has little effect on the spectra. These water molecules probably coordinate to the sodium ions and to the lattice. When about ten water molecules per large cavity are adsorbed the spectra of the two types of ferrous ions begin to merge into one. Complete hydration of the zeolite causes a further change with an increase in both the i.s. and the q.s. of the ferrous ions though no change is seen in the absorption area. The parameters derived from these spectra are listed in Table II. It is probable that the addition of a third water molecule forms the complex $(Z-O)_3Fe^{2+}(H_2O)_3$. The formation of an octahedral complex by transition metals in hydrated zeolites is commonly found (15) though in most studies the octahedral species is assumed to be $Fe(H_2O)_6^{2+}$. However, as no decrease in

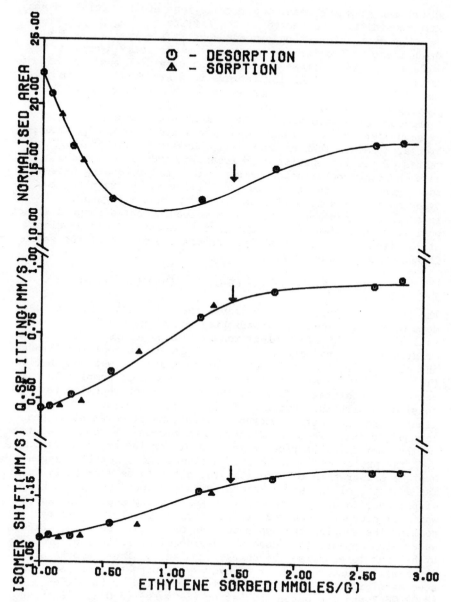

Figure 6. *Effect of adsorbed ethylene on isomer shift; quadrupole splitting and absorption area of spectra taken at room temperature. Arrows indicate one ethylene per Fe^{2+} cation.*

absorption area was seen, the ferrous ions are not solvated and must remain firmly attached to the lattice.

The most interesting process seen in the spectra is that which leads to two ferrous sites in the partially hydrated zeolite. As the incease in absorption area upon addition of water indicates that all the ferrous ions are firmly attached to the zeolite lattice, a mechanism in which the ferrous ions are jumping between different sites can be ruled out. Similarly a process which involved the water molecules jumping between adjacent ferrous ions would also give a decrease in the absorption area rather than an increase. The similarity of the i.s. values of the two species of ferrous ion indicates that both have the same coordination number with similar bonding between the ligands and the ion. The q.s. difference between the two types must derive from effects due to the water ligands. Slow rotation or vibration of the water molecules between two distinct positions could lead to changes in the electric field gradient at the iron nucleus. Possibly dissociation of one of the water ligands could occur, viz.

$$(Z-O)_3Fe^{2+}(H_2O)_2 = (Z-O)_3Fe^{2+}(H_2O)(OH)^- + (Z-O)-H^+.$$

The larger electric field gradients resulting from the replacement of the more equally distributed univalent ions by half the equivalent number of divalent ions would encourage such a dissociation. The production of hydroxyl groups in zeolites through the electric fields acting on water molecules is well known (16) (17) though these are usually observed in almost completely dehydrated zeolites. However, the dissociation reaction is consistent with the evidence of tight bonding between the water ligands and the lattice whereas suggestions of movement of the ligands is not. Thus the dissociation process is considered to be the most feasible process to account for the spectra obtained.

The proton movement between a ligand and a lattice oxygen must be slower than 10^{-7}s for the two sites to be seen individually. The addition of excess water can evidently speed up the exchange of protons, to account for only one site being seen when excess water is present. However, if the process is slowed down by cooling the sample, the two sites can once again be seen. Excess water also appears to reduce the degree of dissociation, probably by helping to equalise the charge imbalance. Thus removal of water will cause the number of hydroxyl groups to increase. The species giving the inner doublet is identified as $(Z-O)_3Fe^{2+}$ $(H_2O)(OH)^-$ and the outer doublet as $(Z-O)_3Fe^{2+}(H_2O)_2$. Such an assignment is as expected with the more symmetrical species having the larger q.s.

In most studies on the dehydration of zeolites (16)(17)the production of hydroxyl groups is found to occur in the final stages of dehydration. The water molecules dissociating are considered to remain even after prolonged dehydration at high

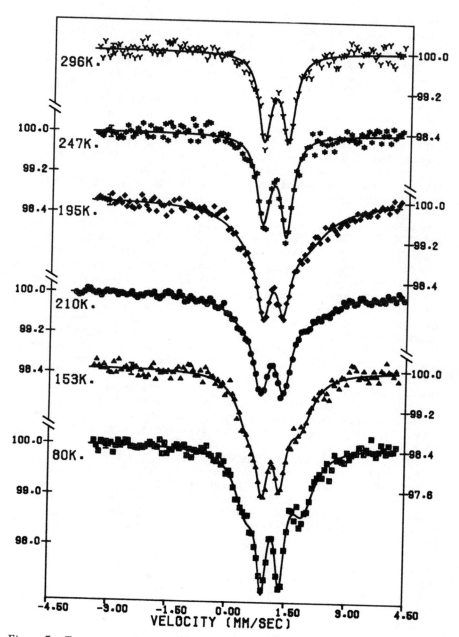

Figure 7. Temperature dependence of spectrum of ferrous–A containing 0.8 mmol of ethylene per gram of zeolite

temperatures though some of the cations could lose water by the process $2M(OH) \rightarrow MOM + H_2O$. However, in ferrous-A the ferrous ions sit alone in the window and there is no evidence of a $Fe^{2+}-O-Fe^{2+}$ complex which would involve two different ferrous sites unless the complex were formed in the sodalite cage. This is ruled out from the results obtained on the adsorption of the organic gases. This implies that in ferrous-A the mobile protons must recombine with the hydroxyl group during dehydration to facilitate full dehydration. The dehydration temperature of 100°C is surprisingly low although the 3-fold geometry for the ferrous ion is very unusual. It is, however, comparable to the dehydration temperature found for cobalt-A. (8) The electric fields in the zeolite evidently can help the dehydration and stabilize the cations in their unusual geometries.

Adsorption of Ethylene. The spectra of Figure 5 are readily interpreted as resulting from the mobility of the ethylene molecules. First the dramatic change in the recoilless fraction of the ferrous ions is readily explained. When an ethylene molecule is attached to a ferrous ion, the ion is pulled slightly out of the plane of the window and when the ethylene molecule jumps away the ion relaxes to its original position. Hence the ions are forced to oscillate over distances larger than those associated with simple thermal vibrations and the recoil free fraction decreases, decreasing the absorption area on adsorption of ethylene. As more ethylene is added, the number of unoccupied ferrous sites decreases and the residence time of an ethylene molecule on any one site is increased. Hence the recoilless fraction should increase again as the sites are filled with a minimum absorption area occurring at 50% coverage of the ferrous ions. The experimental minimum occurs at one ethylene molecule to 2 ferrous ions, indicating that ethylene molecules prefer the ferrous ions to any other type of site. If the sodium and ferrous cations were equivalent sites, the minimum would occur at 4.7 ethylene molecules per large cavity.

The effect of a fluctuating environment on the shape of the spectra of the ferrous ions can be obtained by the method used to describe the phenomenon of chemical exchange in n.m.r (18) and its extension to describe paramagnetic relaxation in Mössbauer spectroscopy.(19) The ethylene molecules are assumed to jump in a random manner but the time spent by the ferrous ions between being free and complexed (referred to as states A and B) is assumed negligible. If τ_A^{-1} is the probability of state A changing and τ_B^{-1} that of state B, a steady state is reached when $P_A/\tau_A = P_B/\tau_B$ where P_A and P_B are the fractional populations of states A and B respectively. The density matrix formalism developed by Wickman (19) is used to describe the changes in the spectra.

For long lifetimes ($\tau \rightarrow \infty$) two lorentzian shaped doublets with the parameters of states A and B respectively are obtained.

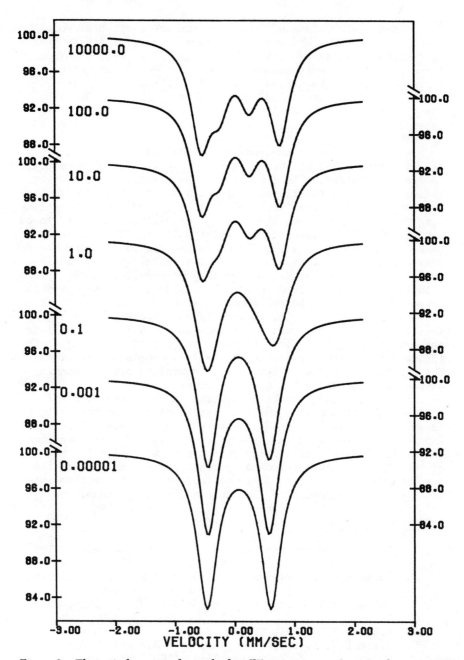

Figure 8. *Theoretical spectra obtained when* [57]*Fe environment fluctuates between two states* A *and* B. *The lifetime of state* A *is quoted in units of lifetime of the Mössbauer event.*

For intermediate values of τ the spectra are complex as some examples calculated and given in Figure 8 show. In these spectra P_A (the inner site) is 0.35. The spectra show that deviations between the two special cases of $\tau \to 0$ and $\tau \to \infty$ occur in the range $10.0 \tau_N$ to $0.1 \tau_N$ where $\tau_N = 9.55 \times 10^{-8}$s.

Comparison of the calculated spectra in Figure 8 and the spectra of Figure 7 show extensive similarities. The spectra taken at around 200 K show a state similar to that expected if $\tau_A \approx 10^{-7}$s.

The ferrous ion in the sodalite window has C_{3v} symmetry and this is not altered by the sorption of ethylene if this molecule is freely rotating. The d orbital energy levels will, however, change from those for the uncomplexed ion. The ferrous-ethylene complex shows a temperature dependent q.s. from which the splitting between the d_{z^2} and the d_{xz}, d_{yz} doublet is estimated to be around 400 cm^{-1}. The lattice contribution to the q.s. will also be reduced. However, without knowing how far the ferrous ion is drawn from the sodalite window, it is difficult to accurately quantify these changes. The room temperature i.s. of the ferrous-ethylene complex of 1.20 mm s^{-1} is similar to that observed for ferrous ions in tetrahedral geometry with oxide and halide ions (20) but in these the q.s. is usually much larger. This suggests that the lattice component still remains quite large. The ferrous ions cannot, therefore, be moved far out of the plane of the window as this would decrease the lattice component significantly. These facts are consistent with the assumption of weak bonding of the ethylene molecules to the ferrous ions.

The observation of mobile sorption through the effect of the sorbate molecules on the characteristics of the surface cations is an unusual use of Mössbauer spectroscopy. There is obviously further scope to study the dynamic properties of ferrous-sorbate complexes in more detail using this technique. Also the rotation of the ligand could be studied by measuring the spectra of the complexes at down to liquid helium temperatures where this rotation would be expected to slow down and finally cease.

Literature Cited.

1. Morice J.A. and Rees L.V.C. Trans. Faraday Soc., (1968) 64, 1388
2. Delgass W.N. Garten R.L. and Boudart M. J. Chem. Phys. (1969) 50, 4603
3. Delgass W.N. Garten R.L. and Boudart M. J. Phys. Chem. (1969) 73, 2970
4. Dickson B.L. Ph.D Thesis (University of London, 1973)
5. Breck D.W. Eversole W.G. Milton R.M. Reed T.B. and Thomas T.L. J. Amer.Chem.Soc., (1956) 78, 5963
6. Barrer R.M. and Meier W.M. Trans. Faraday Soc., (1958) 54, 1074
7. Sherry H.S. and Walton H.F. J. Phys. Chem. (1967) 71, 1457

8. Gal I.J. Jankovic O. Malcic S. Radovanov P. and
    Yodorovic M. Trans. Faraday Soc. (1971) 67, 999
9. Housley R.M. Erickson N.E. and Dash J.G. Nuclear
        Instr. Methods (1964) 27, 29
10. Smith J.V. and Dowell L.G. Z Krist. (1968) 126, 135
11. Yanagida R.Y. Vance T.B. and Seff K. Chem. Comm.
                                            (1973) 383
12. Dickson B.L. Erickson G.A. Rees L.V.C. and Fitch F.
                in preparation
13. Riley P.E. and Seff K. Chem. Comm. (1972) 1287
14. Clark M.G. Bancroft G.M. and Stone A.J. J. Chem.
                        Phys. (1967) 47, 4250
15. Mikheikin I.D. Zhidomirov G.M. and Kazanskii U.B.
                    Russ. Chem. Rev. (1972) 41, 468
16. Ward J.W. Molecular Sieve Zeolites, Part I; Adv.
                    Chem. Series (1971) 101, 380
17. Ward J.W. Trans. Faraday Soc. (1971) 67, 1489
18. Lynden-Bell R.M. Progr. N.M.R. Spectr. (1967) 2, 163
19. Wickman H.H. Mössbauer Effect Methodology, Vol. 2,
            ed. I.J. Gruverman (Plenum, New York, 1966)
                        p.39
20. Greenwood N.N. and Gibb T.C. "Mössbauer Spectroscopy"
            (Chapman and Hall, London, 1971)

27

Structural and Kinetic Studies of Lattice Atoms, Exchangeable Cations, and Adsorbed Molecules in the Synthetic Aluminosilicate, Y Zeolite

EDWARD E. GENSER

Department of Chemistry, California State University, Hayward, Calif. 94542

The application of NMR to the study of surfaces and adsorbed species is one of those rare instances in which weakness may be regarded as a virtue. The nucleus, since it is buried deep inside the atom or ion of interest, suffers relatively minor perturbations from intermolecular interactions and conversely, the nucleus exerts a negligible influence on the interaction itself. The perturbations on the nuclear magnetic levels are manifested as a change in the NMR spectrum relative to that of the unperturbed species. The objective is then to interpret this change in terms of the structural and dynamic features of the surface-adsorbate interactions. Thus the magnetic nucleus may be useful as a nuclear probe or "spin label" in the study of interfacial and surface phenomena.

This article does not pretend to be a comprehensive review of the applications of NMR to such systems (1,2,3) nor of the NMR technique itself (4,5,6). It is intended rather to illustrate the method by reference to a specific system which has been extensively used as a catalyst (7), viz., the crystalline Aluminosilicate, Y Zeolite.

Y Zeolite is a synthetic material which is structurally related to the natural mineral Faujasite (8). It is especially interesting from the point of view of NMR for several reasons. First, it consists of large pores, or supercages 13.7 Å in diameter which are connected by channels, or windows with a diameter of 7.5 Å. As a result of its very open, porous structure, Y Zeolite has the capacity to adsorb a significant fraction of its own weight of various molecules. For example, about 25% of the weight of hydrated Y is present as water, a figure which corresponds to nearly 30 water molecules per supercage. Furthermore, this water is quite mobile, having a diffusion coefficient comparable to that in liquid water (9). Thus in this rather unique system a mobile phase is superimposed on a rigid Aluminosilicate lattice framework. Since the NMR spectra are quite different and characteristic for each, one may study the phases separately and attempt to correlate the results.

A second important feature of the Y Zeolite is that it is basically a defect structure in which AlO_4^- are occasionally present as replacements for SiO_4 tetrahedra. (For Y Zeolite the Si/Al ratio covers the range from about 2-3.) As a consequence, positive ions must be incorporated in the lattice to preserve electrical neutrality. These ions are conveniently capable of replacement with other cations by exchange from aqueous solution (10). The cation type and location in the lattice are crucial factors in the function of Y Zeolites as active catalysts. Thus there are four basically distinct environments for magnetic nuclei in the Y Zeolites. They are:

1. In a rigid lattice position. These include the magnetic nuclei ^{27}Al, and ^{29}Si (7.5% isotopic abundance).

2. In a charge compensation cation. Here one may take advantage of cation exchange and choose an ion whose nucleus possesses convenient properties for use as a spin label. Some examples are ^{23}Na, ^{7}Li, and ^{1}H. (The latter are usually attached to framework oxygen atoms.)

3. Adsorbed molecules. Here one can think of observing the nuclei in a molecule which is chemically or physically adsorbed on the interior surface of the lattice framework. Some particular isotopes of interest are ^{1}H, ^{13}C, ^{19}F, and ^{31}P.

4. Liquid-like molecules. While these may interchange with those described in part 3, it is convenient to consider molecules in the interior of the supercages as a distinct category which behaves like ordinary liquids in many respects. The nuclei of interest as labels include those listed in part 3 and addition of other isotopes to this list depends on details of the nuclear site in the molecule of interest. The discussion of Nuclear Quadrupole interactions which follows will help to clarify this point.

A third advantage which makes these Y Zeolites attractive systems for study by NMR is that crystal structures of a number of these have been determined by X-ray diffraction (8,11). In some cases the cation positions and those of the adsorbed species (e.g., H_2O) have also been located (12). This is important because the unequivocal interpretation of the NMR results cannot usually be made in the absence of such information. This is one of the main difficulties in interpreting the NMR spectra in amorphous Silica-aluminas.

The NMR Method. As a general rule the NMR spectra of nuclei in solids are broad lines and usually have little or no structure. The same nuclei in liquids may give rise to narrow lines and a relatively complex structure. The relative sensitivity depends on the ratio of the resonance field and the line width, i.e., on the ratio: $Q = H_{0/\Delta H} = \nu_{0/\Delta \nu}$, where H_0 and ΔH are the resonance field and line width expressed in gauss and ν_0 and $\Delta \nu$ are the resonance frequency and line width expressed in hz. As an example, the ^{23}Na resonance in aqueous solution in a field of 10^4 gauss has a line width of about 0.03 gauss, so that Q_{liq} = 3.3 X 10^5. On the other hand, the ^{23}Na resonance arising from the ion in the solid (hydrated) Y Zeolite has a width of 6 gauss and a corresponding Q_{solid} = 1.6 X 10^3. This is a reduction of about 200, a very conservative figure since the ion in the solid is more dilute than in the solution. This accounts for the characteristically low S/N (signal-to-noise) ratio in solids as seen in Figure 1 where the ^{23}Na and ^{27}Al resonances of hydrated Y Zeolite are shown. If one wishes to place a truly quantitative interpretation on the experimental results, it is usually necessary to resort to the method of signal accumulation with the aid of a computer. In this method the spectrum is repeatedly scanned and each successive spectrum is added to the resultant of all previously accumulated spectra. The theoretical improvement in the S/N ratio after p scans relative to that for a single scan is equal to $\sqrt{p}$. Figure 2 shows the accumulated result of 32 scans of the spectrum in Figure 1. The result is somewhat better than theory because a slower sweep and increased electronic filtering were used.

Nuclear Interactions in Solids.

a. Magnetic interactions. These consist of the coupling of the observed nucleus with those of its neighboring nuclei. The magnetic dipole-dipole interaction is orientation dependent and is an important contributor to the line width in solids. Since it depends on the internuclear separation (more correctly, on r_{ij}^{-3}) it can frequently be used to confirm a postulated structural configuration of nuclei by the theory which relates the second moment of the resonance line to the internuclear distances and magnetic moments (13). Moreover, when relative motion of the nuclei occurs it can modify the second moment in a predictable manner which reveals the frequency of the motion involved. The rigid lattice theory has been applied to LiY in which the Na$^+$ ions have been replaced by Li$^+$ from solution exchange (14). Details of this application in Reference 14 shows that if one assumes a distribution of Li$^+$ ions in the dehydrated NaY (11), the calculated theoretical second moment is 0.64 gauss^2. The experimental result obtained from a time averaged spectrum (128 scans) is 1.06 gauss^2. The proximity of theory and experiment establish the fact that magnetic dipolar broadening dominates the line width. If quadrupole broadening is negligible, the result suggests that the Li$^+$ ions may approach one another more closely (perhaps in pairs) than the Na$^+$ ions do in NaY. It was observed also that the results for LiY were the same at room temperature and at -102°C indicating that the Li$^+$ are held in fixed positions in the lattice relative

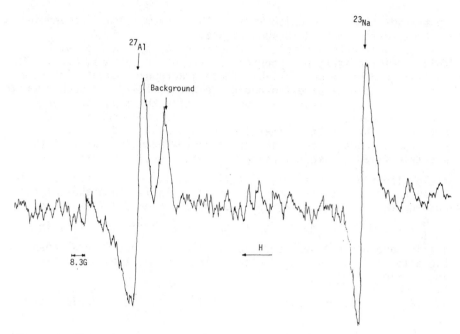

Figure 1. ^{27}Al and ^{23}Na *NMR resonances in NaY (No. 2) at 10.4 MHz. Background is due to ^{27}Al in probe.*

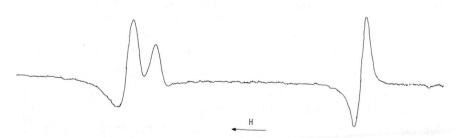

Figure 2.   *Noise averaged total spectrum of the ^{23}Na and ^{27}Al resonances from hydrated Y zeolite including probe background signal. Total scan width = 250 gauss.*

to the NMR time scale which implies that any motional frequencies must be small compared to about 2 Khz.

   b. Nuclear Quadrupole interactions (4,15). Nuclei with spins I > 1/2 possess nonspherical charge distributions, i.e., nuclear quadrupole moments, Q. These quadrupole moments interact with the gradients of extra-nuclear electric fields in such a way that for fields of less than cubic symmetry the magnetic energy levels are perturbed and hence the magnetic resonance is also affected. The result is that the energy levels and thus the spectrum depends on the orientation of the principal axes of the electric field gradient tensor V_{kl} with respect to the magnetic field direction. In a polycrystalline solid one sees the spatially averaged result which depends on the relative magnitude of the quadrupole interaction energy E_Q, and the magnetic Zeeman energy, E_M. There will be one value of E_Q for each distinct site in the lattice and hence a superposition of spectra where several possible sites exist as in a Y Zeolite. The form of the quadrupole interaction is:

$$E_Q = (1/6)(1 + \gamma)\Sigma_{kl}Q_{kl}V_{kl} \qquad \text{k,l = x,y,z in the principal axis system.}$$

In this formula, γ is the Sternheimer anti-shielding factor (15). It is produced by polarization effects in the atom or ion due to the nuclear quadrupole moment. It effectively acts to amplify the magnitude of the quadrupole interaction. The net effect of all this is that one usually expects that the quadrupole interactions for nuclei will, at the very least produce some resonance line broadening. It is not at all uncommon for the quadrupole effects to completely dominate the magnetic effects and prevent the observation of any NMR at all. For example, the ^{27}Al (I = 5/2) NMR is observable in NaY only in the hydrated form. The same is true of the [23]Na resonance (I = 3/2). It was suggested (14) that the reason for this was that the water molecules in the hydrate act as an effective dielectric by interposing themselves between the nucleus and the surrounding charges thereby reducing the field gradient by a factor $1/\varepsilon$ where ε is an apparent dielectric constant. To check these notions the theoretical second moment calculated for [23]Na in NaY due to magnetic dipolar broadening alone is 0.42 gauss^2, compared to the experimental result of 53.9 gauss^2. The large discrepancy is attributed to the fact that the primary broadening mechanism arises from quadrupole interactions. A similar conclusion was reached in the case of [27]Al for which the experimental second moment was 40.4 gauss^2. As a consequence of these findings, it may be concluded that in both cases one is observing only the $m_I = 1/2 \rightarrow m_I = -1/2$ transition in the spectrum. ^7Li appears to be an exception to this behavior even though it too has a quadrupole moment ($I = 3/2$). A possible explanation for this may be due to the relatively small quadrupole moment of ^7Li and to the low value of the anti-shielding factor γ due to the small size of the Li^+ ion. A comparison of the three nuclear isotopes follows.

	$^{23}Na^+$	$^{27}Al^{+3}$	$^7Li^+$	
Q*	0.1	0.149	-4.2×10^{-2}	*(in multiples of $e \times 10^{-24}\ cm^2$)
γ	4.53	2.59	-0.256	

Apparently then, the magnitude of E_Q is about 40 times smaller for $^7Li^+$ than it is for the other two isotopes (assuming the field gradients are similar for all three), thus permitting one to observe the magnetic dipolar interaction in the dehydrated LiY. Thus $^7Li^+$ is a plausible choice as a nuclear probe when a monovalent cation is required.

One further piece of information was obtained from the NaY spectrum (14). According to theory, the relative intensities of the observed resonances $+1/2 \rightarrow -1/2$ should be $^{23}Na/^{27}Al = (4/10)/(9/35) = 1.56$. The experimentally determined figure is 0.64. If one assumes that all of the ^{27}Al are observed, this would imply that only 41% of the ^{23}Na nuclei are observable. Since there are about 52% of the Na^+ ions located in the super-cages in the dehydrated NaY (and perhaps more in the hydrated NaY) it was suggested that those were the Na^+ ions observed by NMR.

NMR of Adsorbed Species.

a. Effects of motion on the resonance (4,17). Since adsorbed molecules in the pores are capable of motion we antici-pate that the orientational effects encountered in solids for both magnetic dipolar interactions and nuclear quadrupole interactions will tend to average out to the degree that the motional frequen-cies involved are large compared to the static broadening in the solid. The nuclear interactions affect the resonance in liquids mainly through their influence on the relaxation times T_1 and T_2.

For nuclei with quadrupole moments in asymmetric environments, this may shorten the relaxation times to such an extent that no resonance absorption is observable. If $I = 1/2$, the magnetic dipolar interactions may be virtually eliminated by the rapid tumbling of the molecules with the result that sharp lines are observed in non-viscous liquids. The situation in Y Zeolite is complicated by the fact that some molecules may be adsorbed in the smaller sodalite cages (12) and are likely to be less mobile than

those in the supercages. In the latter situation one can expect
that some molecules near the "walls" of the cages will be less
mobile than those in the supercages near the center. These molecules will be surrounded by others of the same species in a
liquid-like environment. The interesting question of mobility of
these adsorbed molecules in Zeolites has been successfully studied
by NMR pulse techniques (2,18) and will not be discussed here.

  b. The effect of paramagnetic ions on the resonance of
adsorbed species (4). The nucleus is subject to dipolar broadening in the presence of electronic magnetic dipoles. However, due
to their large magnitude, the electronic dipoles associated with
transition metal ions, for example, may have significant effects
on the NMR of the neighboring nuclei. The effects are complicated
and depend on the details of the ion and its crystal field, the
ion-nuclear separation, the relative ion-nuclear motion, etc.
Recently, in this laboratory NMR studies have been made of various
adsorbed molecules in Y Zeolites which have been "doped" (exchanged) with selected Lanthanide ions. The object was to determine the feasibility of observing chemical shifts or line broadening in the NMR of the adsorbed species, which are due to the
presence of the paramagnetic Lanthanide ion. Such shifts and/or
broadening might be capable of interpretation in terms of the
structure of the surface complex and the exchange kinetics in the
supercages. Some ions which have been employed include Dy, Yb,
Ce, Eu, and Pr (all +3 ions). Some molecules which have been
studied as adsorbates include H_2O, CH_3OH, C_2H_5OH, $C_2H_5NH_2$, C_3H_8
and C_2H_4. The method and its possible applications will be
discussed along with the results of work currently in progress.

Literature Cited

1. Haneman, D., Resonance Methods, in "Characterization of Solid
    Surfaces," Kane, P.F. and Larabee, G.B., eds., pp. 337-377,
    Plenum Press, New York (1974).
2. Pfeifer, H., Nuclear Magnetic Resonance and Relaxation of
    Molecules Adsorbed on Solids, in "NMR-Basic Principles and
    Progress," Vol. 7, pp. 53-153, Springer, Berlin (1972).
3. O'Reilly, D.E., Magnetic Resonance in Catalytic Research,
    Adv. Catalysis (1959), 12, 31.
4. Abragam, A., "The Principles of Nuclear Magnetism," Clarendon
    Press, Oxford (1961).
5. Slichter, C.P., "Principles of Magnetic Resonance," Harper
    and Row, New York (1963).
6. Carrington, A., and McLachlan, "Introduction to Magnetic
    Resonance," Harper and Row, New York (1967).
7. Venuto, P.B., Advan. Chem. Ser. (1971), 102, 260.
8. Baur, W.H., Am. Mineralogist (1964), 49, 697.
9. Karger, J., Z. Phys. Chem. (Leipzig), (1971), 248, 27.
10. Sherry, H.S., Advan. Chem. Ser. (1971), 101, 350.

11. Eulenberger, G.R., Shoemaker, D.P. and Keil, J.G., J. Phys.
     Chem. (1967), 71, 1812.
12. Smith, J.V., Advan. Chem. Ser. (1971), 101, 171.
13. VanVleck, J.H., Phys. Rev. (1948), 74, 61.
14. Genser, E.E., J. Chem. Phys. (1971), 54, 4612.
15. Cohen, M.H. and Reif, F., Solid State Physics (1957), 5, 321.
16. Das, T.P. and Bersohn, R., Phys. Rev. (1956), 102, 733.
17. Bloembergen, N., Purcell, E.M. and Pound, R.V., Phys. Rev.
     (1948), 73, 27.
18. Resing, H.A., Adv. Mol. Relax. Proc. (1972), 3, 199.

28

NMR Investigations on the Behavior of Water in Starches

H. LECHERT and H. J. HENNIG

Institute of Physical Chemistry, University of Hamburg, Laufgraben 24,
2000 Hamburg 13 W. Germany

The application of NMR-methods on problems of the behaviour of water in biological material has its origin
in the very beginning of the use of this method for
the study of phenomena of molecular structure and mo-
bility(1).The starting point of our investigations on
starches of different origin had been the problem of
"bound" and "free" water at the starch which had been
studied in the literature with varying success and results using NMR(2,3)and a lot of other experimental
methods(4-8).The interest in these questions has its
origin in the significance of the knowlegde of the water bond in biological material at all and also in
problems of the technical application. The state of
swelling and the kind of water bonding is, for instance, an important quality determining factor of food
products and their behaviour in industrial processes
and storage. Especially show products with starches as
binding agents often an undesired separation of water
at freezing and subsequent thawing, a problem familiar
to every cook or housewife.Applying NMR on these problems,above all,the mobilities of the water molecules
are studied in the different sorption states reflecting different bond strengths in these states.As is
well known,the mobility of a nucleus,being incorporated in a moving molecule,is measured by the correlation time τ_c,giving the time interval,in which fixed
spatial relations of the nuclei can be expected.This
correlation time is connected with the relaxation times T1 and T2 of the nuclear magnetization,obtained by
a pulsed NMR-experiment,by the wellknown BPP-theory(9)
the results of which are presented in the Figure 1,
and need not to be discussed in detail.The value of
T1/T2=1.6 at the T1-minimum is characteristic for isotropic motion of the molecules and one correlation ti-
me.Any deviation from this value in a mixture contai-

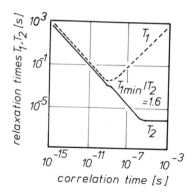

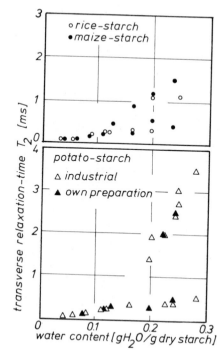

Figure 1. Dependence of the nuclear relaxation times T_1 and T_2 on the correlation time τ_c, according to the theory of Bloembergen, Purcell, and Pound (9)

Figure 2. Dependence of the transverse relaxation time T_2 on the content of water in different starches

ning,for instance,starch,points to a more or less
specific interaction of the water with the polymer
matrix.This interaction may cause a distribution of
correlation times,or also special kinds of reorientation of the water molecules,deviating from the isotropic motion.Under favourable conditions,from the magnetization decays,giving the relaxation times,more
than one exponential can be drawn,corresponding to two
ore more relaxation times belonging to regions with
different molecular mobility.Exchange effects between
these regions and the life times of the molecules in
one of them are described by the theory of Zimmerman
and Brittin(10)and Woessner and Zimmerman(11,12).
Now to the starch.Firstly,we have measured the transverse relaxation time T2 of the adsorbed water in dependence on the water content.The results for the
maize-and rice-starch are demonstrated in the upper
half of the Figure 2.The lower half contains the respective curves of potato starch.One sample has been
of technical origin,whereas the other has been isolated from potatoes under careful conditions.Up to about
15%of water only a single relaxation time occurs.Above
this water content two T2can be detected.The first appearance of the more mobile kind of water lies for all
starches near the water content,where by sorption experiments the socalled "capillary water" has been de-
tected(4,5).Because the effects for the potato starch
are more distinct than for the other starches,in the
following experiments preferably this starch has been
studied.Figure 3 shows the temperature behaviour of T2
for different water contents of the potato starch.
Striking in these diagrams is the decrease of the relaxation time of the component with the better mobility at higher temperatures.According to the theoretical
results of Woessner and Zimmerman(11,12)this behaviour
is characteristic for an exchange of protons between
regions of different mobility of the molecules.For a
slow exchange,the relaxation times as well as the probabilities of the occupation of the different regions
can be taken undisturbed from the magnetization decays
For exchange rates of the order of magnitude of the
relaxation rates the time constants of the decay and
the relative intensities depend in a quite complicated
way on the real relaxation times,the exchange rate and
the probabilities of occupation of the different re-
gions.For the conditions of our measurement these relations can be simplified to

$$\frac{1}{T_{2a}'} = \frac{1}{T_{2a}} + \frac{1}{\tau_a}$$

where $T2_a'$ is the relaxation time,taken from the decay and τ_a is the life time of the molecule in the state a.For the other state with the lower molecular mobility holds a respective relation.Sufficiently above the maximum, the real T2 of the more mobile protons is quite long and the corresponding relaxation rate can be neglected.The apparent relaxation rate,taken from the magnetization decay is then equal to the exchange rate given by the reciprocal life time of the molecules in the state a.To give an impression of the amounts of the life times the τ_a have been evaluated for the 20%sample.At $53^{\circ}C$ one observes $\tau_a=3.7ms$,at $93^{\circ}C$ $\tau_a=1.2ms$.For the sample with 33%water, a further kind of water can be observed which has a very good molecular mobility.This water proved to be freezable in calorimetric experiments,carried out comparatively. It can be assigned as "free" water in terms of the definition used by Kuprianoff(6).Figure 4 shows the rerults of the measurements of the longitudinal relaxation times for a sample of dehydrated starch and two samples with 20% water and 33% water respectively.The minimum of the protons of the starch framework is shifted out of the range of measurement of our apparatus towards higher temperatures.The course of the longitudinal magnetization shows only one exponential in the whole range corresponding to a uniform behaviour of T1 Looking at the life times of the molecules in the range of the better mobility, it can be seen that these times are distinctly shorter than the measured T1. Therefore,the exchange is slow against T2 and fast against T1,which means that in T1 only an averaged value can be observed from the NMR-measurements.The value of T1/T2 at the T1-minimum is quite large,indicating strong deviations of the mobility from the isotropic case or a broad distribution of correlation times.Looking at the dielectric relaxation,such a broad distribution can be observed, and seems to be plausible in such a complicated system.On the other hand,however,some samples of the potato starch of technical origin showed decay shapes of the transverse magnetization which could not be splitted into exponentials. One of these decays is shown in the Figure 5 in comparison to the respective wide line spectrum.It can be seen that the decay as well as the resonance line can be splitted into three components. A broad line is superimposed by a splitted line and a narrow component.The broad line is due to water molecules sorbed at the starch structure carrying out an anisotropic motion. In the decay function,correspondingly,a fast decay at the beginning is followed by

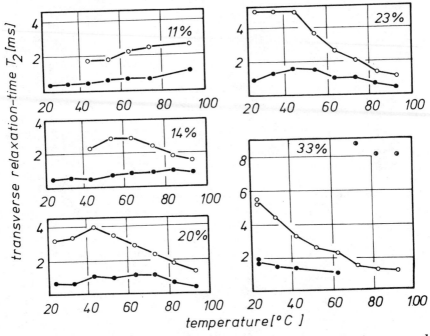

Figure 3. Temperature behavior of the transverse relaxation time T_2 of potato starch for different water contents

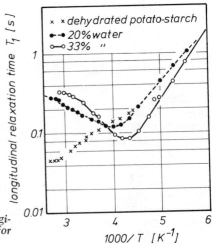

Figure 4. Temperature behavior of the longitudinal relaxation time T_1 of potato starch for different water contents

some kind of beat.The narrow component can be seen in
the decay function by the asymmetry of the beat with
respect to the base line.To have a measure for the a-
mount of the splitted component present in such a decay, it has proved to be practical, to define the difference between the minimum a and the maximum b as an
auxiliary parameter k, with respect to the total intensity which will be used in some tables later on.
The splitted component in the spectrum can be explained by residuals of the dipole dipole interaction of
the water protons left by an anisotropy of the motional process of the water molecules.The anisotropic
motion of water molecules on biological structures
like,for instance,membranes or fibres can be expected
to be a rather common process possibly responsible
for a lot of special properties and processes of these structures in the organism.Thorough studies of this
phenomenon may give deeper insights into the function
of water and also of the respective structure.In the
following experiments on potato starch shall be,there-
fore,demonstrated at least some considerations on the
detectability of this process of anisotropic motion
by NMR.For the special case of the molecular structure of water, proton resonance as well as deuteron resonance can be used to get information on the mentioned motional process.For the case of the proton-resonance water can be regarded as a system of two spins
coupled by the interaction of its magnetic dipoles.
This system adopts three energy levels in a magnetic
field, causing a splitted line, the components of
which have the distance

$$\Delta\omega = \frac{3\gamma^2 h}{r^3}(3\cos^2\Theta - 1)$$

where γ is the gyromagnetic ratio of the protons and
r their distance in the water molecule. Θ is the angle between the proton-proton-vector and the applied
magnetic field.The deuteron has spin 1 and its energy
is splitted into three levels,too,if a system with D_2O
molecules is brought into a magnetic field. The cause
of this splitting is given,however,by the interaction
of the quadrupole moment Q of the deuteron with the
electric field gradient q in the field of the oxygen
ion.The resulting line in the spectrum is again a
doublet,with a distance of its components given by an
expression quite similar to that of the dipole-dipole
interaction, where Θ is in this case defined as the
angle between the direction of the OH bond,which is
identical with the direction of the symmetry axis of

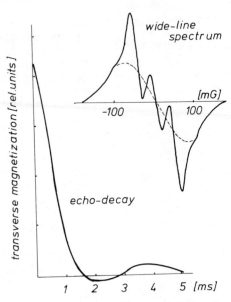

Figure 5. Comparison between the echo-decay and the wide-line spectrum of the Ca-form of a potato starch with 20% water content

the electric field gradient q, and the applied magnetic field.

$$\Delta\omega = \frac{3}{4} \frac{e^2 qQ}{h} (3\cos^2\Theta - 1)$$

For a reorientating water molecule in both expressions the angular term can be splitted into

$$(3\cos^2\Theta - 1) = (3\cos^2\Theta' - 1)(3\cos^2\Theta'' - 1)/2$$

where Θ' is for both types of interaction the angle between the applied magnetic field and the preferred axis of reorientation.It contains the structural information on the reorientation process.The other term, containing the angle Θ'' describes the time behaviour of the reorientating molecule.For powders or possibly fibres the first term has to be integrated over all angles Θ',the axis of reorientation can assume.For powders results from this integration the two-peak distribution,well known from the powder spectrum of the two spin system with fixed molecules.The second term has to be time averaged.The averaged value vanishes,if the motion is isotropic and can be generally taken as a measure of the anisotropy of the mobility.

This holds regardless of the slightly different definition of Θ'' in the two cases,which gives the angle between the proton-proton-vector and the preferred axis of reorientation for the proton resonance and the angle between the OH-vector and the mentioned axis in the case of the quadrupole interaction.If,now,an exchange of the water molecules occurs,the energy levels of a proton,left at the molecules are changed for the dipole-dipole-interaction,not,however,for the quadrupole interaction of the deuteron.The occurrence of the splitted component in the spectrum is,therefore,for the protons quite sensitive against exchange effects, which can be,on the other hand studied preferably by the proton resonance.The anisotropic motion itself is preferably studied with the deuteron resonance.Another limit of detection of an anisotropic motion by the discussed splitting is given,mostly at lower temperatures,if the mobility of the molecules in the state of anisotropic motion is too slow,so that the line widths of the components cover over the splitting. Table I shows the results of studies on potato starch samples,in which the phosphate groups have been neutralized with different cation.The amount of the splitted component present in the spectrum,is given by the parameter k,mentioned above.

Table I.Values for the auxiliary parameter k for different ions on the starch and different temperatures.

temperature $^\circ C$	Li	Na	Cs	Ca	Ba
21				3.2	0.0
16				7.4	3.5
11	1.8	1.0		11.1	5.9
1	6.2	2.8	0.4		7.0
-4		0.0	1.6		

pH=8.1 watercontent 20 %

It can be seen that the temperature range of the intensity of the splitted component depends strongly on the kind of cation present in the starch.An explanation of these effects seems to be possible by an assumption of a cation dependent exchange of entire water molecules(13),as well as by an exchange of the protons of the water molecules catalyzed by these ions. Effects of this kind have been observed by Hertz and Kluthe(14)in concentrated salt solutions.The amount of splitting measurering the anisotropy,is neither dependent on temperature nor on the kind of the cation in the limit of accuracy.This leads to the suggestion that the regions of anisotropic motion are not the cations.The intensity of the narrow component increases

with increasing cation radius,suggesting that this
component belongs to the water molecules at the cations,involved in the exchange process.An acid treated
sample showed no splitted component in the whole temperature range.Therefore,the dependence of the amount
of the splitted component on the pH of pretreatment
has been studied on a Na-starch.The results are demonstrated in the Table II.
Table II.Dependence of the auxiliary parameter k measurering the amount of splitted component of a Na-
starch at 1°C and a water content of 20% on the pH-
value.

pH	7.3	7.5	7.8	8.0	8.2	8.4	8.6	9.0
k	0.0	1.3	2.2	2.2	2.8	1.8	2.2	2.2

It can be seen that the auxiliary parameter k increases rather steeply up to pH = 8.2 and remains then
nearly constant.This result may explain that in a lot
of substances,where an anisotropic motion of adsorbed
water can be expected,no splitting of the proton resonance has been observed.According to the above discussion,the phenomenon of the anisotropic motion has
been studied on starch samples which have been deuterated by a treatment with heavy water.This treatment

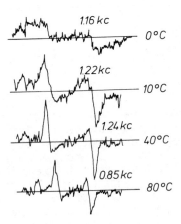

*Figure 6. Deuteron resonance
spectra of a deuterated potato
starch sample with a D_2O content of 24%. The amount of
splitting is given beside each
spectrum.*

leads to a complete exchange of the water in the starch,and the protons in the OH-groups,not,however,of the protons of the CH-groups in the glucose units.This fact will be used later on for quite detailed studies of the swelling mechanism.Figure 6 shows the resonance lines for the deuterons of a common potato starch with 24% D2O at various temperatures.The spectra show the expected splitting.The values for the amount of splitting are given beside each resonance line.In contrast to the proton resonance in the respective H2O containing sample no central narrow line can be observed in the deuteron resonance.This suggests that for the exchange process at the cations,mentioned above,the exchange of protons should be favoured against the exchange of entire water molecules.At lower temperatures the loss of mobility of the water molecules in the state of anisotropic motion can be seen in the broadening of the components of the deuteron spectrum.At temperatures above about $50^{\circ}C$ a slight decrease in the splitting can be observed.This decrease is observable up to temperatures above $100^{\circ}C$ where water distilles already from the sample.The loss of anisotropy,which corresponds to the decrease of the splitting may be possibly caused in a rearrangement of the starch structure,consisting in an irreversible expanding.This process reduces the degree of anisotropy and diminishs the amount of the ordered molecules, which can be also seen from the broadening of the components above $40^{\circ}C$.Another hint to that explanation is given by the fact that a sample which had been at $80^{\circ}C$ for some time shows a reduced value of the anisotropy cooling back to room temperature.Further it is a common experience from the technical use of the potato starch that at temperatures above $50^{\circ}C$ changes in structure can be obtained by treatment with steam or water.In the Figure 7 the respective spectra for a sample with a D2O-content of 39% are demonstrated.It can be seen that the splitting for a given temperature is smaller than for the sample with the lower water content,which has been principally observed already in the proton-resonance experiments.Further the decrease of the splitting with increasing temperature is distincly stronger.Obviously,the anisotropy of the reorientation is reduced with increasing water content,which may be due to a changed probability of the occupation of an ordered state of molecules(15) as well as in an extension of the starch structure,which is observed in the X-ray data.In this connection it should be mentioned that in contrast to the starch,in the cellulose no extension can be observed in the

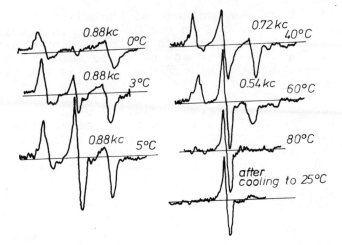

Figure 7. Deuteron spectra of a deuterated potato starch sample with a D_2O content of 39% at different temperatures. The amount of splitting is given beside each spectrum.

23 °C

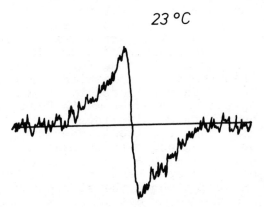

Figure 8. Deuteron resonance spectrum of a deuterated maize starch sample with a D_2O content of 30%

structure for water contents and correspondingly no
decrease in the anisotropy of the water mobility(17).
Effects comparable with those, observed in the starch,
have been measured in a row of papers in collagen fib-
res(16)which show both,the extension of the structure
and the decrease in anisotropy of the water mobility.
The most striking difference in the spectra of Figure
6 and Figure 7 is given by the occurrence of the narrow component in the middle between the components of
the splitted line.It can be seen that this line grows
in intensity between 0°C and 5°C, which is near the
melting point of the pure D2O at 4°C, and remains then
at constant intensity over a large range of temperatu-
res.This behaviour differs from that of the narrow
line in the proton resonance,mentioned above.Calorime-
tric experiments carried out parallel to the NMR-mea-
surement show that above about 30% H2O a small amount
of freezable water appears in the starch,which is
called "free" water according to the definition used
by Kuprianoff(6)as has been already mentioned.Obvious-
ly,the narrow line in the spectra of Figure 7 is due
to this "free" water.Proceeding with the temperature
above 80°C,the splitted line vanishes in the 39%D2O
sample and cannot be restored by cooling back to room
temperature.This indicates that even a small amount of
"free" water causes some kind of internal swelling,de-
stroying the structure responsible for the anisotropic
motion of the water. For a further investigation of
these effects a sample of maize starch with a content
of 30% D2O has been studied.As can be seen from the
Figure 8 in this starch no indication of a splitting
can be observed.Apart from the protein content of the
native maize-starch the crystal structure of the
starches are different.We have,therefore,isolated the
socalled B-type of the potato starch and measured the
deuteron resonance on a deuterated sample.No detectable splitting of the resonance could be observed.The
phenomenon of the anisotropic motion of the water molecules on the potato starch seems,therefore,to be a
peculiarity of the intact granule of the potato
starch.The quite sensitive reaction of the deuteron
spectrum offers a lot of possibilities of investigations on changes in the granular structure under the
influence of cooking or heat treatment in technical
processes.Finally,some experiments on possibilities
for studies of the swelling process shall be reported.
It has been mentioned already,that the OH-protons and
the protons of the water of the starch can be easily
replaced by deuterons without exchanging any of the
CH-protons.In the most interesting problems,concerning

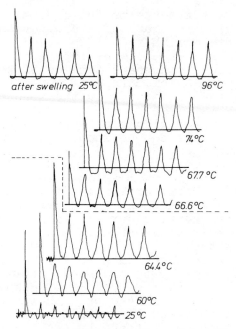

Figure 9. *Magnetization decays of the CH–protons of a deuterated potato starch in the course of swelling in D_2O. The picture in the upper left corner results after cooling the sample from 96°C to room temperature.*

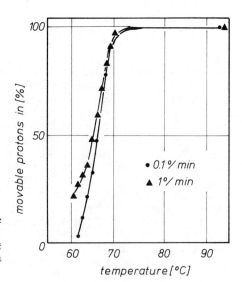

Figure 10. *Dependence of the amount of swollen potato starch in a mixture of 3% deuterated starch in D_2O for different heating rates between 55° and 60°C on temperature.*

swelling,water is present in a large excess.The influence of the swelling process on the protons of the water is,therefore,difficult to measure.Measurering, however,the proton resonance of the CH-protons of a starch deuterated in the described manner and swollen in D2O,the swelling can be studied on the high-polymer itself,because the T2 ofthe high-polymer in the swollen state is some orders of magnitude larger than that in the solid state.Figure 9 shows Carr-Purcell-Experiments at different temperatures of a mixture with only 3% potato starch.The portion of the mobile starch chains grows near the temperature of swelling with increasing rate until at 96°C only mobile starch chains are present in the mixture.Cooling back to room temperature and leaving overnight the signal of the solid is partly restored.This method seems to be extremely suited for studies of retrogradation and,for instance, freeze-thaw phenomena,the studies of the molecular processes of which are quite difficult.In the Figure 10 the course of the swelling process for two samples of a 3 % potato starch mixture is shown for different rates of heating in the temperature range below the swelling temperature.It can be seen that the sample heated more slowly,swells at a distinctly higher temperature,a behaviour which is also known from the technical behaviour of steam treated starch.At 70°C the swelling process is finished and all protons are in the movable state.In the Figure 11 some experiments

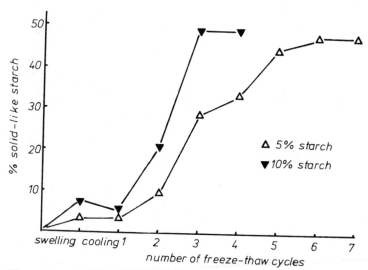

Figure 11. Dependence of the amounts of solid starch in two potato starch gels on the number of applied freeze–thaw cycles

are summarized describing the forced retrogadation by
subsequent freeze-thaw processes.For these experiments
two samples with starch contents of 5 and 10% had been
swollen and left at room temperature for about 24
hours.In this time,in the mixture with 10% starch 7%
of the total signal appears in the signal belonging
to the solid phase and in the sample with 5% starch
3%.The samples have then cooled to -20°C with a defined rate.After thawing again the amount of solid
was determined.Repeating this procedure some times
this amount of solid increases as is demonstrated in
the Figure 11.The reported experiments show,that NMR-
methods can give quite detailed information on molecular processes in colloid and interface systems,if it
is possible to get sufficiently defined substances,as
it is often the case in the range of biological material,and if it is further possible to prepare the
question on the system with sufficient clearness.
Wewant to thank the "Arbeitsgemeinschaft für Indus-
trielle Forschungsvereinigungen" and the "Forschungs-
kreis der Ernährungsindustrie" for the excellent support of our work.

Literature Cited

1. Shaw,T.,and Elsken,R.,J.Chem.Phys.(1953)$\underline{21}$ 565
2. Toledo,R.,Steinberg,M.P.,and Nelson,A.I.,
                     J.Food Sci. (1968)$\underline{33}$ 315
3. Sudhakar Shanbhag,Steinberg,M.P.,and Nelson,A.I.,
                     J.Food Sci. (1970)$\underline{35}$ 612
4. Schierbaum,F.,   Stärke-Starch
                                      (1960)$\underline{12}$ 257
5. Schierbaum,F.,and Täufel,K.,
                        Stärke-Starch
                                      (1962)$\underline{14}$ 234
6. Kuprianoff,J.,   Soc.Chem.Ind.(London),(1958)
7. Riedel,L.,Dechema-Monographien,Band 63,Verlag
                        Chemie Weinheim 1969
8. Duckworth,R.B.,   J.Food Technol.
                                      (1971) $\underline{6}$ 317
9. Bloembergen,N.,Purcell,E.M.,and Pound,R.V.,
                        Phys.Rev. (1948)$\underline{73}$ 679
10. Zimmerman,J.R., and Brittin,W.E.,
                        J.Phys.Chem.(1957)61 1328
11. Woessner,D.E.,   J.Chem.Phys.(1961)$\underline{35}$ 41
12. Woessner,D.E., and Zimmerman,J.R.,
                        J.Phys.Chem.(1963)$\underline{67}$ 1530
13. Eigen,M.,   Pure Appl.Chem.
                                      (1963) $\underline{6}$ 105
14. Hertz,H.G., and Klute,R.,
                        J.Phys.Chem.NF
                                      (1970)$\underline{69}$ 101

15. Chapman,G.E.,Danyluk,S.S.,and Lauchlan Mc,K.A.,
    Proc.Roy.Soc.London B
    (1971)178 465
16. Migchelsen,C., and Berendsen,H.J.C.,
    J.Chem.Phys.(1973)59 296
17. Dehl,R.E.,   J.Chem.Phys.(1968)48 831

29

NMR Relaxation in Polysaccharide Gels and Films

S. ABLETT and P. J. LILLFORD
Unilever Research Laboratories, Colworth House, Sharnbrook, Beds., U.K.

S. M. A. BAGHDADI and W. DERBYSHIRE
Department of Physics, University of Nottingham, Nottingham, U.K.

Introduction

The family of gels formed from the polysaccharides, agarose, kappa carrageenan and iota carrageenan provide an interesting and relatively simple series for investigation. The mechanism of gelation in this series is thought to involve the formation of junction zones, and perhaps even super junction zones, involving double alpha helices (1,2). Unlike the alginates the presence of divalent cations is not necessary for gelation. Each of the polysaccharides studied contains sequences in which residues of 3 linked β D galactose alternate with residues of 4 linked 36 anhydro α galactose. Whereas agarose possesses no sulphate groups and iota carrageenan has every residue sulphated, kappa carrageenan provides an intermediate example. The presence of charged sulphate groups would be expected to have a significant effect on many of the observable properties, influencing polysaccharide solubility in the aqueous sol, and the ease of association or aggregation to form a gel, and in turn the gel strength, it's stability to freeze-thaw cycles and possibly ageing effects.

Most previous NMR investigations have been concerned with the properties of the water in agarose gels (3-16). In general the relaxation rates of the water protons are considerably enhanced over bulk water values. Explanations have generally been based upon a crude two phase model, where water is considered to be either free, when it behaves as bulk water, or to be associated in some manner with the polysaccharide chains, when it is termed 'bound'. Exchange between bound and free phases is normally rapid on an NMR time scale, and averaged molecular properties are observed. Although the amount of bound water is limited, estimates ranging from 0.1 to 3.6 grams H_2O/gram. of agarose (9,17) an inhibition of molecular reorientation rate with a consequent enhancement of the bound water intrinsic relaxation rate causes a marked effect on a population weighted average relaxation rate. The molecular reorientation rate within the bound phase is directly related to the mobility of the polysaccharide to which the water molecule is bound, and thus the nuclear spin spin relaxation rate is a sensitive indicator of the state of gelation, changes on gelation typically being in excess of an order of magnitude. The corresponding change in spin lattice relaxation on gelation is in general much less marked. Changes in relaxation during thermal cycles which result in gelation and subsequent gel melting, demonstrate the large thermal hysteresis observed in any measurable parameter sensitive to the gelation state of the agarose, presumably indicating that a high degree of cooperativity is involved in it's gelation and gel melting. The prediction of the simple two phase model of a linear relationship between relaxation rate and polysaccharide concentration is confirmed for both spin lattice and spin spin relaxation, but for the T_2 relaxation the intercept at zero concentration is not the predicted spin spin relaxation rate of bulk water and it's temperature dependence is incorrect (15). At all gel concentrations studied the ^1H spin spin relaxation rate has a maximum value in the neighbourhood of 300 K. This is attributed to the onset of a rapid exchange with a more rigid species, which consequently has a greater relaxation rate. This, as will be demonstrated later is an additional component to the bound water in which the water protons are less mobile than in the bulk. As temperature is increased, and the rapid exchange condition is more nearly satisfied, this additional phase makes a greater contribution to the averaged relaxation rate, thereby increasing it, and offsetting the negative temperature dependence normally observed. Above 300 K exchange is rapid and the usual negative temperature dependence is resumed. The temperature dependence is positive under conditions of intermediate exchange. On cooling agarose gels freezing occurs before the slow exchange condition can be attained. The validity of an explanation of the minimum in the observed T_2 value in terms of the model outlined above, involving an increasingly effective exchange process with a more rigid proton species, instead of in terms of a model invoking a temperature dependent conformational change of the polysaccharide, is demonstrated by the observation that the temperature of the minimum in the proton T_2 value is dependent upon the isotopic composition in a suite of samples of uniform water content, but variable H_2O/D_2O ratio (9). Additional confirmation is provided by the absence of a T_2 minimum in the 1% relaxation and a shift in the temperature of the minimum in the ^2H T_2 value from 300 K. The shifts in the temperatures of the T_2 minima are of a magnitude and in a direction consistent with the theory. There is some speculation on the nature of the more rigid species, suggestions have been made that the species comprises very tightly bound water molecules (8) or hydroxyl groups on the agarose molecules (6).

The two phase model is certainly an oversimplification. If the bound water phase is identified with a component that remains unfrozen below the bulk freezing point, it is necessary to invoke a distribution of correlation frequencies to describe the motions responsible for relaxation within this non freezing phase. The distribution necessarily involves a significant contribution from low frequency motions. These might occur as a consequence of anisotropic motion, the component of motion about one molecular axis being slow compared with that about another, or alternatively a very tightly bound water species might be involved similar to that reported from experiments on water adsorption on some celluloses (18,19). If this latter interpretation is correct this is presumably not the component responsible for the exchange T_2 minimum observed at 300 K, because that exchanging species would be in a slow exchange condition at temperatures below the freezing point.

Freezing and subsequent thawing causes agarose gels to synerese and the water proton relaxation to become non exponential, with the relaxation rate of the longer component becoming more similar to that of bulk water.

This paper is concerned with an extension of the measurements on agarose gels to water adsorbed on agarose films in order to further test the validity and limitations of the two phase model, and to attempt to clarify some of the outstanding problems, for example the nature of the species responsible for the exchange T_2 minimum at 300°C, the anomalous intercept in the $1/T_2$ concentration plots, and to seek some explanation for the wide diversity in estimated hydration values. It might be hoped that any successful model would have a wider applicability, and extension of the studies from gels to films provide an opportunity of developing such a model. The extension to kappa and iota carrageenan systems permits a study of the effects of a progressive introduction of sulphate groups on the polysaccharide.

Comparison of the gelation of agarose, kappa carrageenan and iota carrageenan.

On being heated at temperatures in excess of 373 K and subsequently cooled to 293 kappa and iota carrageenan, like agarose, readily formed gels. Within a given sample batch, data was reproducible provided that these two temperatures were enclosed in any thermal cycle. Although small variations in detailed behaviour were observed in samples from different batches, the relaxation behaviour did not seem to depend upon mechanical treatment, for example the relaxation data on iota gels were similar in two samples, one of which had been forced many times through a 5μ filter. In some respects the ^{1}H relaxation behaviour was similar to that observed in agarose gels, relaxation rates were enhanced over bulk water values, the spin spin relaxation rate exceeding the spin lattice rate by a considerable factor. The relaxation rate enhancements for both gel and sol had the linear concentration dependence expected on the basis of the two

phase model. These dependences can be related to the number of
water molecules bound to each disaccharide repeat unit on the
assumption of a rapid exchange between bulk and bound phases.

$$\frac{n}{T} = \frac{n_a}{T_a} + \frac{n_b}{T_b} = \frac{n}{T_a} + n_b \left[\frac{1}{T_b} - \frac{1}{T_c}\right]$$

where n is the total number of
water molecules in the sample, n_a the number in the bulk and n_b
the number in the bound phase. T is the observed relaxation time
and Ta and Tb the intrinsic relaxation times in the bulk and bound
phases. If the concentration of solute = c grams of polysaccharide
/gram of water, and Nb is the number of bound water molecules per
solute unit

$$\frac{1}{Ma} \left[\frac{1}{T} - \frac{1}{Ta}\right] = \frac{Nb.c}{Ma} \left[\frac{1}{Tb} - \frac{1}{Ta}\right]$$

Ma and Ms are the molecular weights of water and the disaccharide
repeat unit. If the assumption is made that Tb is independent of
polysaccharide type comparison of the gradients of the concentration plots of relaxation rate allow relative values of Nb to be
determined. To avoid any possible complications occurring as a
consequence of additional contributions to spin-spin relaxation
as a result of the onset of an exchange process, this procedure
was only adopted for spin lattice relaxation with results shown
in table 1.

Table 1 Estimated Relative Values of Nb (the number of bound water
molecules/disaccharide repeat unit).

	Agarose	Kappa Carrageenan	Iota Carrageenan
[2]H relaxation in the sol	4.4	2.4	2.0
[2]H relaxation in the gel	4.8	3.1	2.0
[1]H relaxation in the sol	2.5	2.7	2.0
[1]H relaxation in the gel	7.0	2.9	2.0
No. of OH groups per disaccharide repeat unit	4	3	2

The assumption of a uniform T_{1b} in the series agarose, kappa
and iota carrageenan is obviously a gross one, although the
relaxation behaviour of the non freezing component in a 10% gel of
iota carrageenan is very similar to that reported in a 10% gel of
agarose (16). In addition it is known that the bound phase is
complex, and hence the consistency of these results and the similarity with the number of hydroxyl groups per disaccharide unit
is encouraging.

As in agarose a minimum was observed in the T_2 value for the kappa carrageenan gels, but at a somewhat higher temperature 323K instead of 303K, whereas that for the iota carrageenan gels was lower at 293K. As in agarose the temperatures of the T_2 minima were independent of gel concentration, although on some occasions some variability between batches was observed, which might reflect slight differences in material, or some undetected difference in procedure. If the observation of a T_2 minimum is explained in terms of the onset of a rapid exchange with a more rigid species, then at the minimum the intrinsic relaxation time of that species is comparable to the lifetime of the exchangeable proton in that species. A shift in the temperature of the minimum in kappa carrageenan to a higher value implies that either the intrinsic relaxation time is less or that the exchange is slower than in agarose. In iota carrageenan a shift of the minimum to a lower temperature implies the opposite.

The magnitude of the thermal hysteresis observed in a gel-sol cycle decreased in the sequence agarose,- kappa carrageenan ,- iota carrageenan. If this behaviour is attributed to the increasing presence of charged sulphate groups whose electrostatic interactions modify aggregation behaviour, it might be expected that an increase in the ionic strength of the solvent would, by itself modifying the electrostatic interactions between the sulphate groups, cause changes in the gelation behaviour. Incorporation of potassium chloride into kappa carrageenan gels increases the size of the thermal hysteresis observed in a sol-gel cycle. The other noteworthy feature is that the exchange minimum became much more pronounced in a gel containing potassium chloride.

Freeze-Thaw Stability

When an agarose gel was frozen and subsequently thawed the ^1H spin-spin relaxation became non exponential and the gel exhibited syneresis, whereas the ^1H solvent relaxation of a gel of iota carrageenan remained single component and no syneresis was observed. The values of the relaxation rates and the temperature of the T_2 minimum were reproducible over a number of freeze-thaw cycles for the iota gels. Again the presence of charged sulphate groups would seem to be implicated in the mechanism for the reduction or prevention of freeze-thaw damage. It might be expected that electrostatic repulsion between the charged sulphate groups on the surfaces of the helices of iota carrageenan either prevents their aggregation on freezing, or their disruption when the gel thaws. In contrast the extensive aggregation of the uncharged agarose on freezing is not reversed on thawing. Non exponential relaxation might arise from water attached to or trapped by the aggregates, and water in the interstices which is essentially free. The physical separation of these regions must be sufficiently large for diffusion between them to be slow on an NMR timescale to explain the observation of non exponential relaxation.

Agarose Gels and Films.

Below the bulk freezing point the ^1H relaxation of the non-freezing water in an agarose gel was indistinguishable from that in a sample of dry agarose to which an equivalent amount of water had been added, 0.59 grams H_2O/gram of agarose (15), implying that the molecular properties of the bound water were independent of any differences in the conformation and packing of the polysaccharide in the frozen gel and wet powder. In both situations the polysaccharide molecules were relatively immobile, whereas in the sol the agarose molecules exhibited a considerable mobility which modulated the low frequency components of the bound water molecules modifying their intrinsic spin spin relaxation rates. Thus the use of water adsorbed onto a dry powder to an extent equivalent to the non freezing component in the gel offers a method of extending direct observations on the bound phase to temperatures above 273K, without the complication of an intervention by a bulk phase. When such studies were undertaken both a maximum and a minimum were observed in the spin-spin relaxation time. This is consistent with the behaviour of a system where a rapidly relaxing phase is passing from a slow, through an intermediate to a fast exchange regime, with a more abundant slower relaxing component (20). The observation of a single component relaxation at the lower temperatures indicates that the relaxation rate of the rapidly relaxing phase is possibly unsuitably large for the instrumental settings, or that it is insufficiently different from the bulk rate, or that it's population is too small for direct observation. The observation of relaxation over the complete range from slow, through intermediate to fast exchange does confirm the assignment of the T_2 minimum to an exchange process. However, as the exchange behaviour was observed in agarose films containing no freezable, that is bulk water, the exchange process is not between free and bound water and occurs within the bound water-polysaccharide system.

Labelling of the minimum of four separate phases now involved can be a problem. In general the most mobile phase is free water, which has intrinsic relaxation rates $1/T_{1a}$ and $1/T_{2a}$, equivalent to those of bulk water. The fractional population P_a is a function of the water content in a gel or a sol. This water is considered to behave as bulk water, and to freeze to bulk ice on reduction of temperature below 273K. On freezing the T_2 value is reduced from seconds to microseconds, exchange with the other phases becomes slow, and the spectrometers are usually operated in such a manner that this phase is not detected. This phase is largely absent in the hydrated films. The second phase, labelled bound, is identified as that component which remains unfrozen. Whilst it is necessary to invoke a distribution of correlation times, and therefore of environments to explain the frequency and temperature dependences of the relaxation rates,

the relaxation behaviour is often simple exponential and averaged
relaxation rates $1/T_{1b}$ and $1/T_{2b}$ are observed, implying that
exchange between the separate environments is rapid. Exchange
between the a and b phases is considered to be rapid at all
temperatures above, and to be slow at all temperatures below the
bulk freezing point. The third phase whose exchange with the
previous two components was deemed responsible for the maximum
and minimum T_2 values is an evenly more rapidly relaxing species
with intrinsic rates $1/T_{1c}$ and $1/T_{2c}$. As mentioned in the introduction the nature of this phase, hydroxyl protons or very tightly
bound water is still in doubt. The fourth phase, d, comprises
protons in the polysaccharide molecule. The location of the
hydroxyl protons is uncertain, phases b, c and d being possible
candidates.

In order to provide further information on the properties of
the bound phase, and on the nature of the exchanging species,
films of agarose, hydrated to different extents, were prepared.
The relaxation of the most moist film, which had a water content
corresponding to the non freezing component, had a relaxation
behaviour identical, within experimental error, to that reported
for the non freezing fraction in an agarose gel (16). The drier
films exhibited higher relaxation rates (figure 1)

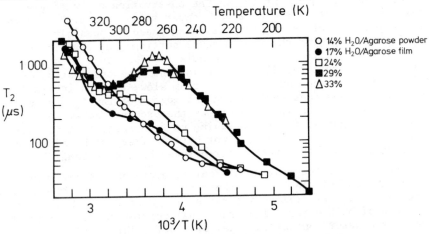

Figure 1. Proton T_2 in hydrated agarose films

consistent with the preferential removal on dehydration, of the
more mobile constituents of the distribution of the bound water
phase, instead of an equivalent reduction over the complete range.
The driest sample used, with a 14% water content, was a powder
instead of a film. It's relaxation, whilst being of the correct
magnitude and having an appropriate temperature dependence, was
not completely consistent with that of the films, suggesting that
the bound water relaxation was slightly different in the powders

and films. The exchange T_2 maxima and minima were less marked in the drier films, being barely observable in the 17% film and undetectable in the 14% powder. This suggests that the additional species whose exchange is passing from the slow to the fast regime, between 270 and 303K has a motional rate comparable to a film with a water content the order of 15% which corresponds to 3 water molecules/disaccharide residue. This number is very similar to the estimate of the amount of water that is very tightly bound to cellulose and does not contribute to the narrow water signal superposed on a broad cellulose signal (18, 19). This also corresponds to the more rigid region of the distribution investigated in the frozen gels. A film with such a water content would not show a T_2 maximum and minimum as there would be no, or very little, difference in the observed relaxation rate whatever the exchange rate. The values of the observed spin spin relaxation rates are in good agreement with the cw linewidths reported by Aizawa et al (3).

At the higher temperatures in the rapid exchange condition, the observed relaxation rates were independent of water content, indicating that the relaxation was dominated by the component, whether hydroxyl protons or very tightly bound water molecules whose exchange with the adsorbed water phase was responsible for the exchange T_2 maxima and minima. At temperatures below the intermediate exchange region the spin spin relaxation was, as described earlier, dependent upon water content in a manner consistent with the adsorbed water having a broad distribution of motional rates and the more mobile species being removed preferentially.

These observations in the film have not provided any information on the identity of the very tightly bound species, the c phase. Agarose gels prepared at a number of pH values exhibited similar relaxation behaviour within experimental error, for example all exhibited a minimum in the T_2 value at 303K irrespective of pH value (figure 2).

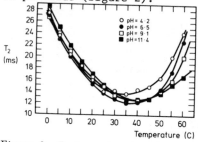

The observation of a pH independent exchange rate would seem to indicate that the process does not involve the exchange of hydrogen atoms between hydroxyl groups and water molecules, or between water molecules, that is unless this exchange were not the determining step. For example it might be speculated that the hydroxyl hydrogen atoms are involved in H bonds in maintain-

Figure 2. Proton spin-spin relaxation in 10% agarose gels

ing a polysaccharide helical structure, and that exchange of the hydroxyl hydrogen might well involve a local rupture of the helical structure. However, this behaviour is not observed in aqueous solutions of poly L lysine and poly L glutamic acid, where exchange of the peptide NH proton is pH dependent (21). The other explanation entails the exchange of complete water molecules, the number of very tightly bound water molecules constituting the c phase inevitably being the order of, or less than 10% by weight, that is two or fewer water molecules per residue. Some careful adjustment of relative amplitudes and relaxation rates would be required to explain the apparent single component relaxations observed in the films under slow exchange conditions. At the high water contents the relative c population is low, whereas at the lower water contents the b and c relaxation rates become more nearly equivalent rendering resolution difficult. In principle ^{17}O relaxation can be used to distinguish between the exchange of protons and intact water molecules. A 5% agarose gel was prepared in $H_2^{17}O$ (22), the T_2 relaxation time had the normal positive temperature dependence showing no evidence of intermediate exchange. Unfortunately this does not preclude the exchange of intact water molecules as conditions are such that the system would be in the slow exchange regime for ^{17}O relaxation, the observed relaxation time is the order of 1 ms which taken with a bulk phase relaxation time of 10 ms predicts a b phase relaxation of 30μs. If water molecules were involved in the c phase their relaxation times would be even shorter, and hence very much smaller than the values of τ_c necessary to fit the ^1H and ^2H data in the vicinity of the T_2 minima.

Woessner and Snowden (9) and Andrasko (13) have varied a further parameter, the isotopic composition. Woessner and Snowden assumed two contributions to the ^1H intrinsic relaxation rates, β from interactions with protons capable of being exchanged with deuterium in D_2O, and α from all other sources. By varying the isotopic composition the term β could be scaled by a factor W. Clearly the model adopted earlier in this paper, involving four proton fractions, bulk, bound, very tightly bound and non exchangeable, can be tested using the Woessner and Snowden published isotopic dilution data.

$$\frac{1}{T} = \frac{Pa}{Ta} + \frac{Pb}{Tb} + \frac{Pc}{(1-Pc)(Tc+\tau c)}$$

It is assumed that the non exchangeable protons are not observed, but that they may make contributions to the relaxation of the non exchangeable protons. Woessner and Snowden considered only one gel concentration

$$\frac{1}{T} = \frac{Pa*}{Ta*} + \frac{Pc}{(1-Pc)(Tc+\tau c)}$$

x is the fraction of exchangeable protons remaining as protons and L is the factor representing the reduction in the ^1H relaxation as a result of a dipolar coupling to a deuteron instead of to another proton

$$\frac{1}{T} = \alpha a + \beta aW + \frac{Pc}{(1-Pc)\left[\dfrac{1}{\alpha c+\beta cW} + \tau c\right]}$$

Following the customary procedure three temperature regions may be distinguished

(i) High where $\dfrac{1}{\alpha c+\beta cW} \gg \tau c$

$$\frac{1}{T} = \alpha a + \beta aW + \frac{Pc(\alpha c+\beta cW)}{1-Pc}$$

$$= A + BW$$

where $A = \alpha a + \dfrac{Pc\ \alpha c}{1-Pc}$

and $B = \beta a + \dfrac{Pc\ \beta c}{1-Pc}$

(ii) Intermediate where $\tau c \gg \dfrac{1}{\alpha c+\beta cW}$

$$\frac{1}{T} = \alpha a+\beta aW + \frac{Pc}{(1-Pc)\tau c} = A + BW$$

where $A = \alpha a + \dfrac{Pc}{(1-Pc)\tau c}$

and $B = \beta a$

(iii) Low where $\alpha a + \beta a\ W \gg \dfrac{Pc}{(1-Pc)\tau c}$

$$\frac{1}{T} = \alpha a + \beta aW \qquad \text{where}\quad A = \alpha a$$

$$B = \beta a$$

Within the three temperature regions the dependence of relaxation rate upon W is predicted to be linear, but linearity would not be expected to occur at the junctions between regions. The separate α and β rates decrease with increasing temperature, whereas τc increases. Because of the onset of freezing Woessner and Snowden did not report measurements in the low temperature regime. However, with the exception of the lowest temperature studied, 283K graphs of $\dfrac{1}{T2}$ against W were linear within experimental error.

At 283K there was a small curvature corresponding to a decrease in gradient at the higher W values. This is the appropriate direction but it is somewhat surprising that departures from linearity occured at 283K and not at the higher temperatures, at the transition between the intermediate and high temperature regimes.

The similar temperature dependences of the gradient and intercept indicate that the rates α_c and β_c have similar activation energies. It is to be noted that the T_2 minima observed by a number of workers are consistently asymmetric implying that the activation energies for Tc and τc are, perhaps not surprisingly different.

As in the Woessner-Snowden analysis the type of analysis presented here provides information on the quantities $Pc\tau_c$, $Pc\alpha_c$ and $Pc\beta_c$. The data on the films suggests that the c phase corresponds to six or fewer protons per disaccharide unit. At a gel concentration of 7.85% agar/gram of water this corresponds to a Pc value $\approx$ 1% giving α_c and β_c values of $2 \times 10^3 s^{-1}$ and $3 \times 10^3 s^{-1}$ at 318K, corresponding to a ^1H relaxation time of 200μs, comparable to that recorded directly from the films.

In summary the fit is encouraging, the data reported in this and our previous papers is consistent with that reported by other workers. Our estimate of Pc corresponding to 0.1 gram H_2O/gram of agarose is identical to that estimate of the bound phase by Woessner and Snowden using a different series of assumptions.

Iota carrageenan gels and films.

With the previously mentioned exception that gels of iota carrageenan did not exhibit a thermal hysteresis in ^1H relaxation behaviour on undergoing a sol-gel cycle, the general ^1H relaxation behaviour in iota carraggenan gels was similar to that of the agarose gels. Both the spin lattice and spin spin relaxation rates were linearly dependent upon concentration, but unlike the agarose gels the intercepts of the graphs of relaxation rate against concentration were similar to the relaxation rates of bulk water for both spin lattice and spin spin relaxation instead of being an order of magnitude too large for spin spin relaxation. Furthermore, the temperature dependences were of the appropriate form. As in the agarose gels the spin spin relaxation rate had a stronger concentration dependence, but for both spin lattice and spin spin relaxation the gradients of the concentration plots were reduced from the agarose values to values more characteristic of sols, for the spin spin relaxation the reduction was in excess of an order of magnitude. Two explanations might be sought, a reduction in the amount of bound water in the iota carrageenan gels, or alternatively a reduction in the relaxation rates of the bound water. Again the bound water is identified as that remaining unfrozen below the bulk freezing point. Like the agarose gels the amount of non freezing water is almost independent of temperature below the freezing point, and is proportional to the concentration of iota carrageenan over a concentration range from 0.02 to 1.6 grams of iota carrageenan per gram of water, the lower values corresponding to gels and the higher ones to films. Notwithstanding the wide concentration range and the different physical state of the system the amount of non freezing, that is bound water, was constant at 0.43±0.03 grams of water/gram of iota carrageenan. This is less than the 0.59 value obtained for agarose but not sufficiently so to explain the large reductions observed in the concentration dependences of the observed relaxation rates. This corresponds to 200 grams of bound water per disaccharide repeat unit, which is similar to the 180.5 grams determined for agarose.

In a previous section the concentration dependences of the spin lattice relaxation rates of sols and gels of the three polysaccharides were used to derive relative values of $N_b \left[\dfrac{1}{T_b} - \dfrac{1}{T_a} \right]$ where N_b was the number of bound water molecules per disaccharide residue. If it was assumed that the relaxation rate of the bound phase was independent of the polysaccharide the relative values of N_b were in most cases seemingly related to the number of hydroxyl groups suggesting that these provided the primary binding sites. In fact the comparibility of the estimates of 200 and 180.5 grams of bound water per disaccharide unit in agarose and iota carrageenan gels, that is 11 and 10 water molecules/residue, suggest that whilst encouraging the analysis was invalid and that the similarity of the apparent N_b values to the ratio of the number of hydroxyl groups was fortuitous.

The spin lattice and spin spin relaxation behaviours of the non freezing component are almost identical to those reported in agarose at the same frequency (16), and correspond to the bound phase possessing a complex distribution of correlation frequencies. A determination of the complete distribution of spectral frequencies would strictly necessitate measurements of T_1 as a function of frequency and $T_{1\rho}$ as a function of H_1 field value over a range of temperatures. Because of the similarity in relaxation behaviour to the non freezing component in agarose gel this has not been undertaken, it might be expected that similar distributions might occur.

Extrapolations of this relaxation behaviour to temperatures above the freezing point together with an assumption that the bound phase is represented by 0.43 grams of water/gram of iota carrageenan allows the concentration dependences of the two relaxation rates to be calculated. The results are shown in figure 3, for clarity only the 278K results are displayed.

Whilst the fit to the spin lattice relaxation data was not particularly good, that to the spin spin relaxation was excellent considering the errors inherent in any extrapolation.

Thus we have estimated that the amount of bound water expressed in molecules of water per disaccharide unit is similar in gels of agarose and iota carrageenan. We have also determined that the relaxation rates of the bound phase are similar in the

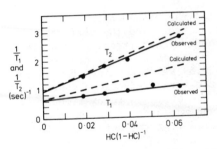

Figure 3. Concentration dependence of relaxation in iota carrageenan gels at 5°C

two systems and yet the observed concentration dependences of relaxation rate in the agarose gels exceed those in iota carrageenan. A partial explanation is that the molecular weight of iota carrageenan is considerably greater, 464 instead of 306. A second factor is that the model was only partly successful in each case in predicting relaxation rates above 273K. In agarose the spin lattice relaxation was predictable but the observed spin spin relaxation rates were an order of magnitude larger than expected, in iota carrageenan the spin spin relaxation rates were predictable and the spin lattice relaxation rates were lower than expected.

In an attempt to clarify the situation measurements were made on hydrated films of iota carrageenan. A film having a moisture content of 37.9% equivalent to 15.7 H_2O molecules/disaccharide residue, exhibited a 26% reduction in signal intensity on freezing, corresponding to a non-freezing component of 0.45 grams of water/gram of agarose. The T_1 and T_2 values exhibited a small discontinuity on freezing. Analysis below 273K was difficult, because although the observed spin spin relaxation was single component, spin lattice relaxation was non exponential, and although resolved into two components, the relative amplitudes varied with temperature. With this exception, the general behaviour was similar to that observed in the non freezing fraction of the iota gels. Like the T_1 relaxation on the non freezing fraction of the gel, the long T_1 component passed through a minimum, it was of similar value and occurred at a similar temperature. The ratio of the T_1 to T_2 at the T_1 minimum temperature was similar as was the dependence of spin spin relaxation upon temperature. These similarities are again strongly indicative that the relaxing species in the non freezing component of the gels is similar to that occurring in the films.

A film with a reduced water content, 15.8%, 4.8 molecules/residue, showed no discontinuities in relaxation on passing through 273K. The relaxation times were shorter than in the wet film, corresponding again to a selective removal of the more mobile fraction of the water molecules in the bound phase on dehydration. Here analysis was further complicated by an observed reduction in signal intensity with decreasing temperature. As in the film with higher water content spin lattice relaxation was non exponential.

Above 273K the T_2 value of the wet film decreased with increasing temperature consistent with the onset of an exchange process similar to that observed in agarose and in the iota gel, but unlike these other examples no minimum in T_2 was observed, the T_2 value decreased slowly with increasing temperature and the spin spin relaxation became non exponential at temperatures greater than 333K. The film also showed a thermal hysteresis, the T_2 relaxation times were lower on the cooling part of the cycle, and the non exponentiality persisted down to 323K. In the dry film the T_2 values were of the order of 1 ms instead of the 10 ms recorded for the wet film. The T_2 values increased with increasing temperature, but always remained below the wet film values. Hence, as in the drier films of agarose, and unlike the wetter films and the gels of agarose and iota carrageenan no evidence of a T_2 exchange minimum was observed. Again a thermal hysteresis and non exponentiality were observed in the relaxation behaviour, but in the dry film this continued down to 290K and persisted over a period of several days.

The obvious interpretation of the occurrence of thermal hysteresis and non exponentiality in the relaxation behaviour is that some change, not necessarily permanent, occurred in the water binding capacity of the films at the higher temperatures. A limited number of rate studies were undertaken, for example in a dry film left for two hours at 323K, the T_2 value decreased from 2.86 to 2.42 ms, whereas the moist film relaxation time decreased over a period of 18 hours at 328K from 8.3 ms becoming non exponential with component times of 6.6 and 3.1 ms. Decreases in relaxation times such as these occurring over a period of hours will have affected the values of relaxation times recorded as a function of temperature, they would also explain the occurrence of thermal hysteresis. In a moist film suddenly placed at 313K, the T_2 value decreased from 8.3 to 6.0 ms in the first 100 minutes. The 8.3 ms value is greater than T_2 values recorded at lower temperatures in the normal measurements and suggests that but for this additional effect a normal T_2 minimum corresponding to the completion of an intermediate exchange process, would have been observed, although it would appear that the T_2 minimum in the film would have been more shallow than in the gel.

A film was prepared containing D_2O instead of H_2O at a sufficiently low concentration that freezing did not occur. Both of the observed ^2H relaxations were, within experimental error, single component exponential, with T_1 independent of temperature at 2.2 ms and T_2 changing smoothly from 60 μs at 263K to 130 μs at 293K.

Conclusions

It would appear that the model described in this paper is adequate to explain the gross features of the water NMR relaxation in gels and hydrated films of the family of polysaccharides including agarose, kappa carrageenan and iota carrageenan, instead of merely the agarose gels considered previously. The model involves three proton phases, bulk water, bound water and a very tightly bound species. The non-exchangeable polysaccharide protons are not considered to be involved directly. Identification of the bound water with the non freezing component predicts a bound water component of ten or eleven water molecules per disaccharide residue, almost irrespective of polysaccharide species. Hydrated films of polysaccharide containing the same water content had identical relaxation behaviour to the non freezing component, and offered a method of extending direct measurements on the bound species to temperatures greater than 273K without the intervention of a bulk water phase. Extrapolation of the relaxation behaviour of the non freezing components to temperatures above 273K yields calculated relaxation rates consistent with observed values for spin lattice relaxation in agarose gels and spin spin relaxation in iota carrageenan gels, but incorrect values for spin-spin relaxation in agarose gels and spin lattice relaxation in iota carrageenan, the discrepancies being in opposite directions. No explanation has been found and these observations probably indicate a limitation in the somewhat naive model. The nature of the very tightly bound species whose exchange with the bulk and bound water changes from slow to rapid as temperature is increased is still uncertain, experiments intended to distinguish between alternative suggestions of hydroxyl groups and very tightly bound water molecules were unfortunately not unambiguous. These experiments involved studies of ^{17}O relaxation and the pH dependence of the exchange T_2 minimum. However, experiments on films indicated that the very tightly bound phase, if water, corresponded to 0.1 grams of water/gram of agarose, or alternatively would be consistent with the number of hydroxyl groups. It is suggested that this very tightly bound phase is that identified by Woessner and Snowden as bound water, and using a different method of analysis also given a concentration of 0.1 grams of water/gram of agarose. The model and data presented here are consistent with the isotropic dilution studies of Woessner and Snowden.

References

1. D.A. Rees, Adv. Carbohyd. Chem. Biochem. 24, 267, (1969).
2. D.A. Rees, Chem. and Ind. 630, (1972).
3. M. Aizawa, J. Mizuguchi, S. Suzuki, S. Hayashi, T. Suzuki, N. Mitomo and H. Toyama; Bull Chem. Soc. Japan 45, 3031, (1972).
4. M. Aizawa, S. Suzuki, T. Suzuki and H. Toyama; Bull. Chem. Soc. Japan 46, 116 (1973).
5. J. Clifford and T.F. Child; Proc. 1st. Eur. Biophys. Congr., 461 (1971).
6. T.F. Child and N.G. Pryce; Biopolymers 11, 409, (1972).
7. T.F. Child, N.G. Pryce, M.J. Tait and S. Ablett; Chem. Comm. 1214 (1970).
8. D.E. Woessner, B.S. Snowden Jr. and Y.E. Chiu; J. Coll. and Inter. Sci. 34, 283, (1970).
9. D.E. Woessner and B. Snowden Jr.; J. Coll. and Inter Sci. 34, 290, (1970).
10. O. Hechter, T. Wittstruck, N. McNiven and C. Lester; Proc. Nat. Acad. Sci. 46, 783, (1960).
11. G. Sterling and M. Masuzawa; Makromol. Chem. 116, 140, (1968).
12. R.K. Outhred and E.P. George; Biophys. J. 13, 83, (1973).
13. J. Andrasko; Biophys. J. 15, 1235, (1975).
14. J. Andrasko; J. Mag. Res. 16, 502, (1974).
15. W. Derbyshire and I.D. Duff; Disc. Far. Soc. 57, 243, (1974).
16. I.D. Duff and W. Derbyshire; J. Mag. Res. 17, 89, (1975).
17. A.G. Langdon and H.C. Thomas; J.Phys. Chem. 75, 1821 (1971).
18. E. Forslind; NMR Basic Principles and Progress 4, 145, (1971).
19. M.L. Froix and R. Nelson; Macromolecules 8, 726, (1975).
20. M.A. Resing; Adv. Mol. Relax. Proc. 1, 109, (1967).
21. S. Capelin and W. Derbyshire, unpublished results.
22. S. Ablett and P.J. Lillford, unpublished results.

30

Ions of Quadrupolar Nuclei as NMR Probes of Charged Surfaces

ROBBE C. LYON, JÜRGEN WECKESSER, and JAMES A. MAGNUSON

Program in Biochemistry and Biophysics, Washington State University, Pullman, Wash. 99163

During recent years nuclear magnetic resonance of ^{23}Na ions has been used to probe various biological membrane surfaces and model membrane systems. Cope recognized that the continuous wave sodium resonances generated from samples of biological tissue were characterized by integrated signal intensities reduced to 30-40% of that expected from the sodium content (1).

He concluded that 60-70% of sodium ions in tissues was broadened to nmr invisibility by complexing with tissue macromolecules, while the remaining sodium ions were free in solution in tissue water and contributed to the visible signal (2). This explanation supported the association-induction hypothesis (3). An alternate interpretation was presented by Shporer and Civan (4). They suggested that the reduction in the signal intensity of the ^{23}Na resonance was due to a single population of sodium ions experiencing nuclear quadrupole interactions. If a sodium nucleus experiences an anisotropic electric field the dipolar energy transitions will be perturbed due to the interaction of the nuclear electric quadrupole moment and this electric field gradient. This results in three resonance lines characterized by a central resonance line contributing 40% to the total intensity and two satellite lines which each contribute 30%. This effect has been observed for liquid crystals of sodium linoleate in water by Shporer and Civan (4), lecithin and sodium cholate in water by Lindblom (5) and sodium decyl sulfate and decyl alcohol in water by Chen and Reeves (6). Although the satellite lines have been observed using liquid crystalline model membranes, this splitting of the resonance lines has not been detected in biological samples.

A detailed theoretical treatment of quadrupolar interactions has been presented by Berendsen and Edzes (7). Because of random orientations of large local electric field gradients in biological material the quadrupole splitting becomes widely distributed resulting in satellite lines which appear extremely broad and become indistinguishable from the background noise. The satellite lines were detected indirectly by pulsed nmr of muscle tissue by resolving the non-exponential decay of the transverse relaxation into two exponential decays with relaxation times T_{2s} and T_{2f}. T_{2s} and T_{2f} denote the slow fraction corresponding to the narrow central resonance and the fast fraction corresponding to the broad satellite lines. T_{2s} is the relaxation time which is measurable from the observed line width in biological samples. The interpretation that there is only one species of sodium ions in biological tissue rather than the existence of 60% bound and 40% free has been generally accepted (8). This species of sodium ions is associated with an anisotropic environment and experiences nonzero electric field gradients over the diffusion ranges covered by the ions during their correlation times. This situation would require that the biological tissue is heterogeneous over distances of the order of 100 A (6), which is usually the case.

The existence of a free species of sodium ions along with an associated species of sodium ions was first observed by Magnuson and Magnuson (9) using cellular dispersions of gram-negative bacteria. By varying the concentration of cells from low to high the integrated signal intensity varied from 100% to 40%. This suggested the presence of a free species of sodium ions which contributes 100% of its signal to the intensity in conjunction with the associated species which contributes 40% of its signal. This effect has also been observed by Monoi (10) by varying the concentrations of the particulate fraction of rat liver tissue in saline solution.

If there exists only two populations of sodium ions in cellular dispersions, the observed line width, $\Delta\nu_{obs}$, of the central resonance should be the sum of the contribution from the free species and the contribution from the associated species

$$\Delta\nu_{obs} = f_i\Delta\nu_i + f_a\Delta\nu_a \qquad (1)$$

where f_i and f_a are the fractions of free and associated species, respectively, $\Delta\nu_i$ is the free sodium ion line width (standard line width of a NaCl solution), and $\Delta\nu_a$ is the average line width of the associated sodium ion which is characteristic of the magnitude of the electric field gradient within the anisotropic environment.

The fractional population of each sodium ion species can be determined from the integrated intensity. Since the free species and the associated species contribute 100% and 40% of their signals, respectively, to the observed intensity, I,

$$I = f_i(100\%) + f_a(40\%). \qquad (2)$$

Using $f_i + f_a = 1.0$, leads to the following expressions for the fraction of ions in each environment in terms of the measurable quantity, intensity,

$$f_i = \frac{I-40}{60} \qquad (3)$$

$$f_a = \frac{100-I}{60} \cdot \tag{4}$$

To support the model presented by equation (1) the observed line width should be directly and linearly proportional to the integrated intensity. Substituting equations (3) and (4) into equation (1) yields

$$\Delta\nu_{obs} = -\left(\frac{\Delta\nu_a - \Delta\nu_i}{60}\right)I + \left(\frac{100 \, \Delta\nu_a - 40 \, \Delta\nu_i}{60}\right). \tag{5}$$

This relationship is characterized by the associated line width, $\Delta\nu_a$, which is dependent on the biological organism. This parameter will be calculated for each system investigated.

This study is focused on the examination of the relationship expressed by equation (1) by altering the two sodium ion populations. The distribution of sodium ions between these two populations is dependent on: (1) the number and magnitude of the charged sites interacting with sodium ions, which is dependent on the concentration and type of bacteria employed; (2) the total concentration of sodium ions; and (3) displacement of sodium ions from negatively charged sites by competing cations. Each of these situations has been monitored by continuous wave ^{23}Na nmr for cellular dispersions of Pseudomonas aeruginosa and Proteus vulgaris.

Methods and Materials

Instrumentation. Continuous wave nmr absorption spectra were obtained with a modified Varian DP-60 nmr spectrometer by direct sweeping of the radio frequency. The V-4210 radio frequency unit was locked to a General Radio Model GR-1164 frequency synthesizer which was swept with a voltage ramp supplied by a Nicolet Fabri-Tek 1072 time averaging computer. Time averaging was necessary to increase the signal-to-noise ratio. The magnetic field was locked at 14,000 G to an external water proton reference signal with a resonance frequency of 60.0 MHz. Audio modulation of 2000 Hz was used in conjunction with a Princeton Applied Research Model 121 lock-in amplifier for base line stabilization. The line widths were determined by direct reading of the radiofrequency. The time averaging computer was utilized to perform signal integrations. Errors in these three types of measurements are estimated to be $\pm5\%$. The central resonances of ^{23}Na ions were observed at 15.872 MHz.

Preparation of Bacterial Dispersions. The gram-negative bacteria, Pseudomonas aeruginosa and Proteus vulgaris, were grown by inoculation in 3.7% Brain Heart Infusion Broth (which had been autoclaved at 15 psi at 121°C for 15 minutes for sterilization) and incubation for 18 hours (late log phase of growth) at 32°C on a shaker. The bacteria were harvested in polyethylene bottles by centrifugation at 12,000 g for 20 minutes, resulting in a thick pellet of bacteria. The growth medium was removed by suction and the pellets were washed by resuspension in 0.16 M NaCl, 1 mM MgCl$_2$ and centrifugation as before. The pellets were washed twice by this procedure and then suspended in various NaCl, MgCl$_2$ solutions for observation by nmr. The bacterial concentrations are described as dry weight of bacteria per volume of solution.

Results and Discussion

Effects of Bacterial Concentrations. The effects of varying the bacterial concentration of cellular dispersions of Pseudomonas aeruginosa in 0.16 M NaCl at two different magnesium concentrations, 0.001 M Mg^{++} and 0.01 M Mg^{++}, are illustrated in Figures 1 and 3, respectively. The observable line widths increased in a fairly linear fashion with the increase in cellular concentrations, as expected. This reflects an enhancement of the negative charge density and a corresponding increase in the size of the anisotropic environment with respect to each sodium ion. The decrease in integrated intensity is due to a shift of sodium ions from the population free in solution to the population associated with the increasing electric field gradients. The ten-fold increase in Mg^{++} concentration resulted in a shielding of this negative charge density with respect to the sodium ions. At the higher Mg^{++} concentration the line widths are reduced and a greater fraction of sodium ions are free in solution. This tendency for magnesium ions to displace sodium ions from the anisotropic electric fields will be examined later in greater detail.

Figures 2 and 4 illustrate the linear dependency of the observed line width on the integrated intensity. The use of linear regression by the method of least squares generated the empirical relationships

$$\Delta\nu_{obs} = -0.862\ I + 95.1 \quad (r^2 = 0.963) \tag{6}$$

and

$$\Delta\nu_{obs} = -0.715\ I + 83.1 \quad (r^2 = 0.982) \tag{7}$$

for the variation in bacterial concentration at 0.001 M Mg^{++} and 0.01 M Mg^{++}, respectively. The statistical value of r^2, the coefficient of determination, indicates how closely the equation fits the experimental data (maximum $r^2 = 1.0$) and is defined by the expression

$$r^2 = \frac{\left[\sum xy - \dfrac{\sum x \sum y}{n}\right]^2}{\left[\sum x^2 - \dfrac{(\sum x)^2}{n}\right]\left[\sum y^2 - \dfrac{(\sum y)^2}{n}\right]} \tag{8}$$

for the general equation $y = a_1x + a_0$. From these empirical relationships the associated line widths (at 40% integrated intensity) were calculated to be 60.6 Hz and 54.5 Hz, respectively.

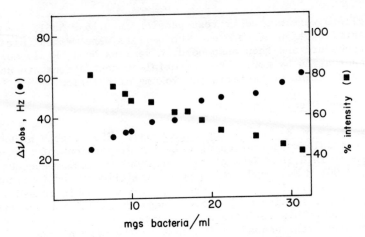

Figure 1. Variation in bacterial concentration of Pseudomonas aeruginosa in 0.16M NaCl at 0.001M MgCl₂

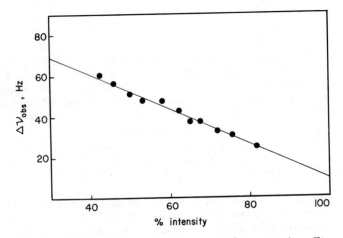

Figure 2. Observed linewidth vs. integrated intensity from Figure 1

Effects of Total Sodium Concentration. The effect of increasing the total sodium ion concentration at 0.001 M Mg^{++} is illustrated in Figures 5 and 6 for cellular dispersions of Pseudomonas aeruginosa at a bacterial concentration of 35 mg/ml and Proteus vulgaris at 100 mg/ml, respectively. The increase in integrated intensity and corresponding decrease in observed line width as the total sodium ion concentration increases, indicates a reduction in the fraction of sodium ions experiencing an electric field gradient. This redistribution of sodium ions in favor of the free population appears to obey a rectangular-hyperbolic relationship for both bacterial systems. Linear plots of the intensity versus the reciprocal sodium concentration (M^{-1}) supported this assumption. The empirical equations

$$I = 81.5 - \frac{5.17}{[NaCl]} \qquad (r^2 = 0.973) \qquad (9)$$

and

$$I = 78.0 - \frac{5.09}{[NaCl]} \qquad (r^2 = 0.975) \qquad (10)$$

were calculated for Pseudomonas aeruginosa and Proteus vulgaris, respectively. At high sodium concentrations the intensities approach 81.5% and 78.0%, or in terms of the associated fraction of ions, utilizing equation (4), f_a reaches minimums of 0.31 and 0.37, respectively.

It appears that the addition of sodium ions results in the distribution of ions between the two populations rather than the sodium ions strictly adding to the free population. A substantial fraction of the ions remain associated with the electric field gradients. These minimum associated fractions of ions may represent the fractional volume of solution within the bacteria if the sodium ions are distributed to produce equimolar concentrations inside and outside of the cells (except within the region near the outside charged surface). From the dry weights, the total fractional volume occupied by the bacteria compared to the total solution volume can be approximated to be 0.1-0.2 for a dry weight of 35 mg/ml and 0.3-0.4 for 100 mg/ml, depending on the organism. Dividing the associated fraction of sodium ions into two subpopulations, f_{a1} for the fraction near the charged surface and f_{a2} for the fraction inside the bacterial cells, equation (2) can be expressed as

$$I = (1-f_{a1}-f_{a2})(100\%) + f_{a1}(40\%) + f_{a2}(40\%). \qquad (11)$$

The intracellular associated fraction, f_{a2}, should be a constant proportion of the free fraction of sodium ions. Since the free fraction reaches a maximum at high sodium concentrations the relationships of

$$f_{a2} = 0.45 \ f_i \qquad (12)$$

and

$$f_{a2} = 0.59 \ f_i \qquad\qquad (13)$$

can be estimated from the f_a minimums of 0.31 and 0.37. Substituting equation (12) for f_{a2} into equation (11) the expression

$$f_{a1} = \frac{0.086 \ M}{[NaCl]} \qquad\qquad (14)$$

satisfies the empirical equation (9) for Pseudomonas aeruginosa at 35 mg/ml. Substituting equation (13) for f_{a2} into equation (11) the expression

$$f_{a1} = \frac{0.085 \ M}{[NaCl]} \qquad\qquad (15)$$

satisfies the empirical equation (10) for the Proteus vulgaris at 100 mg/ml. The consistency of these results supports the model presented by equation (11). This indicates that the total number of ions near the charged surface remains constant. The fraction, f_{a1}, merely decreases on addition of sodium ions due to the increase in the free population and the associated population within the cells.

The dependency of the observed line widths on the integrated intensities yielded the linear empirical relationships

$$\Delta\nu_{obs} = -0.933 \ I + 110.4 \qquad (r^2 = 0.988) \qquad (16)$$

and

$$\Delta\nu_{obs} = -0.486 \ I + 66.8 \qquad (r^2 = 0.955) \qquad (17)$$

from the data presented in Figures 5 and 6, respectively. Associated line widths of 73.1 Hz and 47.4 Hz were calculated from equations (16) and (17).

Effects of Magnesium Concentration. Competition for negatively charged sites by divalent metal ions is an effective means of altering the distribution of sodium ions between the associated and free states. By using magnesium ions, which are generally impermeable to gram-negative bacteria, it was anticipated that only the associated sodium ions near the outside charged surface of the cells would be displaced into solution. At high magnesium concentrations the remaining associated species of sodium ions should be located within the cells.

The effect of varying the magnesium ion concentration at 0.16 M NaCl for Pseudomonas aeruginosa at 40 mg/ml and Proteus vulgaris at 40 mg/ml are shown in Figures 7 and 8, respectively. Due to the large range of magnesium ion concentration utilized during these titrations, the magnesium content is expressed as $pMg = -\log [Mg^{++}]$, where the $[Mg^{++}]$ is given as a molar concentration. The displacement of sodium ions is clearly displayed for both systems. At high magnesium concentrations the intensities can be roughly estimated to approach the values of 90% for the Pseudomonas aeruginosa and approximately 93% for the Proteus vulgaris. These

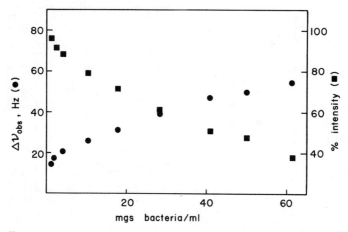

Figure 3. Variation in bacterial concentration of Pseudomonas aeruginosa in 0.16M NaCl at 0.01M MgCl₂

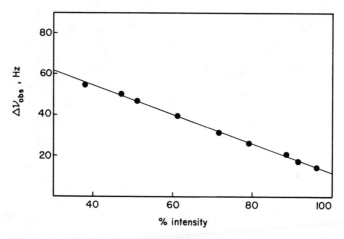

Figure 4. Observed linewidth vs. integrated intensity from Figure 3.

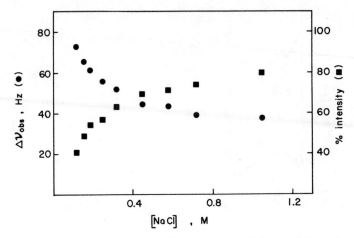

Figure 5.   Variation in total sodium ion concentration for Pseudomonas aeruginosa *in 0.001M MgCl₂ at 35 mg/ml dry weight*

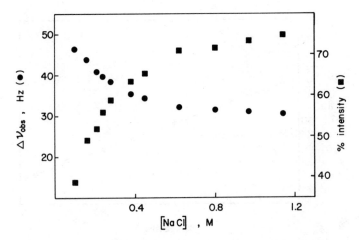

Figure 6.   Variation in total sodium ion concentration for Proteus
vulgaris *in 0.001M MgCl₂ at 100 mg/ml dry weight*

values correspond to minimum associated fractions of 0.17 and 0.12, respectively, and fall into the range estimated for the total fractional volume occupied by the bacterial cells at a concentration of 40 mg/ml dry weight. Adjusting the Proteus vulgaris value for 100 mg/ml would yield a value of 0.30. For both organisms these minimum associated fractions are smaller than the values obtained by varying the sodium concentration. But these are closer to the estimates of the fractional volume occupied by the cells. It appears that only the associated sodium ions located outside of cells are displaced by magnesium.

The observed line widths were again linearly dependent on the integrated intensity as shown by the empirical relationships

$$\Delta\nu_{obs} = -0.967\ I + 107.0 \qquad (r^2 = 0.961) \tag{18}$$

and

$$\Delta\nu_{obs} = -0.541\ I + 72.5 \qquad (r^2 = 0.902) \tag{19}$$

determined from the data illustrated in Figures 7 and 8, respectively. From these equations the associated line widths were calculated to be 68.3 Hz for Pseudomonas aeruginosa and 50.9 Hz for Proteus vulgaris.

Conclusions

From these results it is clearly demonstrated that the distribution of sodium ions between a free state and an associated state in bacterial dispersions is dependent on the concentrations of bacteria, sodium ions, and magnesium ions. In each case the decrease in observed line width was linearly dependent on the increase in integrated intensity, as described by equation (5). The consistent correlation between the equations presented and the experimental data directly supports the model represented by equation (1). The observed line width consists of contributions from sodium ions distributed between two general environments, each characterized by an average line width. The associated line width reflects the magnitude of the interaction of the sodium ions with the electrostatic fields produced by the bacterial cells. These values of the associated line width appeared to be independent of the concentrations of bacterial cells, sodium ions, and magnesium ions, but dependent on the organism. The average associated line widths for Pseudomonas aeruginosa and Proteus vulgaris were calculated to be 64.1 Hz with a standard deviation of 7.1 Hz and 49.2 Hz with a standard deviation of 1.8 Hz, respectively. Since the free line width and the associated line width remains constant for a given organism, the variation in observed line width is solely dependent on the distribution of sodium ions.

Distinguishing between the intracellular and extracellular associated fractions of sodium ions has been a problem. A separ-

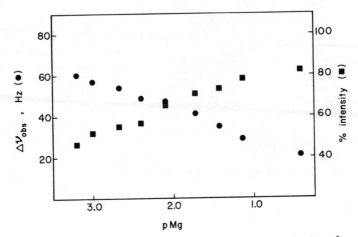

Figure 7. *Variation in magnesium ion concentration for* Pseudomonas aeruginosa *in 0.16M NaCl at 40 mg/ml dry weight*

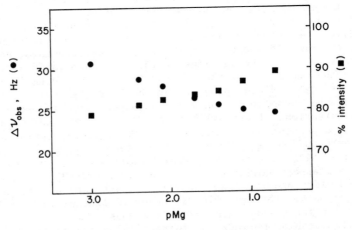

Figure 8. *Variation in magnesium ion concentration for* Proteus vulgaris *in 0.16M NaCl at 40 mg/ml dry weight*

ate line width cannot be resolved for each from our experimental results. A single average line width for the entire associated fraction is consistent with the model presented. During the sodium titration experiment the extracellular associated fraction appeared to be inversely proportional to the total sodium concentration. The addition of magnesium ions reduced the extracellular associated fraction by displacing the sodium ions associated with the extracellular cell surface. In both cases the effects on the sodium ions interacting with the external cell surfaces could be monitored. This appears to be an effective method of examining the environment near the charged cell surfaces. (Supported in part by funds provided for biological and medical research by State of Washington Initiative Measure No. 171 and by USPHS grant CA 14496).

Literature Cited

1. Cope, F. W., Proc. Natl. Acad. Sci. U.S.A. (1965), 54, 225.
2. Cope, F. W., Biophys. J. (1970), 10, 843.
3. Ling, G. N., "A Physical Theory of the Living State: The Association-Induction Hypothesis." Blaisdel Publishing Co., Waltham, Mass. (1962).
4. Shporer, M. and Civan, M. M., Biophys J. (1972), 12, 114.
5. Lindblom, G., Acta Chem. Scand. (1971), 25, 2767.
6. Chen, D. M. and Reeves, L. W., J. Am. Chem. Soc. (1972), 94, 4384.
7. Berendsen, H. J. C. and Edzes, H. T., Ann. N. Y. Acad. Sci. (1973), 204, 459.
8. Edzes, H. T. and Berendsen, H. J. C., Ann. Rev. Biophys. Bioeng. (1975), 4, 265.
9. Magnuson, N. S. and Magnuson, J. A., Biophys. J. (1973), 13, 1117.
10. Monoi, H., Biophys. J. (1974), 14, 645.

31

The NMR Quadrupole Splitting Method for Studying Ion Binding in Liquid Crystals

GÖRAN LINDBLOM, HÅKAN WENNERSTRÖM, and BJÖRN LINDMAN

Division of Physical Chemistry 2, Chemical Center,
P.O.B. 740, S-220 07 LUND 7, Sweden

Charged interfaces greatly influence the properties of many technical amphiphile systems and have a
wide functional role in living systems. It is,
therefore, not surprising that the interest for eluci-
dating ionic interactions at interfaces in colloidal
systems is rapidly growing. Of particular significance
would be the development of experimental methods which
permit a direct and specific probing into the binding
state of small ions. Interest is here naturally focus-
sed on nuclear magnetic resonance. This method permits
the specific observation of many simple ions of technical and biological interest, such as the alkali and
halide ions as well as Mg^{2+}, Ca^{2+}, PO_4^{3-} and NH_4^+.
Different NMR parameters provide insight into diffe-
rent aspects of the interaction and dynamic state of
the ions. Furthermore NMR is non-perturbatory and the
state of the sample as regards consistency, turbidity
etc. causes no fundamental difficulties. Many of the
ions mentioned are in addition to their magnetic
moments characterized by electric quadrupole moments
and these have through their interactions with electric field gradients a dominating influence on the
appearance of the NMR spectrum. The reduction of the
quadrupole interactions due to rapid molecular motion mani-
fests itself in terms of relaxation effects. For an
environment which is anisotropic on the appropriate
time-scale, only partial averaging occurs and the
residual quadrupole interaction shows up in the spectrum as a quadrupole splitting.

The first report of counterion quadrupole splittings for lyotropic mesomorphous systems was made as
late as 1971 (1) but since then the quadrupole splitting method has become widely applied to study ionic
interactions in amphiphilic systems in our own laboratory as well as in others. The present article is

concerned with giving a general review of the method,
surveying the theoretical basis and describing different types of information provided as well as pitfalls
and difficulties in the interpretation. We will be
mainly dealing with the magnetic resonance aspects of
the problem while detailed accounts of studies of
counterion binding in specific systems will be given
elsewhere.

Basic Theory

The theory required for dealing with quadrupole
splittings in mesomorphous systems ($\underline{2}$) is obtained by
introducing into the general theory of quadrupolar
effects in solids ($\underline{3}$) the effects of partial orientation as given for example in the treatments of Bucking-
ham and McLauchlan ($\underline{4}$), of Diehl and Khetrapal ($\underline{5}$) and
of Luckhurst ($\underline{6}$). In the presence of quadrupole interactions the spin hamiltonian may be written as a sum
of the nuclear Zeeman interaction and the quadrupole
coupling term

$$H = H_Z + H_Q = -\nu_L I_z + \beta_Q \sum_{q=-2}^{2} (-1)^q V_{-q} A_q \qquad (1)$$

The hamiltonian is expressed in frequency units. ν_L is
the Larmor frequency, the V_q's are the irreducible components of the electric field gradient tensor (of
second rank) and the A_q's the standard components of a
second rank spin tensor operator. β_Q is defined by
$\beta_Q = \frac{eQ}{2I(2I-1)\hbar}$ where eQ is the quadrupole moment. In
Eq. (1) chemical shift anisotropies have been neglected.
For an isotropic liquid the mean value of H_Q is
zero and the quadrupole interaction influences only
the relaxation. For an anisotropic medium, however,
the mean value of H_Q is nonzero and a quadrupole splitting appears in the NMR spectrum. The quadrupole hamiltonian in Eq. (1) may be evaluated in any coordinate
system but it is convenient to express the spin opera-
tors in a laboratory-fixed coordinate system and the
electric field gradients in a principal axes coordinate
system fixed at the nucleus. The hamiltonian may then
be rewritten as

$$H_Q = \beta_Q \sum_{q,q'} (-1)^q V_{-q}^M A_{q'}^L D_{q'q}(\Omega_{LM}) \qquad (2)$$

$D_{q'q}$ is a second rank Wigner rotation matrix element
and Ω_{LM} signifies the three eulerian angles specifying
the transformation from the molecular coordinate system
(M) to the laboratory system (L). In Eq. (2) the V_q's
are to be taken in the molecular frame and the A_q's in

the laboratory frame. Molecular motion is described as a time-dependence of Ω_{LM}.

If we consider a uniaxial liquid crystal, like a nematic mesophase or a lamellar or hexagonal amphiphilic mesophase, the system possesses cylindrical symmetry about an axis called the director. For this case it is convenient to perform the transformation from the molecular frame to the laboratory frame via the director coordinate system (D). A schematic illustration of the three coordinate systems is given in Figure 1. For a macroscopically aligned sample having the same director orientation throughout the sample, the quadrupolar hamiltonian may now be written

$$H_Q = \beta_Q \sum_{qq'q''} (-1)^q V_{-q}^M A_{q''}^L D_{q'q} (\Omega_{DM}) D_{q'q''} (\Omega_{LD}) \qquad (3)$$

The mean value of H_Q is given by Eq.(4) for the case where a nucleus can be considered to remain within a region of a given director orientation over a time which is long compared to the inverse quadrupole interaction.

$$\overline{H_Q} = \beta_Q \sum_{qq''} (-1)^q V_{-q}^M A_{q''}^L \overline{D_{0q}(\Omega_{DM})} D_{q''0}(\Omega_{LD}) \qquad (4)$$

(The mean value of $D_{q'q}(\Omega_{DM})$ is zero for $q' \neq 0$ if there is a threefold or higher symmetry around the director axis.)

The quadrupole term is generally small compared to the Zeeman term and, to first order, it is only the secular part of $\overline{H_Q}$ that contributes to the time-independent hamiltonian, H_0, i.e.

$$H_0 = -\nu_L I_z + \beta_Q V_0^M S D_{00} (\Omega_{LD}) (3I_z^2 - I^2) \qquad (5)$$

The order parameter S is given by

$$S = \frac{1}{2} \overline{(3 \cos^2\Theta_{DM} - 1)} + \eta \overline{\sin^2\Theta_{DM} \cos 2\phi_{DM}} \qquad (6)$$

where η is the asymmetry parameter ($\eta = \sqrt{6} V_2^M/V_0^M$) and Θ and ϕ are the eulerian angles β and γ, respectively. For the zeroth component of the electric field gradient tensor we have

$$V_0 = \frac{1}{2} \frac{\partial^2 V}{\partial z^2} .$$

The hamiltonian in Eq.(5) gives $2I + 1$ energy levels and the NMR spectrum consists of $2I$ equally spaced peaks. A schematic diagram is given in Figure 2 for $I = 3/2$. The distance between two peaks, which will be referred to as the (first-order) quadrupole split-

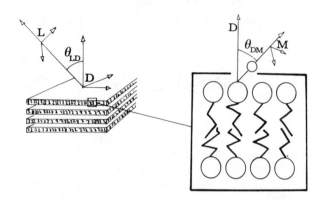

Chemica Scripta

Figure 1. Schematic of the mesomorphous structure in a lamellar phase. The different coordinate systems used in the text are outlined in the figure: laboratory frame (L), director frame (D), and molecular frame (M). θ_{LD} and θ_{DM} are angles between z-axes in laboratory—director systems and director—molecular systems, respectively (2).

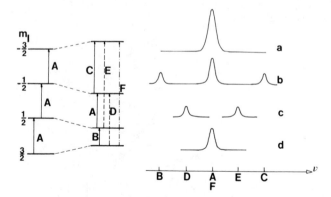

Figure 2. Schematic showing NMR spectra for spin $- 3/2$ nuclei (right) and corresponding energy levels (left); (a) single quantum transitions, isotropic solution, (b) single quantum transitions, first order quadrupole splitting, (c) double quantum transitions, (d) triple quantum transition

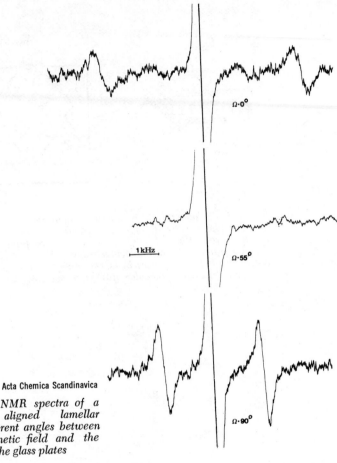

Acta Chemica Scandinavica

Figure 3(a). ²³Na NMR *spectra of a macroscopically aligned lamellar mesophase at different angles between the applied magnetic field and the normal to the glass plates*

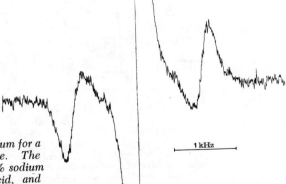

Figure 3(b). ²³Na NMR *spectrum for a corresponding "powder" sample. The sample composition was 48.5% sodium octanoate, 23.9% octanoic acid, and 27.6% water by weight. Temperature 27 ± 2°C (8).*

ting is for a macroscopically aligned sample given by

$$\Delta(\theta) = \left| 3 \beta_Q V_0^M (3 \cos^2\theta_{LD} - 1) \right| \tag{7}$$

From an observed splitting the quadrupole coupling constant e^2qQ/h can be calculated through $(e^2qQ)/h = 4I(2I-1)\beta_Q V_0$.

For a powder sample where all director orientations are equally probable, the distance between the major peaks in the NMR spectrum coresponds to $\Delta(\theta)$ for $\theta_{LD} = 90°$. The powder splitting, Δ_p, is therefore given by

$$\Delta_p = \left| 3 \beta_Q V_0^M S \right| \tag{8}$$

Typical quadrupole-split NMR spectra for aligned and powder samples are given in Figure 3.

For very strong quadrupole interactions also the second-order perturbation term may have to be taken into account. For nuclei with $I = 3/2$ the second-order correction to the energy levels primarily leads to a shift of the central line in the spectrum. For a macroscopically aligned sample the shift is given by

$$\delta\nu_{1/2, -1/2} = \left[\frac{3}{4}\beta_Q V_0^M S\right]^2 \frac{I(I+1) - 3/4}{\nu_L} (\cos^2\theta_{LD}-1) \cdot$$

$$\cdot (9 \cos^2\theta_{LD}-1) \tag{9}$$

For a powder sample, the central line in the NMR spectrum gives, as illustrated in Figure 4, two marked peaks which are separated by

$$\Delta_p^{(2)} = \left[\frac{5}{4} \beta_Q V_0^M S\right]^2 \frac{I(I+1) - 3/4}{\nu_L} \tag{10}$$

General Aspects on the Observation of Counterion Quadrupole Splittings

Experimental Observation. We have earlier reported first-order quadrupole splittings for all the alkali ions (7-11) as well as for the chloride (10) and bromide ions (12). All these studies were performed with continuous-wave techniques using a Varian wide-line spectrometer. Representative spectra showing first- and second-order quadrupole effects are given in Figures 3 and 4 for ^{23}Na and ^{35}Cl, respectively, and spectra for the other ions were given in Refs. 9 and 12. The wide-line spectrometer is still being used for very large splittings but in other cases pulse techniques offer a much higher precision. Using a

Bruker pulse spectrometer the splitting is obtained
from the time-modulation of the free-induction decay
after a 90° pulse. A typical ^{23}Na spectrum obtained
in this way is shown in Figure 5. Alternatively, the
spectra are recorded on a modified Varian XL-100
spectrometer using the Fourier transform technique. A
typical ^{133}Cs spectrum obtained in this way is shown
in Figure 6.

As was mentioned above the central line may be
affected by second-order quadrupole interactions for
strong quadrupole couplings and this has been observed
for ^{35}Cl, ^{37}Cl (1) and ^{85}Rb (9,10). In the absence of
second-order effects the width of the central line is
generally determined by quadrupole relaxation. We have
as yet not been able to detect any effects due to chemical shift anisotropy. To distinguish between broadening due to relaxation and second-order quadrupole
effects, which is not always straight-forward purely
from spectral shape, an examination of the variaion of
spectrum with magnetic field strength should normally
be conclusive (1,10). (Cf. Eqs.(9) and (10).) If not
otherwise stated the splittings reported below refer
to first-order splittings of powder samples at probe
temperature (25-30°C).

Macroscopic Alignment. All the observations mentioned above refer to powder samples where all values
of $\cos\theta_{LD}$ are equally probable. However, a macroscopic
alignment of the sample can be achieved by orienting
the sample with a solid surface (13) or with the magnetic field (14,15). The effect of the orientation on
the spectrum gives information on the mode of aligne-
ment but the signal to noise ratio for the satellite
peaks can also be improved as their intensities become
restricted to a small frequency range. The ^{23}Na spectrum for a lamellar mesophase oriented in thin layers
between glass-plates is shown in Figure 3. By investigating the splitting as a function of the orientation
of the glass plates relative to the magnetic field,
the director was found to be directed perpendicularly
to the glass plates, i.e. the sample is oriented with
the lamellae along the glass surfaces (8). The magnitudes of the quadrupole splittings of powder and
aligned samples are found to correspond to each other
according to Eqs.(7) and (8) indicating essentially
planar lamellae even in the powder sample.

Multiple Quantum Transitions. The NMR of quadrupolar nuclei in liquid crystals provides unusually
favorable conditions for the observation of multiple
quantum transitions. This is due to a combination of
small quadrupole splittings and short T_1 values. It

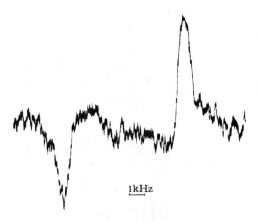

1kHz

Chemical Physics Letters

Figure 4. Experimental chlorine-35 magnetic resonance spectrum of a sample with the composition 50% octylammonium hydrochloride, 30% H_2O, and 20% decanol by weight. The resonance frequency was 5.78 MHz (1).

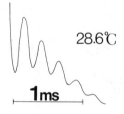

28.6°C

1ms

Figure 5. ^{23}Na resonance free induction decay after a 90° radio frequency pulse for a hexagonal mesophase sample composed of 51.0% sodium octanoate and 49.0% water

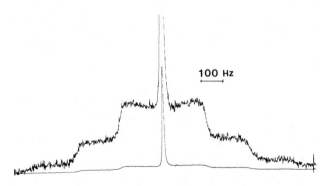

100 Hz

Figure 6. ^{113}Cs FT-NMR spectrum at 30°C for a lamellar mesophase sample with the composition, 42.2% cesium octanoate, 29.6% octanoic acid, and 28.2% water by weight. Accumulations, 8000. (By the courtesy of Hans Gustavsson.)

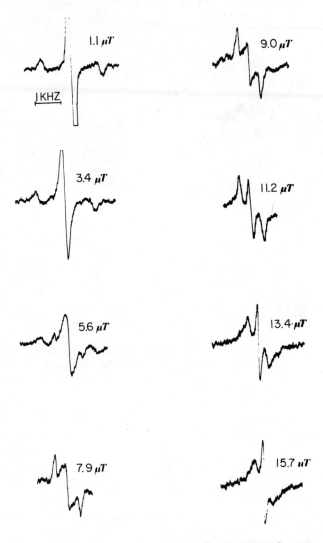

Figure 7. ^{23}Na NMR spectra at different RF field amplitudes for a lamellar liquid crystalline sample composed of 39.1% sodium octanoate, 33.3% decanol, and 27.6% water by weight (16).

was recently demonstrated (for ^{23}Na) that both double and triple quantum transitions are easily obtainable under typical experimental conditions (16). As may be inferred from the energy diagram in Figure 2 the double quantum transitions appear at the frequencies $\nu_L \pm \Delta/2$ while the triple quantum transition appears at the Larmor frequency. The experimental ^{23}Na spectra shown in Figure 7 are illustrative: At low rf-field strengths only the single quantum transitions are observed but with increasing rf-field these transitions are saturation broadened while the double quantum peaks become visible. At even stronger rf-fields also the double quantum peaks are saturation broadened while the central peak increases in intensity. This is due to the triple quantum transition. Obviously, the possibility of misinterpreting the double and triple quantum peaks must be carefully considered but studies at variable rf-field strength should be helpful in this respect. For pulse experiments, where the NMR signal is observed in the absence of rf-field, multiple quantum transitions do not appear.

Effect of Phase Structure. The order parameters of hexagonal and lamellar mesophases of the same system are directly related to each other if the microscopic structure at the water-amphiphile interface can be assumed identical for the two phases. Thus if the molecular motion around the rods of the hexagonal structure proceeds in a time short compared to the inverse splitting, the splitting of the hexagonal phase should be half that of the lamellar phase (2). The observations for the sodium octyl sulfate-decanol--water system (Figure 8) are consistent with this model and the ionic interactions appear to remain essentially unchanged at the phase transition. For other systems, for example the sodium octanoate--decanol-water, sodium octanoate-octanoic acid-water and sodium octanoate-pentanol-water systems (17), the variation of counterion splitting with phase structure cannot be explained simply in terms of phase anisotropy.

Theoretical Considerations of Counterion Quadrupole Splittings

Multi-Site System. The counterions in amphiphile--water liquid crystals can reside in different bonding environments which are characterized by different quadrupole splittings. It seems reasonable to assume that in most cases the rate of exchange between the different binding positions within a homogeneous phase is much more rapid than the difference in quadrupole splitting between the different sites. In this case, which is supported by experimental findings, the observable first-order quadrupole splitting is given by

$$\Delta(\Theta) = \left| (3 \cos^2\Theta_{LD}-1)3\beta_Q \sum_i p_i V_i S_i \right| \qquad (11)$$

for a macroscopically oriented sample and by

$$\Delta_p = \left| 3\beta_Q \sum_i p_i V_i S_i \right| \qquad (12)$$

for a powder sample (2). p_i denotes the fraction of counterions in sites $\underline{i}$ which are characterized by the field gradient V_i and the order parameter S_i.

The absolute signs of Equations (11) and (12) may be noted since they imply that the contributions of different sites may have opposite signs and thus a partial or total cancelling out may occur as a result of chemical exchange. As will be seen below, both V and S may have either positive or negative signs.

It can be seen from these expressions that the quadrupole splittings are determined by three factors, i.e. the distribution of counterions over different binding sites and the intrinsic electric field gradients and order parameters of the sites. These three quantities are not easily separated and certain assumptions are required to proceed in the analysis.

The Order Parameter. The order parameter, characterizing the partial orientation of the field gradients sensed by the counterions, can vary between 1 (for a rigid case with $\Theta_{DM} = 0^\circ$) and $-\frac{1}{2}$ (for a rigid case with $\Theta_{DM} = 90^\circ$) assuming $\eta = 0$. It is determined by both the average Θ_{DM} value and by its range of fluctuation according to Equation (6). (For certain counterion interactions, the asymmetry parameter is zero for symmetry reasons; for example, a counterion located symmetrically with respect to the three oxygens of $-SO_3^-$ or $-OSO_3^-$. Even with a sizeable η value it will generally only have a small influence on the magnitude of the splitting. Noting that the η term of Equation (6) may appreciably influence the splitting only for certain field gradient orientations, we will neglect it in the following discussion. As some misunderstanding seems to persist in the literature, it may also be recalled that the asymmetry parameter may affect the magnitude of the quadrupole splitting but not spectral shape (8).) The order parameter can be zero for two situations: The counterion is moving freely in the aqueous layers so that all directions of the field gradients are equally probable or Θ_{DM} equals the magic angle $(54^\circ44')$, where $3 \cos^2\Theta_{DM} - 1 = 0$.

It may often be a natural starting-point in the interpretation to assume a two-site model with the counterions considered to be either free (f) or bound to the amphiphile aggregate (b). It is usually a good approximation to assume that the free ions are unaffected by the anisotropic environment and put $S_f = 0$. In this case Equation (12) reduces to

$$\Delta_p = \left| 3 \, \beta_Q \, p_b \, V_b \, S_b \right| \tag{13}$$

However, it should be noted that for instance the lamellar mesophases of many amphiphile systems exist down to low water contents; as an example, the thickness of the water layers of the lamellar mesophase of the sodium octanoate-decanol-water system may fall below $8 \cdot 10^{-10}$ m (18). Under such conditions a neglect of S_f seems doubtful.

A schematic drawing illustrating some different situations at an amphiphile-water interface is given in Figure 9 (19).

<u>Counterion Association Degree</u>. It is well known that an appreciable fraction of the counterions can be considered to be bound to the amphiphilic aggregates but the meaning of the word bound may not be the same for different experimental approaches (20). A further complication in connection with the present type of study is that the degree of counterion binding is more difficult to obtain for mesophases than for micellar solutions. For the latter case the convention has often been adopted to consider those counterions as bound which are moving with the micelle as a kinetic entity and then an investigation of the translational diffusion of the counterion directly gives the fraction of bound counterions (21). By using the NMR spin-echo diffusion method it should by using macroscopically aligned samples be possible to obtain the same type of information for, for example, Li$^+$ counterions in liquid crystalline phases. Lacking direct information we may obtain reasonable estimates of the fraction of bound counterions by the assumption that it is at least as high in the liquid crystalline phases as in the corresponding micellar solutions. (Mesophases having a very high water content may be an exception.) For the two--site model we should then have $p_b = 0.6 - 0.9$.

A more difficult question relates to the applicability of the two-site approximation. It may well be that we have to consider two or more types of counterion binding sites but no evidence for this seems to be available for simple surfactant-water systems.

<u>Quadrupole Coupling</u>. In contrast to the case of

most covalent compounds, the field gradient for monoatomic ions are of intermolecular origin. As a consequence of this, the quadrupole coupling may vary considerably with composition, temperature etc. and it is in
practice no way of experimentally obtaining the value
of $\beta_Q V_0^M$. Attempts were made in Ref. 2 to use a simple
electrostatic model, where the field gradients are considered to be due to the charges of the amphiphilic
ions and to the dipole moments of water and other polar
molecules. (An analogous approach has been successfully
used by Hertz (22) for quadrupole relaxation of monoatomic ions.) The quadrupole coupling constant arising
from a point charge Z (in atomic units) at the distance
r (in m) from the quadrupole moment Q (in m^2) gives a
quadrupole coupling constant (2,3)

$$\beta_Q \; V_0^M = \frac{0.52 \cdot 10^6 (1+\gamma_\infty)}{3I(2I-1)} \; \frac{2\varepsilon+3}{5\varepsilon} \; \frac{QZ}{r^3} \qquad (14)$$

γ_∞ is the Sternheimer antishielding factor and ε the
dielectric constant of the medium. Assuming the coun-
terions to be hydrated, which has support from quadrupole relaxation (23) and shielding (24) studies, we
have estimated values of $\beta_Q V_0^M$ for a number of systems.
The values are listed in Table I together with the
values of $1+\gamma_\infty$, Q and r which were employed. ε was
assumed>>1.

Another source of field gradients is given by the
dipole moments of the molecules; for example a solubilized alcohol molecule or a nonionic surfactant. For
a dipole moment μ (in Debye) directed parallel(+) or
antiparallel (−) to the dipole-nucleus vector we obtain

$$\beta_Q \; V_0^M = \frac{\mp 1.09 \cdot 10^{-5} (1+\gamma_\infty)}{I(2I-1)} \; \frac{2\varepsilon+3}{5\varepsilon} \; \frac{Q\mu}{r^4} \qquad (15)$$

Quadrupole coupling constants obtained from Equation (15) are given for some typical values of r and
μ in Table I.

The water molecules also give field gradients.
However, for a hydrated ion the effects of the different water molecules in the hydration sheath will largely cancel out and it is only the asymmetry in the
hydration sheath that produces a net field gradient. A
displacement from the symmetrical configuration by
$0.1 \cdot 10^{-10}$ m of one of the water molecules in the first
hydration layer gives for $^{23}Na^+$ a net quadrupole coupling $3 \beta_Q V_0^M = 21.5$ kHz. It seems though very difficult to obtain the required information on the asymmetry of the hydration layer.

As we have observed for a number of mesophases
composed of ionic amphiphile and water that addition of

Table I

Quadrupole coupling at the ionic nucleus caused by the interaction between the hydrated counterion and different polar head groups estimated from a simple electrostatic model (see text)

Counter-ion	Amphiphile end-group	$1+\gamma_\infty$	$Q \cdot 10^{28}$ (m^2)	$r \cdot 10^{10}$ (m)	$\beta_Q \, v_0^M$ (kHz)
^7Li	$\begin{cases} -CO_2^- \\ -SO_3^- \end{cases}$	0.74	−0.042	4.6	0.74
$^{23}Na^+$	$\begin{cases} -CO_2^- \\ -SO_3^- \end{cases}$	5.1	0.11	4.9	−11
$^{39}K^+$	$-CO_2^-$	18.3	0.09	5.2	−27
$^{87}Rb^+$	$-CO_2^-$	48.2	0.31	5.4	−66
$^{133}Cs^+$	$-CO_2^-$	111	−0.004	5.6	0.83
$^{35}Cl^-$	$-NH_3^+$	58	−0.079	6.2	−44
$^{35}Cl^-$	$-N(CH_3)_3^+$	58	−0.079	7.5	−25
$^{81}Br^-$	$-NH_3^+$	100	0.28	6.4	250
$^{81}Br^-$	$-N(CH_3)_3^+$	100	0.28	7.7	140
$^{23}Na^+$	$-OH$	5.1	0.11	5.1	±2.2
$^{35}Cl^-$	$-OH$	58	−0.079	3.2	64
$^{81}Br^-$	$-OH$	100	0.28	3.4	300

a dipolar uncharged compound has only moderate effects
on the counterion quadrupole splitting, it will be a
natural starting-point also for the three-component
systems to attempt to use Equation (14) to estimate
$\beta_Q \, v_0^M$. It is not easy to judge how realistic this
simple electrostatic model is; in any case the considerable uncertainty in some of the quantities needed
should be noted. Comparison with experimental results
will be decisive but it may be noted that for proteins
the model has had marked success (25).

Discussion of Experimental Findings

Cationic Amphiphile Systems. ^{35}Cl and ^{37}Cl quadrupole splittings have been obtained for the lamellar
mesophase region of three-component system octylammonium chloride-decanol-water as a function of sample
composition and temperature (1,10). The ^{35}Cl quadrupole
splittings converted to first-order splittings (second-
-order effects were observed) are in the range 150-500
kHz (see Table II) and not appreciably temperature
dependent. The concentration dependence is moderate,
with a decreased splitting being observed as the water
content is increased. Apparently, counterion binding
is only slightly affected by temperature and composition changes for this system. The quadrupole splittings
are of the same order of magnitude as the value given
by the simple electrostatic model and an order parameter of the order of unity is suggested. This is
supported by the variable temperature and concentration
results. A possible model to explain this would be in
terms of a $N-C_\alpha$ bond directed perpendicularly to the
lamellae, a symmetric location of Cl^- with respect to
the three hydrogens of NH_3 and a limited range of fluctuation of the field gradient direction.

For the hexagonal phase of the dodecyltrimethylammonium chloride-water system (10) the ^{35}Cl quadrupole
splitting is very much smaller (up to a factor of 100
and more) than for the octylammonium chloride-water
system (Table II) suggesting a markedly different
counterion binding in the two cases. Estimating the
order parameter as above gives values below 0.05 and
about the same value is obtained from the ^{81}Br splitting of the hexagonal phase of the system hexadecyl-
trimethylammonium bromide-water (12). It is not possible presently to give a well-founded interpretation of
these observations as well as of the finding of an
increasing ^{35}Cl splitting with increasing water content
for the dodecyltrimethylammonium chloride-water system
(10). It seems though clear that the qualitative difference between the two systems is due to the fact that

the charged nitrogen is much more screened in tri-
methylalkyl ammonium compounds.

Anionic Amphiphiles with a $-OSO_3^-$ or $-SO_3^-$ End-
-Group. ^{23}Na quadrupole splittings were determined
for the systems sodium octyl sulfate-decanol-water,
sodium octyl sulfonate-decanol-water and Aerosol OT
(sodium di(2-ethylhexyl)sulfosuccinate)-water and the
results are exemplified in Figure 8 and Table II. The
splittings (and hence also the counterion binding) can
be seen to depend only moderately on sample composition and were, furthermore, found to be essentially
independent of temperature (Figure 10). Assuming a
two-site model with $S_f = 0$, and taking the quadrupole
coupling constant from the electrostatic model as described above, the order parameter of the bound counterions is obtained to be of the order of unity. As
described above this would correspond to a situation
with Θ_{DM} around 0° and a small range of fluctuation.
Since the symmetrical $-SO_3$ groups pointing out into
the water layers are expected to have their symmetry
axes parallel to the director, the ^{23}Na splittings
suggest the hydrated sodium ions to be symmetrically
located with respect to the three oxygens. In turn
such a location would suggest a simple electrostatic
ion-ion interaction.

Support for the deduced value of S is provided by
the variable concentration and temperature results
since with an average value of Θ_{DM} of zero, changes in
S may result solely from changes in the range of fluctuation of Θ_{DM}. If these fluctuations are not considerable, the changes in the order parameter will be small

Anionic Amphiphiles with a $-CO_2^-$ End-Group. It
was natural to start our investigations of counterion
quadrupole splittings with the systems sodium octanoate-decanol-water and sodium octanoate-octanoic acid-
-water as these have been so well documented as regards phase equilibria and various physico-chemical
properties in the work of Ekwall and coworkers (26).
As it turned out later, this was somewhat unfortunate
as these systems are characterized by a very complex
counterion quadrupole splitting pattern. In the first
report (9) on these systems this complexity was not
fully realized. For the interpretation, the following
three observations of ^{23}Na splittings in the presence
of anionic surfactant with carboxylate end-group are
important (9,11,17,19):
a) The splittings are much smaller than with the sulfate or sulfonate end-groups and also smaller than
predicted by the model described above with a high S
value.

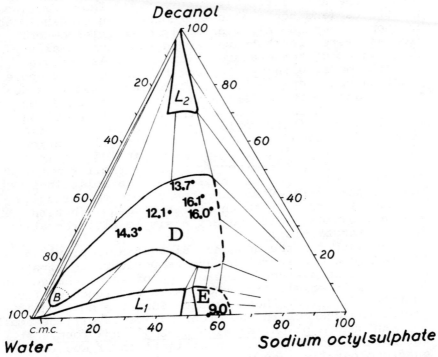

Advances in Liquid Crystals

Figure 8. ^{23}Na quadrupole splittings at 25°C for mesophase sample composed of sodium octylsulfate, decanol, and water. The two mesophase regions studied are denoted D (lamellar) and E (hexagonal) (26).

b) The concentration dependence of the splitting is strong and complex with comparatively large splittings being observed for the water-rich parts of the lamellar phases and with zero splittings at intermediate concentrations.

c) The splittings may be extremely dependent on temperature (Figure 10) and may pass through zero at a temperature well within the stability range of the mesophase (hexagonal or lamellar).

It appears to be relatively unambiguous to interpret the observations of vanishing minimal splittings in terms of a change in sign of the expression within absolute signs of Equation (12). It is, on the other hand, not self-evident why $\Sigma p_i V_i S_i$ becomes zero. There are two possibilities:

i) There may be a sodium ion exchange between two environments having similar magnitudes but different signs of $V_i S_i p_i$ and depending on concentration and temperature either term can be the dominant one.

ii) There may be a change in sign of the order parameter characterizing the bound counterions. A safe distinction between possibilities seems presently impossible but as concerns i) it should be noted that the splitting passes through zero also for binary soap-water mesophases and that, therefore, a second anisotropic site corresponding to interaction with -OH or -COOH cannot explain data. Furthermore, no indication of a second anisotropic binding site is evident in other types of studies. Regarding ii), changes in sign of the order parameter of the bound counterions cannot be due to fundamental changes in the structure of the liquid crystalline phases since corresponding changes are not observed in alkyl chain proton resonance (27) or water deuteron quadrupole splittings (28). Consequently, we attribute in this case the changes in sign of the splitting to a passage of the average θ_{DM} value of the bound counterions through the magic angle. Although ii) seems most plausible, both possibilities appear to be consistent with all our experimental observations of a number of phases. However, as regards the information on the mode of counterion binding, it seems that one is led to consider a partial penetration of the alkali ions between polar head groups irrespectively of which model is chosen.

Based on the assumption that the order parameter of bound counterions passes through zero, a method has been proposed for deducing the sign of the order parameter from variable temperature or concentration studies (19). Furthermore, certain information on phase structure was obtained (19).

390

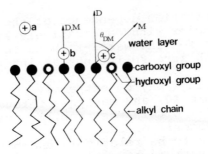

Acta Chemica Scandinavica

Figure 9. Schematic of the counter-ion binding at the lamellar surface according to the discussion in the text. The following three possibilities are shown: (a) The counter-ion is moving freely in the water layer. This location is characterized by S = 0. (b) The counter-ion is located symmetrically with respect to the amphiphile polar end-group. In this case the average angle between the director (D) and the electric field gradient (M) is equal to 0°. (c) The counter-ion is located between amphiphile polar head-groups (19).

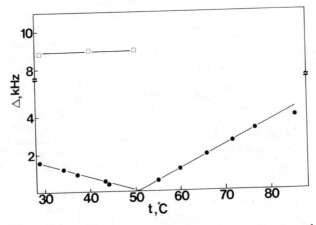

Figure 10. ²³Na quadrupole splitting, Δ, as a function of temperature for a lamellar mesophase sample with the composition 24.3% sodium octanoate, 40.4% decanol, and 35.3% water (●), and for a hexagonal mesophase sample composed of 58.3% sodium octylsulphate and 41.7% water (□) by weight

Nonionic Amphiphile Systems. According to the theoretical considerations, not only charges but also dipoles may be expected to give sizeable field gradients and this is verified by surveying studies of ^{23}Na NMR on mesophase samples composed of nonionic amphiphile, sodium chloride and water (Table II). Apparently, dipolar effects will have to be taken into account in the interpretation of ion quadrupole splittings. The temperature dependence of the splittings has been found to be weak with nonionic amphiphiles.

Zwitterionic Amphiphile Systems. From a biological point of view, studies of mesophases built up of lecithin and other phospholipids are significant since biological membranes contain phospholipid bilayers as an important building-stone. ^{23}Na quadrupole splittings have been obtained, inter alia, for mesophases built up of egg yolk lecithin, NaCl and water, of dimyristoyl lecithin, NaCl and water, of dipalmitoyl lecithin, NaCl and water and of the same components plus cholesterol (29-31). The splitting displays a minimum with varying cholesterol content and increases with increasing temperature and sodium chloride content. The latter observations have led to the suggestion that temperature and electrolyte may induce a conformational change of the phosphorylcholine group (30,31).

Polyatomic Symmetrical Counterions

For polyatomic ions with cubic symmetry such as NH_4^+ or ClO_4^- the interpretation of the quadrupole effects needs a somewhat different approach. Also for these ions the field gradient is zero at the central atom in the absence of intermolecular interactions. The intermolecular forces can create a non-zero quadrupole coupling by two different mechanisms. An external field gradient gives rise to a field gradient at the central atom directly, as with the monoatomic ions. However, the binding of the ion to a lamellar surface can also cause a distortion of the tetrahedral symmetry and the electrons within the ions can then give an effective quadrupole coupling with the nucleus. Which one of these effects that is dominating could be determined by comparison with the quadrupole splitting for a corresponding monoatomic ion.

Quadrupole splittings for symmetrical polyatomic ions have been observed by us for NH_4^+ (11) and by Fujiwara et al. (32) for BF_4^-. However, no systematic studies of these quadrupole splittings have yet been performed.

Finally, it can be mentioned that it is also pos-

Table II

Observed counterion quadrupole splittings of different amphiphilic systems. Data are partly taken from refs. 1, 2, 7, 9-12 and 31. Alkyl chains are written as C_X, where X is the number of carbons in the chain. AOT stands for sodium di-2-ethylhexylsulphosuccinate, EYL for egg yolk lecithin and DML for dimyristoyl lecithin*

Sample composition (%w/w)			Phase	Nucleus studied	Δ_p,kHz
$C_8NH_3Cl-C_{10}OH-H_2O$					
50.0	20.0	30.0	D	^{35}Cl	316
80.0	0	20.0	D	^{35}Cl	446
80.0	0	20.0	D	^{37}Cl	351
$C_{12}N(CH_3)_3Cl-H_2O$					
60.0		40.0	E	^{35}Cl	10.7
70.0		30.0	E	^{35}Cl	7.3
75.0		25.0	E	^{35}Cl	4.3
$C_{16}N(CH_3)_3Br-H_2O$					
65.0		35.0	E	^{81}Br	100
$C_8SO_4Na-C_{10}OH-H_2O$					
34.2	41.3	24.5	D	^{23}Na	16.2
39.5	36.5	24.0	D	^{23}Na	16.0
58.3	0	41.7	E	^{23}Na	9.0
$C_8SO_3Na-C_{10}OH-H_2O$					
43.6	32.0	24.4	D	^{23}Na	12.4
37.8	37.3	24.9	D	^{23}Na	13.9
23.6	38.0	38.4	D	^{23}Na	9.9
$AOT-H_2O$					
25.0	75.0		D	^{23}Na	25.8
37.7	62.3		D	^{23}Na	26.7
50.4	49.6		D	^{23}Na	26.7
89.8	10.2		F	^{23}Na	37.5
$C_7COONa-C_7COOH-H_2O$					
48.1	22.2	29.7	D	^{23}Na	1.1
32.8	36.9	30.3	D	^{23}Na	3.9
45.0	5.0	50.0	E	^{23}Na	1.3

* Phase notations according to Ref. 26.

Sample composition(%w/w)			Phase	Nucleus studied	Δ_p kHz
$C_7COONa-C_{10}OH-H_2O$					
39.2	33.4	27.4	D	^{23}Na	1.0
20.0	44.5	35.5	D	^{23}Na	1.7
31.3	41.0	27.7	D	^{23}Na	2.1
51.0	0	49.0	E	^{23}Na	1.9
42.7	6.9	46.4	E	^{23}Na	1.4
20.2	30.0	49.8	C	^{23}Na	9.4
18.1	27.0	45.9	C	^{23}Na	10.2
9.0	13.0	78.0	B	^{23}Na	not observed
22.2	61.8	16.0	F	^{23}Na	6.6
19.0	63.0	18.0	F	^{23}Na	4.3
16.0	63.0	21.0	F	^{23}Na	2.7
12.0	58.0	30.0	F	^{23}Na	2.6
$C_7COOLi-C_{10}OH-H_2O$					
26.2	35.1	38.7	D	^7Li	0.6
$C_7COOK-C_{10}OH-H_2O$					
30.2	33.2	36.6	D	^{39}K	1.0
$C_7COORb-C_{10}OH-H_2O$					
35.2	30.8	34.0	D	^{87}Rb	18.2
$C_7COOCs-C_{10}OH-H_2O$					
39.6	28.7	31.7	D	^{133}Cs	3.6
Na-Cholate-EYL——H_2O					
16.0	64.0	20.0	D	^{23}Na	4.2
EYL-cholesterol-0.8 M NaCl					
80.0	0	20.0	D	^{23}Na	9.0
55.9	24.7	19.4	D	^{23}Na	12.5
63.6	16.4	20.0	D	^{23}Na	10.5
DML——0.8 M NaCl					
75.0	25.0		D	^{23}Na	5.5
Monooctanoin-0.5 M NaCl					
44.1		55.9	D	^{23}Na	3.1
72.3		27.7	D	^{23}Na	11.3

sible to interpret the quadrupolar coupling of monoatomic ions in terms of distortions (33). We consider,
however, the formalism presented above to be much more
suitable as it allows a direct correlation between the
observed quadrupole splittings and specific molecular
interactions.

Conclusion

It has been our purpose to give a general account
of the quadrupole splitting method for studying ion
binding in mesomorphous systems. Thereby we have presented the relevant theoretical principles and described the general problems of interpretation while the
information provided on ion binding in specific systems is given only in survey. The method is of recent
date and studies performed have had mainly to deal
with the NMR methodological problems. Even so, significant information has been provided on ion binding
phenomena and the quadrupole splitting method shows
great promise in offering a unique possibility of
probing into the geometrical features of the surfaces
of amphiphilic mesophases and biological membrane
systems.

Abstract

Quadrupole splittings in the NMR spectra of counterions may be used to obtain information on the ionic
interactions in surfactant mesophases and model membrane systems. On the basis of a survey of the general
theory, the information provided and types of applications are outlined. Quadrupole splittings have been
obtained for $^7Li^+$, $^{23}Na^+$, $^{39}K^+$, $^{87}Rb^+$, $^{35}Cl^-$, $^{37}Cl^-$
and $^{81}Br^-$ using either continuous-wave or pulse NMR
methods. In the case of strong quadrupole interactions,
second-order effects may dominate the appearance of
the spectrum. Multiple quantum transitions may often
be observed under typical experimental conditions and
constitute a possible source of misinterpretation.
Experimental counterion quadrupole splittings were
obtained as a function of sample composition and phase
structure for different surfactant systems and their
implications on counterion binding are discussed. ^{23}Na
quadrupole splittings vary considerably both in their
magnitude and in their temperature and concentration
dependences with the amphiphile polar head group demonstrating a marked specificity in the counterion
binding. On the basis of previous work deductions are
made about the values of the order parameters charac-

terizing the partial orientation of the field gradients of bound counterions. Tentative conclusions about the geometrical structure of the surfaces of amphiphilic aggreates are presented.

Literature Cited

1. Lindblom, G., Wennerström, H. and Lindman, B., Chem. Phys. Lett. (1971) **8**, 849.
2. Wennerström, H., Lindblom, G. and Lindman, B., Chem. Scr. (1974) **6**, 97.
3. Cohen, M.H. and Reif, F., Solid State Phys. (1957) **5**, 321.
4. Buckingham, A.D. and McLauchlan, K.A., Progress NMR Spectroscopy (1967) **2**, 63.
5. Diehl, P. and Khetrapal, C.L. in "NMR,Basic Principles and Progress" P. Diehl, E. Fluck and R. Kosfeld, Ed., Vol. 1, p. 1, Springer Verlag, Berlin 1971.
6. Luckhurst, G.R. in "Liquid Crystals and Plastic Crystals" G.W. Gray and P.A. Winsor, Ed., Vol. 2, p. 144, Ellis Horwood Ltd., Chichester, U.K., 1974.
7. Lindblom, G., Acta Chem. Scand. (1971) **25**, 2767.
8. Lindblom, G., Acta Chem. Scand. (1972) **26**, 1745.
9. Lindblom, G. and Lindman, B., Mol. Cryst. Liquid Cryst. (1973) **22**, 45.
10. Lindblom, G., Persson, N.-O. and Lindman, B. in "Chemie, Physikalische Chemie und Anwendungstechnik der grenzflächenaktiven Stoffe", Vol. II, p. 939, Carl Hanser Verlag, München, 1973.
11. Gustavsson, H., Lindblom, G., Lindman, B., Persson, N.-O. and Wennerström, H. in "Liquid Crystals and Ordered Fluids" J.F. Johnson and R.S. Porter, Ed., Vol. II, p. 161, Plenum Press, New York, 1974.
12. Lindblom, G., Lindman, B. and Mandell, L., J. Colloid Interface Sci. (1973) **42**, 400.
13. de Vries, J.J. and Berendsen, H.J.C., Nature (London) (1969) **221**, 1139.
14. Black, P.J., Lawson, K.D. and Flautt, T.J., Mol. Cryst. Liquid Cryst. (1969) **7**, 201.
15. Long, Jr. R.C. and Goldstein, J.H., Mol. Cryst. Liquid Cryst. (1973) **23**, 137.
16. Lindblom, G., Wennerström, H. and Lindman, B., J. Magn. Resonance, in press.
17. Rosenholm, J.B. and Lindman, B., J. Colloid Interface Sci. in press.
18. Fontell, K., Mandell, L., Lehtinen, H. and Ekwall, P., Acta Polytech. Scand., Chem. Incl. Met. Ser. (1968) **74**, (III) 1.
19. Lindblom, G., Lindman, B. and Tiddy, G.J.T., Acta Chem. Scand. (1975) **A29**, 876.

20. Mukerjee, P., Mysels, K.J. and Kapauan, P., J. Phys. Chem. (1967) 71, 4166.
21. Lindman, B. and Brun, B., J. Colloid Interface Sci. (1973) 42, 388.
22. Hertz, H.G., Ber. Bunsenges. Phys. Chem. (1973) 77, 531.
23. Lindman, B. and Ekwall, P., Mol. Cryst. (1968) 5, 79.
24. Gustavsson, H. and Lindman, B., J. Amer. Chem. Soc. (1975) 97, 3923.
25. Lindman, B. and Forsén, S., "Chlorine, Bromine and Iodine NMR, Physico-Chemical and Biological Applications", Vol.12 of "NMR, Basic Principles and Progress" P. Diehl, E. Fluck and R. Kosfeld, Ed., Springer Verlag, Berlin, in press.
26. Ekwall, P., Adv. Liquid Cryst. (1975) 1, 1.
27. Tiddy, G.J.T., J. Chem. Soc. Faraday Trans. 1 (1972) 369.
28. Persson, N.-O. and Lindman, B., J. Phys. Chem. (1975) 79, 1410.
29. Persson, N.-O., Lindblom, G., Lindman, B. and Arvidson, G., Chem. Phys. Lipids (1974) 12, 261.
30. Lindblom, G., Persson, N.-O., Lindman, B. and Arvidson, G., Ber. Bunsenges. Phys. Chem. (1974) 78, 955.
31. Lindblom, G., Persson, N.-O. and Arvidson, G., Adv. Chem. Ser., in press.
32. Fujiwara, F., Reeves, L.W. and Tracey, A.S.,J. Amer. Chem. Soc. (1974) 96, 5250.
33. Radley, K. and Reeves, L.W., Can. J. Chem. (1975) 53, 2998.

An Approach to the Quantitative Study of Internal Motions in Proteins by Measurements of Longitudinal Relaxation Times and Nuclear Overhauser Enhancements in Proton Decoupled Carbon-13 NMR Spectra

STANLEY J. OPELLA
Department of Chemistry, Stanford University, Stanford, Calif. 94305
DONALD J. NELSON and OLEG JARDETZKY
Stanford Magnetic Resonance Laboratory, Stanford University, Stanford, Calif. 94305

The increasing use of high resolution NMR for the study of protein structure has drawn renewed attention to the existence of segmental motions in protein molecules, with different domains in the protein structure and different amino acid side chains displaying different degrees of motional freedom. Motions in the nanosecond range, i.e., slow in comparison with bond vibrations, but rapid in comparison with the overall tumbling rate of the macromolecule have been detected early (1,2). Several recent reports (3-9) have focused on specific residues, mostly aromatic, and demonstrated that phenylalanine and tyrosine side chains, even when embedded in the interior of a protein may retain a sufficient degree of rotational freedom to average out the chemical shift nonequivalence arising from secondary and tertiary structure. At least one instance of a motion sufficiently restricted, so that chemical shift averaging does not occur is also known (10).

Most of the early, as well as more recent observations permit no more than the qualitative statement that internal motions which are efficient in averaging out chemical shift, (i.e. $\nu > 10^0\text{-}10^2 \text{ sec}^{-1}$) or more efficient in producing relaxation than overall molecular tumbling (i.e. $\nu > 10^6 \text{ sec}^{-1}$) do occur. The reason for this is that in most instances only limits to the rates of motion can be obtained from chemical shift and line width measurements. Often these limits are several orders of magnitude removed from the actual rate. At the same time calculation of correlation times from a single relaxation time measurement is subject to numerous uncertainties (11).

Our concern has been with the quantitative evaluation of the correlation times characteristic of the different types of motion. To accomplish this we have found it necessary to obtain measurements of at least two different NMR relaxation parameters and an independent measurement of the overall tumbling rate by a different method - e.g. depolarized light scattering (5,10,12). The requirement implicit in this procedure, that two or more measured relaxation parameters be accounted for by the same model and the

same correlation time, severely restricts the number of plausible
models and limits the number of alternative interpretations. An
additional requirement is that the relaxation measurements be of
sufficient accuracy to allow discrimination between different
models.

^{13}C NMR spectroscopy is particularly useful in the approach,
since it offers the possibility of measuring two parameters, the
longitudinal relaxation time T_1 and the Nuclear Overhauser
Enhancement (NOE), whose functional dependence on correlation
times is substantially different (13). Although measurements of
the transferse relaxation time T_2 can in principle be used for the
same purpose, they are experimentally more difficult on large
molecules. T_1 measurements of sufficient accuracy can be obtained
by the inversion recovery method (14), if one carefully guards
against errors introduced by incomplete inversion (low power of
the 180° pulse). The problem of obtaining accurate NOE measurements is more difficult. The requirement for complete decoupling
when observing spectra of large complex molecules with many overlapping resonances usually precludes the use of the most expedi-
ent method of measuring NOE values - i.e. the comparison of line
intensities in the decoupled and the undecoupled spectrum. To
overcome this limitation we have developed a technique for measuring the proton decoupled ^{13}C spectrum without Overhauser effect
contributions to signal intensities (15). The technique is a var-
iant of the well known gated decoupling procedure (16,17) and is
based on the fact that radio frequency irradiation of the protons
has an essentially instantaneous effect on the scalar spin-spin
coupling, while the polarization resulting from the same irradiation develops with a time constant equal to the spin-lattice
relaxation time (T_1). The solution of the applicable equations of
motion (18) demonstrates that with an appropriate choice of delay
times it is possible to observe a fully decoupled spectrum without NOE. The magnitude of the NOE can therefore be evaluated by
simply taking a difference spectrum between two fully decoupled
spectra. The sequence of events, for the gated decoupling experiment as used here is outlined in Figure 1. It consists of turning on the proton radiation simultaneously with a 90° rf pulse
at the carbon frequency. The free induction decay is collected
immediately afterward, with proton decoupling for the shortest
acquisition time consistent with the necessary spectral resolution. Data collection is followed by a delay time during which
the decoupler is turned off.

The rf pulse at time a in the outlined sequence samples the
magnetization of the carbon nuclei by rotating the total magnetization vector from the z direction into the x,y plane, where the
detection of the free induction decay occurs. The signal intensities in the final Fourier transformed spectrum represent the
magnetization at the time of the pulse. The decoupler is on during data acquisition in order to obtain a completely proton decoupled ^{13}C spectrum. The procedure of turning the decoupler off

during the delay time T preceding signal sampling, from b̲ to a̲ in
the diagram, has the function of insuring an equilibrium Boltzmann
distribution at the required time. This gated decoupler experiment requires considerably longer delay periods for nuclear relaxation than experiments involving continuously decoupled systems
because it is limited by spin relaxation when the decoupler is
turned off.

Since the main mechanism of spin-lattice relaxation of carbon
nuclei involves interactions with protons, the observed relaxation
behavior of carbons can be described in terms of a two spin system. In particular for carbons with directly attached hydrogens
the dipolar mechanisms have been demonstrated to be overwhelmingly
dominant (19). The motion-induced fluctuations of the local dipolar magnetic field are the source of the relaxation.

In order to calculate the observable quantities for the nu-
clear spins ($\langle I_z \rangle$ and $\langle S_z \rangle$), the populations of all levels must
be taken into account. With the equilibrium magnetizations
(Boltzmann distributions) defined as I_o and S_o, the time dependence of the observable quantities can be described in terms of the
transition probabilities (W_i).

$$\frac{d\langle I_z \rangle}{dt} = -\rho \, (\langle I_z \rangle - I_o) - \sigma(\langle S_z \rangle - S_o) \tag{1}$$

$$\frac{d\langle S_z \rangle}{dt} = -\rho' (\langle S_z \rangle - S_o) - \sigma(\langle I_z \rangle - I_o) \tag{2}$$

where: $\rho = W_o + 2W_{1_I} + W_2$

$\rho' = W_o + 2W_{1_S} - W_2$

$\sigma = W_2 - W_o$

Steady-state measurements of ^{13}C magnetization (I nucleus)
means that $d\langle I_z \rangle/dt = 0$, while the effect of saturation of the
proton resonances (S nucleus) is to make $\langle S_z \rangle = 0$. Therefore
equilibrium measurements of carbon magnetization under decoupled
conditions gives values for $\langle I_z \rangle$ which are increased over I_o.
Solving equation {1} for $\langle I_z \rangle$ under these conditions yields:

$$\langle I_z \rangle = I_o + \frac{\sigma S_o}{\rho} \tag{3}$$

Equation $\{3\}$ indicates that ^{13}C magnetization ehancement from the nuclear Overhauser effect is a direct consequence of the nuclear spin interactions under steady state decoupled conditions. The magnitude of the Overhauser enhancement is commonly given by y, a normalized ratio of σ and ρ

$$y = \frac{\sigma}{\rho} - \frac{S_o}{I_o} \qquad \{4\}$$

and $NOE = 1 + y$

By defining $I\eta = \frac{\sigma}{\rho} \cdot S_o = I_o \cdot y$ the carbon magnetization can be conveniently expressed as:

$$I_z = I_o + I_\eta \qquad \{5\}$$

The magnetization of a carbon resonance with complete proton decoupling applied continuously for a time long enough to obtain a steady-state level can be described by an expression parallel to equation $\{5\}$.

$$I_o^{\infty} = I_o + I_\eta \qquad \{6\}$$

The magnetization at an arbitrary instantaneous time in the z direction is designated as I_z^i. The time dependenze of the z magnetization of the observed ^{13}C resonance in the proton decoupled experiment can be described by a single exponential function in the integrated form of equation $\{1\}$.

$$I_z^t = I_z^{\infty} - (I_z^{\infty} - I_o^i) \, e^{-t/T_1} \qquad \{7\}$$

The instantaneous magnetization I_z^b at the end of the acquisition time period $(\underline{b})$, of the gated pulse sequence of Figure 1 is given by

$$I_z^b = (I_o + I_\eta) - \{(I_o + I_\eta)\} - I_z^a \, e^{-t/T_1} \qquad \{8\}$$

The magnitude of I_z^b results from an exponential increase in intensity from the level I_z^a, the value immediately after the rf pulse, and including contributions from the nuclear Overhauser effect.

In order to measure NOE a comparison must be made between a spectrum obtained with the decoupler on continuously and a spectrum under gated decoupler conditions. These two cases for the time interval from b to c of Figure 1 must be treated separately. For the decoupled spectrum with full NOE, the decoupler is on during the delay period and the time dependence of the magnetization can be described by the same relationship used for the data acquisition time period. At time c, after the delay T, the magnetization with continous decoupling is given by

$$I_z^c = I_z^\infty - (I_z^\infty - I_z^a)\, e^{-t/T_1} \qquad \{9\}$$

I_z^c may also be computed by considering the magentization as subject to a single relaxation time from the starting value at a to the final value at c, since the nuclear spins are not affected by data acquisition.

$$I_z^c = I_o - (I_o - I_z^a)\, e^{\frac{-(t + t)}{T_1}} \qquad \{10\}$$

When this pulse sequence employs a 90° pulse, $I_z^a = 0$ and equation $\{10\}$ simplifies to

$$I_z^c = I_o \left(1 - e^{\frac{-(t + T)}{T_1}}\right) \qquad \{11\}$$

In order to obtain equilibrium intensities, the pulse repetition interval (t + T) must be sufficiently long to allow complete nuclear relaxation.

When the proton frequency radiation is turned off during the delay period (T), the description of the time dependence of the magnetization of the carbon spins is considerably more complicated than the single exponential functions used in the case above. Because the mutual interaction of coupled ^{13}C and ^1H spins results in non-exponential relaxation behavior, it is necessary to consider the transient magnetizations of both nuclei. The time dependencies applicable are those of equations $\{1\}$ and $\{12\}$. For molecules undergoing rapid reorientation, the frequency terms of ρ, ρ' and σ are negligible, therefore $\rho = \rho'$. Consequently, the relationships can be simplified, and the solution for the ^{13}C nucleus can be written:

$$\langle I_z \rangle = I_o + \tfrac{1}{2}(\langle I_z(c) \rangle - I_o^{eq})\ (e^{-(\rho+\sigma)T} + e^{-(\rho-\sigma)T})$$

$$+ \tfrac{1}{2}(\langle S_z(c) \rangle - S_o^{eq})\ (e^{-(\rho+\sigma)T} - e^{-(\rho-\sigma)T}) \qquad \{12\}$$

I_o^{eq} and S_o^{eq} designate the respective equilibrium values of the magnetizations under the given experimental conditions. Equation $\{12\}$ has been solved previously for a variety of boundary conditions (19,20). The applicable initial and final conditions for the delay period T_1 with no proton irradiation are: carbon nuclei start at $I_z(c) = I_z^b$ and relax to $I_o^{eq} = I_o$ at time $\underline{c}$ with no contribution from I_n, while the hydrogen nuclei start with $S_z(0) = 0$ as a result of the decoupling-induced saturation during the acquisition time period and relax $S_o^{eq} = S_o$. The substitution of these limiting conditions into equation $\{12\}$ gives a value of I_z^c.

$$I_z^c = I_o + \tfrac{1}{2}(I_z^b - I_o)\ (e^{-(1+\eta\frac{\gamma I}{\gamma S})\frac{T}{T_1}} + e^{(1-\eta\frac{\gamma I}{\gamma S})\frac{T}{T_1}})$$

$$-\tfrac{1}{2}(\frac{\gamma S}{\gamma I})\ I_o\ (e^{-(1+\eta\frac{\gamma I}{\gamma S})\frac{T}{T_1}} - e^{(1-\eta\frac{\gamma I}{\gamma S})\frac{T}{T_1}}) \qquad \{13\}$$

The function depends explicitly on t, T, T_1 as well as γ.

The relationships describing explicitly the time dependence of the carbon magnetization after removal of proton irradiation hold without qualification for molecules with short effective correlation times. However, in macromolecular systems nuclear relaxation behavior is determined by relatively slow molecular motions, and ρ no longer is equal to ρ'. Without this simplification the expressions for the time dependence of $\langle I_z \rangle$ following removal of decoupling are more complex. The exponential terms of equation $\{12\}$ are found to be modified, with

$$\rho \pm \sigma = \frac{\rho + \rho'}{2} \pm \tfrac{1}{2}\{(\rho-\rho')^2 + 4\sigma^2\}^{\frac{1}{2}} \qquad \{14\}$$

In the limit as ρ approaches ρ', the expressions for $\langle I_z \rangle$ with rapid and slow motions converge. With ρ much different from ρ', the relaxation behavior cannot be described exactly. This is particularly true for highly anisotropic motions which may be too complex to be described within the existing theoretical framework.

However, qualitative limits can be placed on the deviations from exponential relaxation behavior, since as the Overhauser enhancement becomes very small the strength of the nuclear spin coupling between the two kinds of spins decreases. Without the perturbations of strong nuclear couplings the relaxation becomes exponential in nature. The presence of less than maximal NOE reflects primarily a reduced σ term, since the difference between the transition probabilities W_2 and W_{ij} is less. W_2 is the first transition probability to be affected by a reduction in the rate of molecular motions and W_1 the last (21). Because the sensitivities of the W_{ij} to intermediate rotational correlation times arising from the different frequency dependent terms vary, an appreciable reduction of the NOE results from relatively small deviations from extreme narrowing conditions.

For accurate equilibrium intensity measurements on macromolecular systems the delay time with the decoupler turned off must be of sufficient length. A conservative choice of delay period can be made by considering the requirements of a tightly coupled spin system, since in a "noninteracting" spin system exponential relaxation occurs, which requires less time for the magnetization to reach equilibrium. The longest delay necessary is for rapidly tumbling molecules; if a delay period is calculated for a NOE value according to equation {13}, then it may be considered adequate for the case of a macromolecule. The examples given in this paper illustrate that these requirements are flexible enough to permit accurate measurements. NOE measurements of protons can be routinely performed in a 23.5 kG magnetic field with an acquisition time of 0.4 sec and a delay of 2.0 sec.

Example I: Small Molecules. If the ^{13}C resonance of interest from a molecule undergoing rapid reorientation has a relaxation mechanism resulting from dipolar interactions with protons, the relaxation behavior is independent of resonance frequency and is a function of the effective correlation time. The ^{13}C T_1 is also sensitive to the number of bonded protons. The interpretation of relaxation measurements depends on several assumptions, the most critical of which is that of the dominant relaxation mechanism. The nuclear Overhauser enhancement has one upper limiting value of 3.0 and is independent of both frequency and correlation time. The theoretical maximum value holds for dipolar relaxation. The presence of other contributing mechanisms reduces the NOE. For this reason the parameter has proven useful in detecting the presence of alternative relaxation mechanisms.

It is informative to observe a completely decoupled spectrum without NOE in order to make comparisons with the continuously decoupled spectrum in the case of molecules with simple spectra. The spectrum without NOE can be used to obtain information from integrated intensities for the purpose of assigning resonances and counting the number of nuclei of a particular type. With variable nuclear Overhauser enhancements throughout the spectrum it is not possible to obtain much information from signal intensities. This is a particularly important problem in the study of incorporation of ^{13}C labeled metabolic precursors into natural products.

When making the comparison, the length of time the decoupler must be turned off in order to obtain a spectrum without NOE is prescribed by equation {13}. The usual practice for accurate nuclear resonance relaxation measurements is for equilibrium signal intensities to be within 1% of their theoretical value. For an equilibrium spectrum the repetition rate must be about $5 \times T_1$ of the signal with the longest relaxation time. In the continuously decoupled spectra this is also an appropriate value for $(t + T)$.

The non-exponential character of equation {12} plays a critical role in determining the delay period required to obtain a spectrum without nuclear Overhauser enhancements present. Equation {13} was programmed on an IBM 360 computer with the appropriate values for the physical constants. Figure 2 illustrates the dependence of the value of the sampled magnetization upon the time of turning the decoupler off. The curves are calculated for a resonance which has NOE = 3.0, and the non-exponential relaxation behavior is obvious. The effect of varying the acquisition time is also demonstrated. The concept of a short acquisition time allowing only a small amount of NOE to build up is clearly inappropriate, since in the case where the acquisition period is not sufficient to allow the magnetization to develop beyond the I_0 level, the interaction of the spins causes the magnetization to exceed that value before returning to the equilibrium level. It is also clear from these results that a delay period significantly longer than that used in the inversion-recovery experiment is required to obtain a true equilibrium value of the magnetization. Calculations for a resonance having NOE = 1.5 (not shown) reinforce these conclusions.

An experimental test of these results was made with alanine in solution. In this case the methyl carbon was found to have $T_1 = 1.8 \pm .1$ sec using both inversion-recovery and progressive saturation methods of measurement, in good agreement with a previously reported value (22). This value was used in the program to calculate the ratios as plotted. The experimental value of the NOE is 2.8 for the methyl carbon, while the value for the α carbon is 3.0. The methyl carbon has an NOE slightly reduced from the maximal value; therefore, there is probably a small contribution to the carbon-13 relaxation mechanism from non-dipolar processes. This carbon resonance provides a good test of the theory since the NOE is reduced, illustrating that the equilibrium magnetization levels are affected only by the dipolar part, but that the time course can be determined by normal spin-lattice processes. The fact that it is a methyl group also tests the ideas about equivalent proton spins fulfilling the theoretical requirements for treatment by the two spin approximation.

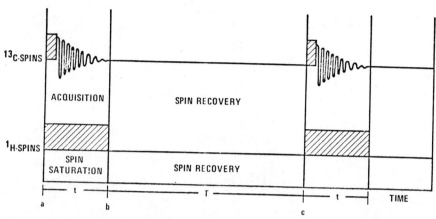

Figure 1. Schematic of the gated decoupler pulse sequence used to obtain carbon-13 magnetic resonance spectra which are proton decoupled but have no intensity contributions from NOE. Acquisition time is t and the added delay with the decoupler turned off is T.

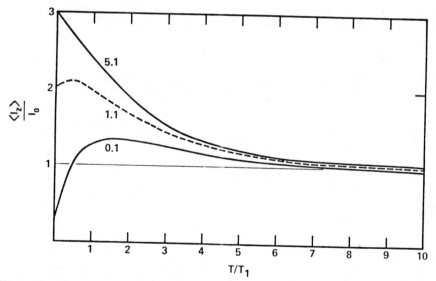

Figure 2. Theoretical plot of the relative intensity of the magnetization ratio to I_o (Boltzmann distribution of spins) vs. the ratio of the delay time with the decoupler turned off to the longitudinal relaxation time. Three different values of the ratio of the acquisition time to the longitudinal relaxation time (t/T_1) are shown as numbers near solid line plots. All lines generated from equation {13} of text. NOE = 3.0.

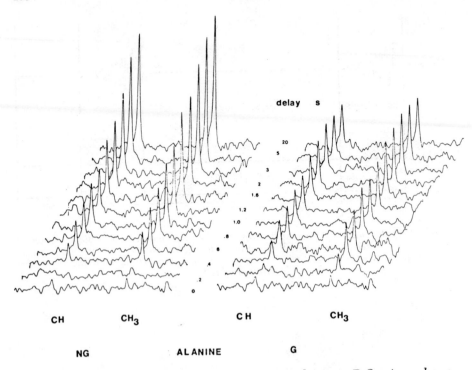

Figure 3. *Stacked plot of the decoupled spectrum of alanine in D_2O using pulse sequence of Figure 4. Comparison of gated and nongated decoupling conditions.*

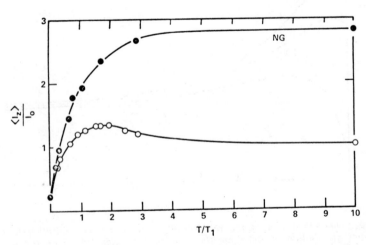

Figure 4. *Comparison of experimental data for the magnetization from Figure 3 with theoretical plots. Circles are experimental values. Lines are generated from equation {13} of text.*

The experimental measurements of the magnetization were made by following the pulse sequence of Figure 1. The delay time, T, was varied while keeping the acquisition time, t, a constant. Figure 3 is a stacked plot of the spectra for both the carbons in the aliphatic portion of the spectrum with the indicated delay times. The dependence of the magnetization upon the ratio of delay to T_1 for the methyl group is presented in Figure 4. The experimental points fit very well to the theoretical curves calculated from equation {13} for the experimental values of NOE and T_1, implying that the complex behavior of the nuclear relaxation with the decoupler turned off during the delay is described well by the theoretical treatment used here. As expected with continuous decoupling, the relaxation is described by a single exponential function with a time constant T_1; the theoretical line is from equation {11}.

    Example II: Large Molecules. In larger, more slowly diffusing, molecules NOE can be reduced from the maximal value even if relaxation is completely dipolar in character. Consequently, those carbon atoms whose motion is restricted to the degrees of freedom given by the diffusion of the protein as a whole may be expected to have low values of NOE. In contrast, carbons in segments with more degrees of motional freedom could approach the maximal value.

    The expectation is generally borne out for globular proteins. Experimental NOE data for one such protein, the muscle calcium binding protein from carp (MCBP) are shown in Figure 5. The carboxyl and carbonyl carbon resonances appear between 125 and 140 ppm. A number of the signals in this region are single carbon resonances. For example, all MCBP's isolated to date contain an invariant arginine residue (Arg-75); the guanido carbon appears as a single carbon resonance at 158.8 ppm. The resonances from the ten phenylalanine residues constitute the major peak in the aromatic region (MCBP contains one histidine residue and no tryptophane). The relatively sharp signals from 135 to 140 ppm arise from the γ-carbons of the phenylalanine and the heterogeneity of these resonances is indicative of the substantial chemical shift nonequivalence induced by the native protein structure. A number of the phenylalanine γ-carbons appear as single peaks. The phenylalnine ring carbons with bonded hydrogens contribute to the broad resonance band centered at 129.5 ppm. The resonance signals from the α-carbons of the protein backbone appear between 50 and 60 ppm. Methylene and methyl carbons occur between 10 and 40 ppm. This upfield region is dominated by the ε-methylene carbons of thirteen lysine residues (39 ppm) and by the β-methyl carbons of twenty alanine residues (∼16 ppm). The resonances at highest field (10 to 14 ppm) are derived from the δ-methyl groups of the five isoleucine residues. Two of these methyl groups appear as single peaks (10.7 and 11.2 ppm).

The integral curves above the spectrum refer to signal intensities obtained under different conditions: (a) with continuous decoupling showing the full NOE, (b) with gated decoupling eliminating NOE. The difference curve (c) therefore gives a measure of NOE for different types of carbons.

Evaluation of Correlation Times from Simultaneous T_1 and NOE Measurement.

A detailed description of the motional state of a protein consists of an assignment of an individual correlation time to every independently movable group or amino acid residue. Such an assignment is possible, at least in principle, because of a direct theoretical connection between the autocorrelation function $G(\tau)$ which describes the time dependence of the nuclear interactions as modulated by motion, and the measurable relaxation parameters (T_1, T_2, NOE). Since the correlation times derived from any set of relaxation measurements depend on the model used in the calculation, a brief statement of the theory as used here is given below. The cases of isotropic and anisotropic relaxation are treated separately.

(a) Isotropic Reorientation. The rotational diffusion coefficient, D, calculated from the Stokes-Einstein equation is appropriate for the description of the reorientation of small globular proteins in solution. This Debye motion is characterized by random steps of small angular displacements. The ^{13}C nuclear relaxation parameters can be readily calculated for this type of motion, since the autocorrelation function is a single exponential with a time constant that is the rotational correlation time $\tau_c = 1/6D$.

The appropriate equations for the interpretation of the data on MCBP are readily derived from the given spectral density

$$\frac{1}{T_1} = \frac{1}{10} \frac{K^2}{r^6} X$$

$$NOE = 1 + \frac{\gamma_H}{\gamma_C X} \left\{ \frac{6\tau_c}{1 + (\omega_H + \omega_C)^2 \tau_c^2} - \frac{\tau_c}{1 + (\omega_H - \omega_C)^2 \tau_c^2} \right\}$$

$$X = \frac{\tau_c}{1 + (\omega_H - \omega_C)^2 \tau_c^2} + \frac{3\tau_c}{1 + \omega_C^2 \tau_c^2} + \frac{6\tau_c}{1 + (\omega_H + \omega_C)^2 \tau_c^2}$$

where $K = \hbar \gamma_H \gamma_C$, r = carbon-hydrogen distance, ω_H = proton resonance frequency, ω_C = carbon resonance frequency, γ_H = proton gyromagnetic ratio, γ_C = carbon gyromagnetic ratio. These functions are plotted in Figure 6 as a function of rotational correlation time. As shown the values are appropriate for a ^{13}C with one bonded proton, like an α-carbon of an amino acid; in order to apply these plots to other groups the T_1, but not the NOE, must be adjusted by dividing by the number of attached hydrogens.

The overall reorientation rate of the native protein can be obtained by considering the α-carbons sterically restricted so that their effective correlation time is the same as that of the protein. Measurement of nuclear relaxation time is the same as that of the protein. Measurement of nuclear relaxation parameters of α-carbons are necessarily approximate because individual signals cannot be resolved and the behavior of an entire spectral region must be evaluated. This approach has considerably greater validity in the present case of a protein with multiple α-helices than in general. Figure 6 shows that the experimental T_1 (55 ms) and NOE (1.2) can only be consistent with a rotational correlation time of about 12 nsec. We have compared the NMR measurements to those of depolarized light scattering (5, 12) and found that the two methods monitor the same motions, with the measured correlation times being within experimental error. In addition, the finding that the relaxation parameters for the α-carbons of this protein are the same as those calculated from the overall tumbling implies that these carbons are indeed rigidly held. On the basis of the fit of the NMR parameters to the data of Figure 6 and the single Lorentzian line shape for the depolarized Raleigh spectrum, the motion of the protein is accurately described as isotropic.

(b) Anisotropic Reorientation: The Intramolecular Motions of Amino Acid Side Chains. The determination of the rates of internal motions of amino acid side chains in a protein structure is a significantly different problem from the calculation of an overall isotropic correlation time. One way of analyzing the motions of side chain groups is to consider each C-H group attached to a rigid body. It is possible to calculate the relaxation parameters for one degree of internal rotation superimposed on the overall isotropic reorientation. However in this case the spectral densities are functions of all the components of the rotational diffusion tensor. Following Woessner (23), the (C-H) spin pair motion can be treated as a random reorientation about an axis attached to a framework which can itself tumble isotropically. The internal rotation is specified by a correlation time, τ_1, and an angle, θ, which the axis of rotation makes with respect to the axis of the body undergoing isotropic rotation. In this model, the motions are considered as stochastic diffusion among a large number of equilibrium positions, so that the diffusion constant for the internal rotation is the inverse of $6\tau_1$. As for the case of isotropic motions, direct calculation of all relaxation parameters is possible:

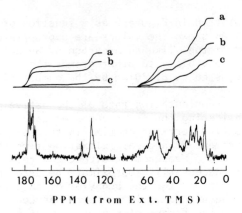

Figure 5. The proton decoupled natural abundance carbon-13 magnetic resonance spectrum of MCBP. (a) The integral of the spectrum with continuous proton noise irradiation. (b) The integral of the gated decoupler spectrum using pulse sequence of Figure 1. It contains no intensity contributions from NOE. The spectrum was obtained with an acquisition time of 0.311 sec followed by a delay of 2.0 sec. (c) The difference spectrum obtained by subtracting b from a.

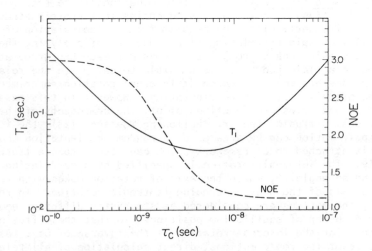

Figure 6. Calculated dipolar T_1 and NOE dependence on isotropic reorientation time τ_c for a ^{13}C nucleus relaxed by a single hydrogen at a distance of 1.09Å in a magnetic field of 23.5 kG

$$\frac{1}{T_1} = \frac{A}{T_{1_c}} + \frac{B}{T_{1_B}} + \frac{D}{T_{1_D}}$$

where $A = \frac{1}{4} + (3\cos^2\theta - 1)^2$

$$B = 3\sin^2\theta \cos^2\theta$$

$$\frac{1}{T_2} = \frac{A}{T_{2_c}} + \frac{B}{T_{2_B}} + \frac{D}{T_{2_D}}$$

$$D = \frac{3}{4}\sin^4\theta$$

$$NOE = 1 + \frac{R(A\phi_c + B\phi_B + D\phi_D)}{(AX_c + BX_B + DX_D)}$$

$$\frac{1}{T_{1_j}} = \frac{1}{10}\frac{K^2}{r^6}X_j$$

$$\frac{1}{T_{2_j}} = \frac{1}{20}\frac{K^2}{r^6}X_j + 4\tau_j + \frac{6\tau_j}{(1+\omega_H^2\tau_j^2)}$$

$$X_j = \frac{\tau_j}{1 + (\omega_H - \omega_c)^2\tau_j^2} + \frac{3\tau_j}{1 + \omega_c^2\tau_j^2} + \frac{6\tau_j}{1 + (\omega_H + \omega_c)^2\tau_j^2}$$

$$\phi_j = \frac{6\tau_j}{1 + (\omega_H + \omega_c)^2\tau_j^2} - \frac{\tau_j}{1 + (\omega_H + \omega_c)^2\tau_j^2}$$

The calculations of these parameters of $\theta = 109.5°$ (the tetrahedral carbon bond angle) and $\theta = 60°$ (for internal reorientation of phenyl side chains about the C_β-C_γ bond axis) are given in Figure 7 for a rigid body with an isotropic reorientation time of 12 nsec.

A carbon-hydrogen pair in an amino acid side chain can generally have several different intramolecular motions. The analysis of motions in terms of one degree of internal rotation is restricted to a few structural groups, where estimates of the specific type of motion can be made and an internal correlation time, τ_1, can provide some insight into the protein structure dynamics. The β carbons of amino acids are expected, in general, to fall into the category of single internal rotation with respect to an adjacent α-carbon undergoing isotropic reorientation. The β carbons of lysine and valine contribute to the resonance band between 24 and 27 ppm. These signals have a relatively homogeneous NOE of 1.4 and an observed T_1 of 50 msec ($2 \cdot T_1 = 100$ msec). There are no isotropic correlation times that simultaneously satisfy these two relaxation parameters. This finding by itself is strong evidence for the motions of these carbons being anisotropic in nature. If the β carbon-hydrogen dipolar vector rotates about the C_β-C_α bond, then the angle of rotation (θ) can be reasonably defined as 109.5°, the value appropriate for a tetrahedral carbon. Referring to Figure 7, the experimental values can be accounted for by this model with an internal rotation of $\tau_1 = 10$ nsec. A similar treatment applied to the resonance band arising from the carbons of asp, leu, and phe results in the finding of a considerably slower rate of internal reorientation.

The β methyl carbons of alanine residues would also be expected to have internal motions that could be reasonably explained by a single internal rotation superimposed on the overall motions of the protein. However, the measured values of $NT_1 = 650$ msec and NOE = 1.8 cannot be accounted for with either the isotropic model or the single internal rotation model. There are several possible sources for this failure of the motional models. The alanine resonance peak at about 16 ppm results from twenty different methyl groups, therefore the relaxation behavior is an average of different motional behaviors, even though the chemical shift is relatively homogeneous. It is also possible that the motions of the methyl groups have more than one additional degree of internal freedom, in which case the appropriate angle for internal rotation is not 109.5°, but some other value.

Most side chain carbon relaxation parameters do not fit at all to the isotropic reorientation model. If the carbons are not in the β position (e.g., carbons further removed from the protein main chain), it is difficult and inappropriate to define an appropriate physical model, with only one degree of internal freedom. For example, the lysine ε carbons in native MCBP have an apparent T_1 of 170 msec ($NT_1 = 340$ msec) and NOE = 2.5. If these carbons were experiencing isotropic reorientation, the T_1 could correspond to $\tau_c = .1$ nsec or $\tau_c = 100$ nsec, however NOE = 3.0 is predicted for the former and NOE 1.15 for the latter. On the other hand, in order to accommodate the experimental NOE value with the isotropic model NT_1 of 70 msec would be necessary, corresponding to an experimental prediction of $T_1 = 35$ msec which is a factor of 5 smaller than the measured value. A more general theoretical treatment of anisotropic motion, with less restrictive models must be applied and is being developed (24).

(b') Special Case: Motions of Phenyl Groups. A specific case in which a more rigorous analysis is possible is that of the phenyl groups. Phenyl groups can undergo reorientation about two different internal axes, C_β-C_γ or C_β-C_α, in addition to the isotropic reorientation of the framework for the molecular axes.

In order to describe the phenyl group motions three physically
distinct correlation times must be considered. Even though the
motions described by these correlation times are independent,
their effects on relaxation are not. Several simplifications are
necessary in order to treat this problem. The most important step
is to employ the measurement of the overall correlation time of
the protein and to use the α-carbon backbone as the isotropically
reorienting framework for the internal axis of rotation. Previous
attempts at describing the intramolecular motions of phenyl groups
in macromolecules by carbon nuclear resonance have been focused on
the model systems of polystyrene (25) and poly (γbenzyl) glutamate
(26). The helical conformation of polybenzyl glutamate may re-
semble a protein structure, however the carbon-13 spectrum suffers
from lack of resolution among the different ring carbons. The
progression of values of relaxation parameters along the side
chain positions indicates that faster motions are present as a
carbon becomes further removed from the α-carbon backbone.

Carbon Position	T_1 (msec)	NOE
CH_2	50	2.0
CH	90	2.1
C_1	--	2.0
$C_{2,3}$	110	2.3
C_4	100	2.1

Table I

Polystyrene Relaxation Data

The intramolecular motions in polystyrene have been described in terms of relatively rapid local segmental motions.
While an effective isotropic rotational correlation time has been
assigned to these rotations on the basis of the T_1 values, a different time is obtained from the NOE measurements from the aliphatic and quaternary phenyl carbons. This situation is analogous
to that for amino acid side chains where no value of τ_c (isotropic) can account for both the T_1 and the NOE data. We have
measured the T_1 and NOE of a 20% solution (W/V) of 25,000 MW
polystyrene. No previous NOE measurements have been made on the
phenyl ring carbons and the results listed in Table I indicate
that there is a differential in NOE among the ring carbon positions. There is no difference among the T_1 values for the positions with bonded protons. The C_4 position has the same value as
the aliphatic carbons, which is a smaller value than the $C_{2,6}$
and $C_{3,5}$ positions. This relaxation behavior can be accounted
for by an internal rotation occurring about the C_1-C_4 molecular

axis. Although a complete analysis of the motions of polystyrene
has not been carried out, this differential NOE shows that the
relaxation methods employed here are sensitive to both the rate
and the angle of internal rotations. The phenyl groups of MCBP,
which have been denatured by guanidine hydrochloride and heat,
may have motions similar to those of the benzyl groups of polystyrene and benzyl glutamate in the random coil form. The relaxation
parameters for the denatured MCBP phenylalanines are T_1 = 200 msec
and NOE = 2.3. The NOE value is similar to the 2.5 found for the
random coil polybenzyl glutamate, however the T_1 is considerably
shorter. These values do not correspond to an isotropic rotational correlation time. In the fully denatured protein many complex
rapid segmental motions are likely to be present, and the resolution of individual phenyl ring carbons is too low to observe differential effects among different positions.

The phenyl ring carbons with directly bonded hydrogens have
a NOE = 1.4, slightly (but significantly) higher than the minimal
value observed for the α-carbon atoms of the protein main chain
(i.e., for the α-carbons, NOE ≃ 1.2). This somewhat unexpected
finding for the carbon atoms forming the tightly knit hydrophobic
core of the parvalbumin molecule can only be explained by assuming that the phenylalanine side chains do in fact possess rapid
intramolecular motion. This motion can be described as follows.
The principal axis for rotation of the phenyl rings can be chosen
to be the C_β-C_γ molecular bond axis for several reasons. It
would be expected that the restricted internal environment of the
hydrophobic core prohibits C_β-C_α axis rotation because of the
large swept volume required. While the differential NOE cannot
be observed in the MCBP spectra because of the lack of resolution
of the ring carbon resonances, the effect found in polystyrene
might serve as a reasonable indication that this method of description is appropriate for phenyl groups. If the predominant
phenyl group rotation is about the C_β-C_γ axis the pertinent C-H
dipolar vectors are at an effective angle of 60° with respect
to the molecular framework. The theoretical plots for the appropriate values of the angle and rate constants are given in
Figure 7. From these an internal correlation time of 4 nsec can
be assigned to the phenyl rings of MCBP.

The general usefulness of this approach is limited only by
the ability to obtain well-resolved, identifiable groups of lines
in the spectrum and by the complexities of the theoretical analysis of anisotropic motions. At the present state of our knowledge
and technology these limitations are not very severe. For a protein of known crystal structure it is therefore possible with a
reasonable investment of time to obtain a rather detailed picture
of its internal dynamics. In the case of MCBP, whose crystal
structure is known (27) we are left with an image of a roughly
cubical cage of six α-helices, rotating as a unit with a time constant of 12 nsec, a quivering core of phenylalanine rings flipping
with a time constant of 4 nsec and a hairy surface of a variety

of side chains bobbing about with time constants of 0.1-10 nsec
and perhaps longer.

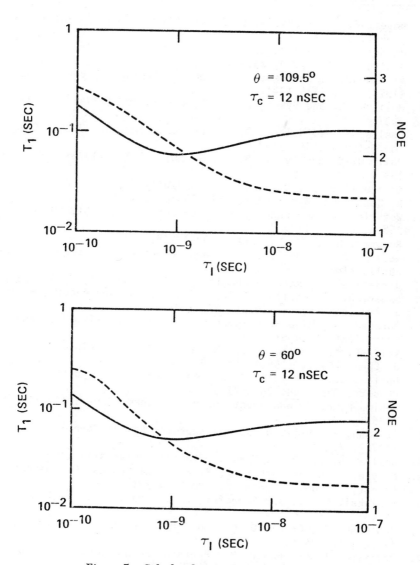

Figure 7. Calculated T$_1$ *and NOE dependence
on the internal reorientation time* τ_1 *for a* ^{13}C—H
*vector inside a rigid protein with an overall iso-
tropic reorientation time* $\tau_c = 12$ *nsec. (A) For
rotation at an angle of 109.5° with respect to the
protein axis. (B) For rotation at an angle of 60°.*

Research supported by National Science Foundation Grant No.
GB 32025.

Literature Cited

1. Bovey, F. A., Tiers, G. V. D. and Filipovitch. J. Polymer Sci.,
   (1959), 38, 73.
2. Jardetzky, O. and Jardetzky, C. D. Methods of Biochemical
   Analysis, (1962), 9, 235-410.
3. Allerhand, A., Childers, R. F. and Oldfield, E., Biochemistry,
   (1973), 12, 1335-1341.
4. Hull, W. E. and Sykes, B. D., Biochemistry, (1974), 13, 3431-
   3437.
5. Opella, S. J., Nelson, D. J. and Jardetzky, O., J. Am. Chem.
   Soc., (1974), 96, 7157.
6. Hull, W. E. and Sykes, B. D., J. Mol. Biol., (1975), 98, 121-
   153.
7. Campbell, I. D., Dobson, C. M. and William, R. J. P., Proc.
   R. Soc. Lond. B., (1975), 189, 503-509.
8. Wuthrich, K. and Wagner, G., FEBS Letters, (1975), 50, 265-
   268.
9. Roberts, G. C. K., Nature, (1975) 258, 112.
10. Nelson, D. J., Opella, S. J. and Jardetzky, O., submitted for
    publication.
11. Jardetzky, O., Adv. in Chemical Physics, (1969), 7, 499-531.
12. Bauer, D. R., Opella, S. J., Nelson, D. J. and Pecora, R.,
    J. Am. Chem. Soc., (1975), 97, 2581.
13. Abragam, A., "The Principles of Nuclear Magnetism," Clarendon
    Press, Oxford, 1961.
14. Vold, R. L., Waugh, J. S., Klein, M. P. and Phelps, D. E.,
    J. Chem. Phys., (1968), 48, 383.
15. Opella, S. J., Nelson, D. J. and Jardetzky, O., J. Chem.
    Phys., (1976), 64, 2533-2535.
16. Kuhlmann, K. F. and Grant, D. M., J. Chem. Phys., (1971), 55,
    2998.
17. Freeman, R., Hill, D. W. and Kaptein, R., J. Magn. Res.,
    (1972), 7, 327.
18. Opella, S. J., Ph.D. Thesis, Stanford University (1974).
19. Noggle, J. H. and Schirmer, R. E., "The Nuclear Overhauser
    Effect," Academic Press, New York, (1971).
20. Stejskal, E. O., and Schaefer, J. J., Mag. Res., (1974), 14,
    160.
21. Kuhlmann, K., Grant, D. M. and Harris, R. K., J. Chem. Phys.,
    (1970), 52, 3439.
22. Armitage, I. M., Huber, H., Pearson, H. and Roberts, J. D.,
    Proc. Nat. Acad. Sci. U.S., (1974), 71, 2096.
23. Woessner, D. E., J. Chem. Phys., (1962), 37, 647.
24. King, R. and Jardetzky, O., to be published.
25. Allerhand, A. and Oldfield, E., Biochemistry, (1973), 12,
    3428.

26. Schaefer, J. and Natusch, D. F. S., Macormolecules, (1972), 5, 427.

27. Kretsinger, R. H. and Nockolds, C. E., J. Biol. Chem., (1973), 248, 3313.

33

Sodium-23 NMR as a Probe for Self-Association of a Guanosine Nucleotide

AGNÈS PARIS and PIERRE LASZLO

Institut de Chimie, Université de Liège, Sart-Tilman par 4000 Liège 1, Belgium

Linewidth of the sodium-23 nuclear magnetic resonance can serve as a probe for molecular aggregation. The sodium-23 nucleus, with a spin number $I = \frac{3}{2}$, is endowed with a quadrupolar moment whose interaction with the local electric field gradient at the site of the nucleus provides an efficient relaxation pathway. Hence, the linewidth will reflect the rate of reorientation of the chemical entity to which the sodium cation is attached, and therefore its degree of aggregation. This correspondence is especially simple if the electric field gradient at the sodium nucleus remains invariant, for instance when the electric field gradient characteristic of a given environment in the monomer is unaffected by polymer formation. This will be the case for an ion pair, such as the phosphate group, with which the sodium cation coordinates on one side, while solvated by water molecules on the other side. Then, the sodium linewidth $\nu_{1/2}$ is simply related to the correlation time τ_c for reorientation of the ion pair, through :

$$\nu_{1/2} = \frac{2\pi}{5} \cdot \left(\frac{e^2 qQ}{h}\right)^2 \cdot \tau_c \qquad (1)$$

where the term between parentheses is the quadrupolar coupling constant, q being the electric field gradient and Q the quadrupolar moment for the sodium nucleus.

The field gradient originates in the electric charges present on the various atoms coordinated to the sodium cation, and surrounding it assymetrically. When the sodium cation is associated to its counter-anion, the ion pair reorients as a whole with τ_c.

Increases in the sodium linewidth are observed as a function of increasing monomer concentration with 5'-adenosine monophosphate (5'-AMP), 2'-guanosine monophosphate (2'- GMP), and with 5'-guanosine monophosphate (5'-GMP). With 5'-AMP and 2'-GMP, these increases are linear with concentration (Figure 1), as expected from the attendant increase in the proportion of the ion pairs. From these, one can evaluate the appropriate values for the quadrupolar coupling constant $\left(\frac{e^2 qQ}{h}\right)$.

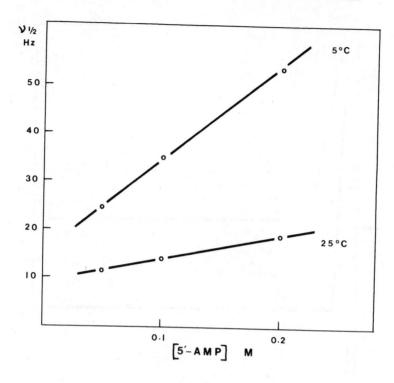

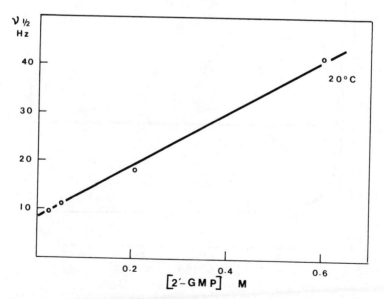

Figure 1.   Band width for the sodium-23 resonance as a function of nucleo-tide concentration (di-sodium salt)

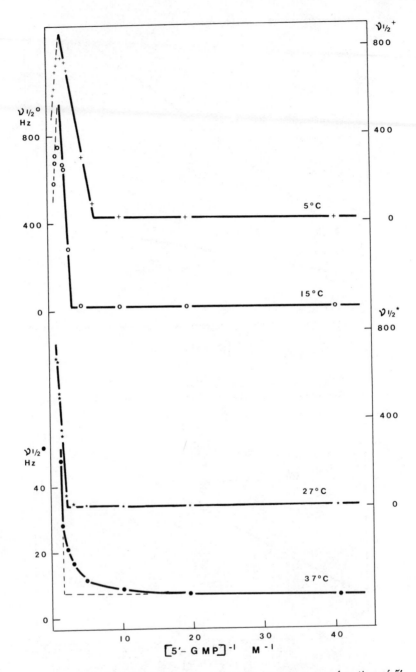

Figure 2. Line broadenings for the sodium-23 resonance as a function of 5'-GMP reciprocal concentration (di-sodium salt) in D_2O. The curves obtained at 0°C and 20°C have been omitted for clarity.

For the sodium salt of 5'-AMP, the stability constant of the sodium-phosphate complex is 2.2 ± 0.2 M^{-1} at 298 K ([1]). We have monitored the sodium-23 nmr linewidth of this complex at the same temperature, for a 0.1 M solution in D$_2$O : using a two-state model, in which the sodium cation exchanges rapidly between the free, symmetrically hydrated, and the phosphate-bound states, one comes up with a characteristic linewidth for the latter of 50 ± 3 Hz. This value, by comparison with the correlation time of $\overline{1}.4 \times 10^{-10}$ s derived from ^{13}C T$_1$ values obtained in the same conditions ([2]), indicates a quadrupolar coupling constant for the sodium ion complexed by a monophosphate in water solution of $0.5^3 \pm 0.08$ MHz. Very similar values are obtained with 2'-GMP. Such values are entirely consistent with findings in our laboratory and in other groups of quadrupolar coupling constants for the sodium cation between 0.5 and 2.0 MHz. These values are diagnostic of an ionic structure with very little covalent character for the coordination bonds.

Turning now to 5'-GMP, a nucleotide known for its aptitude to self-assembly ([3]), one observes considerable increases of the sodium-23 linewidth as a function of concentration at pH = 8.0 ± 0.1. Characteristic "melting" curves are obtained by plotting the linewidth for a 0.26 M concentration against the temperature, with a melting temperature of 10°C and a breadth of 10° (t_m = 21°C, with a breadth of 16°, for a 0.61 M concentration) : these parameters are in excellent accord with those earlier reported by Miles et al. ([3]) from infrared determinations. A solvent isotope effect is apparent : with an 0.45 M solution, the melting temperatures are 16.2°C (H$_2$O) and 19.5°C (D$_2$O). That the melting temperatures depend upon solvent nature and structure points to the importance of hydrophobic forces in the aggregation process - a point further suggested by the entropies reported in the present work (vide infra).

Borrowing a treatment applicable to micelle formation ([4]), we plot the line broadening $\Delta\nu_{1/2}$ as a function of reciprocal concentration (Figure 2) at six distinct temperatures, in D$_2$O solution. In fact, intersection of two linear segments, as observed here, is highly reminiscent of micelle formation. Considering the net equilibrium :

$$n \, B \rightleftarrows B \, n \qquad\qquad (2)$$

with $K = (B_n)/(B)^n$, and setting $(B) = n(B_n) = \frac{1}{2}(B)_0$ at the intersection (or pseudo critical micelle concentration), K is given by :

$$K = \frac{2^{n-1}}{n(CMC)^{n-1}} \qquad\qquad (3)$$

Equation (3) yields values of $\bar{K} = K^{1/n}$ listed below, as a function of the degree of aggregation $\underline{n}$.

n	$\bar{K}$ (27°C; D_2O)
4	2.2
8	2.9
12	3.3
16	3.5
20	3.6
24	3.7

The earlier studies of the Miles group (3) had established presence of planar tetramers, due to hydrogen bonding between the purine bases; these tetramers then stack into an helical arrangement. We have derived the value of $\bar{K}$ separately by monitoring the proton chemical shifts of the H-1' and H-8 resonances in 5'-GMP as a function of monomer concentration in the presence of 2.10^{-3} M EDTA, and 2.10^{-1} M NaCl, in D_2O solution between 0 and 2.10^{-1} M : $\bar{K}$ = 3.5 $\pm$ 1 M^{-1} at 22 $\pm$ 1°C (pH = 7.5), where $\bar{K}$ refers to the series of step-wise equilibria (5) :

$$B + B \rightleftharpoons B_2$$
$$B_2 + B \rightleftharpoons B_3$$
$$-------------$$
$$B_{n-1} + B \rightleftharpoons B_n$$

i.e. without differentiating between tetramer formation by the horizontal hydrogen bonding of the guanosines, and formation of higher aggregates due to the vertical stacking interaction. There is indeed a smooth variation of the proton chemical shifts with monomer concentration. The limiting chemical shift for the dimer B_2 is entirely consistent with a composite of the horizontal and vertical interactions.

Thus, comparison between the two types of data, viz. originating from ^1H and from ^{23}Na nmr, points to a degree of aggregation n = 12-20, with hexadecamer formation a likely candidate.

Sodium-23 nmr is however more informative, since it allows determination of the pseudo critical micelle concentration (CMC), as a function of solvent and temperature. The CMC is determined (Figure 2) within $\pm$ 15 %.

Thermodynamic parameters for the self-association are easily derived, since a plot of ln (CMC) against T^{-1} is linear (ρ = 0.992 for 6 points) yielding :

$$\Delta H^0 = -113 \pm 25 \text{ kcal by mole of hexadecamer}$$
$$\Delta S^0 = -337 \pm 80 \text{ e.u. by mole of hexadecamer in } D_2O$$

and

$$\Delta H^0 = -77 \pm 20 \text{ kcal by mole of hexadecamer}$$
$$\Delta S^0 = +214 \pm 55 \text{ e.u. by mole of hexadecamer in } H_2O.$$

The enthalpy change upon self-association is commensurate with the value of -20 kcal.mol^{-1} observed for tetramer formation with 3'-GMP (6).

With a two-site model, in which the sodium cation is either free in the solution : $(Na^+)_{6H_2O}$, or associated to its phosphate counter-ion, assuming an association constant β independent of 5'-GMP self-association, and using the AMP value (1) of $\beta=2.2$, one obtains the linewidth characteristic of complexed sodium (Table I). The correlation time derived from equation 1 is also indicated : it increases roughly 24-fold in going through the CMC, from a value of ca. 200 ps to a value of 5.900 ps. This is again indicative of a degree of aggregation of the order of 16-20, since the fraction of Na^+ bound above the CMC is assuredly underestimated in Table I by using 5'-AMP as a model, ignoring the fraction of sodium bound in the electrical bilayer. The correlation time for the aggregate, of the order of 5 ns, is consonant with an estimate of 8 ns, based upon the admittedly crude rule of thumb between the correlation time for a macromolecule and its molecular weight : $\tau_c (ns) \sim M/1000$ (7).

The agreement of the melting curves with those reported in the literature from infrared data (3), the agreement of the mean equilibrium constant K derived from ^1H and from ^{23}Na observable, finally the agreement of correlation times for the monomer and the n-mer with expectations, all support the two-state model used here.

The sodium cation reorients with the whole of the aggregate with which it is bound through the phosphate groups : this finding requires that the n-mer itself reorients as a single entity, which is consistent with the observations of Pinnavaia, Miles and Becker (3) of a $\Delta G^{\ddagger} > 15$ kcal.mol^{-1} for exchange of H-8 between four distinct environments. Another requirement, for the sodium cation not to exchange between different phosphate loci on the surface of the aggregate more rapidly than reorientation of the aggregate as a whole, is also met provided that the concentration above the CMC is not too high. The decrease in linewidth, indicated by dashed lines in figure 2, may be attributed to the onset of sodium exchange when adjacent sites become populated to an appreciable extent.

These phenomena may have considerable biological significance. We are currently studying gel formation (8) by 5'-GMP at acidic pH : these may have played a role in the pre-biotic appearance of polynucleotides (9).

In summary, ^{23}Na nuclear magnetic resonance affords a simple and convenient technique for studying micelle-like ordered structures formed by the counter ions in an aqueous environment.

TABLE I

[5'-GMP] M	Association coefficient α	Fraction of Na$^+$ bound B	$\Delta\nu_{1/2}$ obs. Hz	$\nu_{1/2}$ bound ν_B	Correlation time τ_c ($\chi=0.5$MHz)
0.025	0.094	0.047	0.8	24.0	76 ps
0.051	0.170	0.085	2.0	30.6	97 ps
0.097	0.27	0.135	3.7	34.4	109 ps
0.203	0.415	0.21	9.7	53.8	171 ps
0.324	0.51	0.26	15.0	65.3	208 ps
0.45	0.58	0.29	55.0	195.5	0.62 ns
0.58	0.64	0.32	326	1030	3.3 ns
0.60	0.64	0.32	353	1110	3.5 ns
0.74	0.68	0.34	518	1525	4.85 ns
0.75	0.68	0.34	503	1475	4.7 ns
0.90	0.72	0.36	665	1860	5.0 ns
1.00	0.74	0.37	678	1850	5.9 ns

Acknowledgment. We thank Fonds de la Recherche Fondamentale
Collective for a grant towards purchase of the Bruker HFX-90
spectrometer, and IRSIA for award of a pre-doctoral fellowship
to A.P.

LITERATURE CITED

(1) R.M. Smith and R.A. Alberty, J. Phys. Chem., (1956), 60, 180.
(2) T. Imoto, K. Akasaka, and H. Hatano, Chem. Phys. Letters,
(1975), 32, 86.
(3) H.T. Miles and J. Frazier, Biochem. Biophys. Res. Commun.,
(1972), 49, 199.
T.J.Pinnavaia, H.T. Miles, and E.D. Becker, J. Am. Chem. Soc.
(1975), 97, 7198.
(4) H. Gustavsson and B. Lindman, J. Am. Chem. Soc.,(1975), 97,
3923.
J.F. Fendler and E.J. Fendler, "Catalysis in Micellar and
Macromolecular Systems", Academic Press, New York, 1975.
(5) J.L. Dimicoli and C. Hélène, J. Am. Chem. Soc.,(1973), 95,
1036.
(6) J.F. Chantot, Arch. Biochem. Biophys., (1972), 153, 347.
(7) see for instance L.G. Werbelow and A.G. Marshall, J. Am.
Chem. Soc., (1973), 95, 5132.
(8) M. Gellert, M.N. Lipsett, and D.R. Davies, Proc. Nat. Acad.
Sci., USA, (1962), 48, 2013.
H.T. Miles and J. Frazier, Biochim. Biophys. Acta, (1964), 79,
216.
F.B. Howard and H.T. Miles, J. Biol. Chem., (1965), 240, 801.
P.K. Sarkar and J.T. Yang, Biochem. Biophys. Res. Commun.,
(1965), 20, 346.
W.L. Peticolas, J. Chem. Phys., (1964), 40, 1463.
J.F. Chantot, M. Th. Sarocchi, and W. Guschlbauer, Biochimie,
(1971), 53, 347.
J.F. Chantot, Thèse de Doctorat, Université de Paris-Sud,
July 6, 1973 (CNRS registry no. AO 8683).
(9) J.E. Sulston, R. Lohrmann, L.E. Orgel, and H.T. Miles, Proc.
Nat. Acad. Sci. USA, (1968), 60, 409.
B.J. Weimann, R. Lohrmann, L.E. Orgel, H. Schneider-Bernloehr
and J.E. Sulston, Science, (1968), 161, 387.
L.E. Orgel, J. Mol. Biol., (1968), 38, 381.

34

Biological Applications of Spin Exchange

ALEC D. KEITH and WALLACE SNIPES

Biochemistry and Biophysics Department, The Pennsylvania State University, University Park, Pa. 16802

Since nitroxides were introduced as spin labels in 1965 many experiments have been carried out on model membrane and biological membrane systems (1 - 4). Most of these have dealt with some parameter relating to rotational diffusion. However, some have dealt with the measurement of translational diffusion based on the collisional frequency of spin labels (5, 6). One of the great advantages of spin labels for such purposes is the ability to measure events which occur over Angstrom dimensions. Rotational motion experiments can be carried out taking measurements of spin labels dissolved in certain zones of membrane structures. Spin labels, properly designed, localize at specific sites in membrane aqueous dispersions. An appropriate spin label will localize at the hydrocarbon-aqueous interface, totally dissolved in the hydrocarbon interior, or totally in the aqueous medium. The geometry of, and the functional groups on, the spin label determine the properties which lead to zone specificity.

Additional developments have lead to refinements in measuring the properties of specific zones in heterogeneous or biological systems (all biological systems or heterogeneous). The use of paramagnetic ions or chelates of paramagnetic ions make possible the removal of spin label signals which originate from the same environment (7). This technique is useful for systems or conditions where the paramagnetic broadening agent has differential permeability compared to the spin label being used. For example, inorganic salts of transition elements such as Nickel are not normally permeable to intact membrane systems. Therefore, the use of such an impermeable membrane enclosure with TEMPONE and $NiCl_2$ makes it possible to broaden the TEMPONE N-hyperfine lines in the surrounding aqueous medium without broadening the TEMPONE signal originating from the aqueous environment inside the enclosure. Therefore, this approach makes it possible to measure local environmental effects on the spin label population localized within the enclosures. The structures of spin labels and other compounds appropriate for the present report are given in Fig. 1.

The concentration dependency of this line broadening is

426

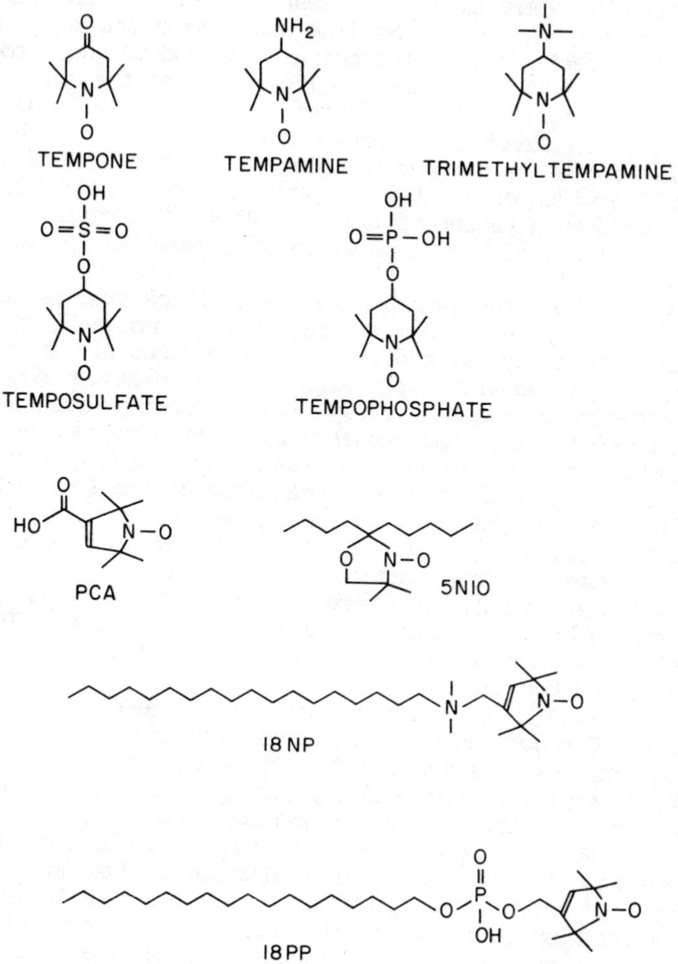

Figure 1. Structure of spin labels. The chemical names of the spin labels shown above are given in the reviews cited and in a variety of other literature.

shown in Fig. 2A (8). The viscosity is held constant so that the
relative efficiencies of different species as line broadening
agents can be compared. Of the compounds shown in Fig. 2, TEMPONE-
TEMPONE and Ni^{++}-TEMPONE are the most effective. The term on the
ordinate of Fig. 2, ΔH_c, is the line width contributed by the concentration effects of the broadening agent and is equal to the
measured line width, H, minus the line width of the spin label at
10^{-4}M concentration or less. Therefore, the minimum line width is
important in considering the minimum concentration of a broadening
agent that can be used. For example, deuterated TEMPONE with a
minimum line width of about 0.25 gauss has its measured line
width lowered by a factor of 112 with 20mM $NiCl_2$ while another
spin label with a minimum line width of 1 gauss is lowered only
by a factor of 11.6.

Fig. 2B shows the same general type of spectral series only
each line now represents a different state of motional freedom (8).
This was achieved by using different proportions of water-glycerol
mixtures. It is apparent from Fig. 2B that collision dependent
line broadening decreases steeply with increasing viscosity. A
TEMPONE concentration that causes 3 gauss line broadening in
water causes only 0.3 gauss line broadening in 75% glycerol. Over
this motional range the minimum line width of the low field and mid
field lines do not change. Decreasing the motion further to
nanosecond rotational correlation times requires 0.5 M TEMPONE to
achieve 3 gauss of line broadening. It is clear from Fig. 2B that
the motional state must be determined before the minimum effective
concentration of a broadening agent can be determined. These concentrations may vary from about 10mM to near 1M depending on the
spin label and the motional state.

Fig. 2C and D show some examples of charge-charge interactions
(8). Fig. 2C shows the decrease in line broadening due to pos-
itive-ion charge-repulsion and Fig. 2D shows the decrease in line
broadening resulting from negative-ion interaction. Both these
examples are illustrative of charge-charge repulsion. Charge
attractions between oppositely charged spin labels and a broadening agent in solution are not easy to detect. However, using
amphiphilic spin labels on phospholipid or membrane interfaces
charge attraction interactions are much easier to demonstrate. For
example the spin label shown as 18PP has its N-hyperfine lines
broadened much more readily by Nickel-Tris than by $K_3Fe(CN)_6$ and
conversely 18NP has its lines broadened more readily by $K_3Fe(CN)_6$.

Fig. 2E shows the concentration dependent line broadening
which occurs in the aqueous channels of porous BioGel P beads (8).
Fig. 2E has the minimum molecular weight denoted on each line
which is excluded from the channels. For example, P100 will allow
a molecule of less than 100,000 MW to pass but will exclude larger
molecules. The beads were tightly packed by centrifugation and as
a result have very little aqueous space between beads. Fig. 2F
shows that rotational diffusion is not drastically modified while
collision dependent line broadening is drastically modified (8).

The results presented in Fig. 2E and F indicate that it may be possible to make measurements in cellular systems which have relevancy to the properties of the diffusion environment.

Fig. 3 shows data about the diffusion constant. The line labeled, $D = 0.22r^2 / \tau_c$, illustrates the relationship between τ_c and D based on the Stokes' Einstein equation for τ_c

$$\tau_c = \frac{4\pi r^3 \eta}{3kT}$$

and Stokes' equation for diffusion

$$D = \frac{kT}{6\pi\eta r^2} \, .$$

Combining these equations and solving for the diffusion constant, D, results in

$$D = \frac{0.22r^2}{\tau_c} \, .$$

The other line is based on the measurement of line broadening due to spin label collision frequency, K.

The actual collision frequency should be twice the measured since only one-half of spin label collisions should be detectible (5, 9, 10, 11); therefore, 2K is a better approximation of the collision frequency. The equation

$$D = 2Ka^2$$

where a is the on center cubic lattice spacing describes the diffusion constant in terms of spin label collision frequency. If less than the expected percentage of collisions result in exchange then the measured line from collision frequency would also lead to a diffusion constant less than expected.

These data form the bases for carrying out biological experiments using spin exchange interactions as a means of localizing spin label signal to one environment of a two or more environment system. They also form the basis for carrying out experiments to determine characteristics of diffusion environments.

Nickel Chloride has been used as a line broadening agent to aid in determining the hindrance to rotational diffusion of a water soluble spin label dissolved in the aqueous cytoplasm of several cell types (7). In general, as the membrane content increased the restriction to rotational diffusion also increased. Subsequent studies were carried out on other systems to determine more about the properties of spin labeled solutes located inside membrane bounded enclosures.

A variety of spin labels were used in an investigation dealing

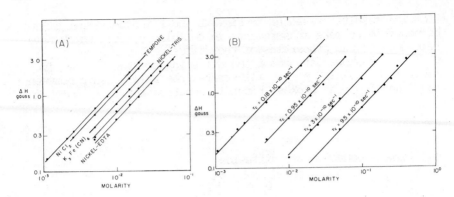

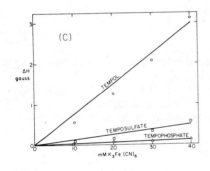

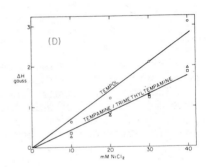

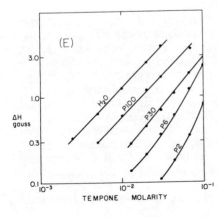

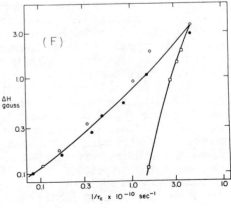

Figure 2. Dependencies of spin-spin interactions. (A) The concentration dependencies of several paramagnetic agents on the line broadening of TEMPONE are shown. (B) The viscosity dependency of spin exchange for the TEMPONE-TEMPONE interaction is shown. Reading the lines from left to right the solvents for spectral measurements are, water, 65% glycerol, 75% glycerol, and 85% glycerol. (C) The effect of a negatively charged broadening agent. Potassium ferricyanide (K₃Fe(CN)₆) is used as a negatively charged broadening agent to determine the efficiencies of broadening on an uncharged, a minus-one-charged, and a minus-two-charged spin label. (D) The effect of a positively charged broadening agent. Nickel chloride is used as a positively charged broadening agent to determine the efficiencies of broadening an uncharged and two positively charged spin labels. (E) The effect of environmental heterogeneity on spin-spin interaction. All BioGel P beads used were 100–200 mesh. BioGel P beads are manufactured by BIO-RAD Laboratories, Richmond, Calif. BioGel P beads are polyacrylamide polymers designed and manufactured for gel filtration chromatography. The P-notation denotes the approximate minimum MW of spherical molecule that is excluded from the pores of polymeric beads. (F) The relationship between collision frequency and rotational correlation time (τ_c). Data from Figure 2(B) and Figure 2(E) are combined and replotted for clarity.

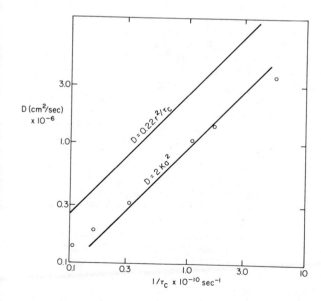

Figure 3. The relationship between the diffusion constant (D) and τ_c. The equations from which each of the above lines were generated are shown in the figure. The collision frequency K or 2K was taken from data such as that presented in Figure 2(A) and (F). One gauss is equal to 2.8 × 10⁶ cps in frequency units and is a necessary conversion factor for this calculation.

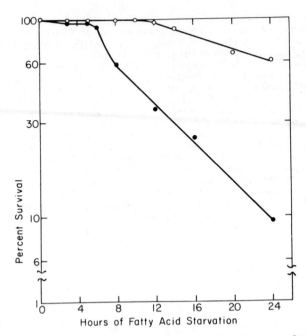

*Figure 4. Viability of fas⁻ mutants under fatty acid deficient conditions. Viabil-
ity of fas⁻ cells supplemented with a saturated fatty acid (open circles). Viability
of fas⁻ cells starved for a saturated fatty acid (closed circles).*

with the viscosity profile of sarcoplasmic reticular (SR) vesicles
(12). In this study it was concluded that the hydrocarbon portion
of the SR membranes were symmetrical with respect to the barriers
that limit spin label motion. An examination of the spectra of
spin labels alkylated onto the outer and inner membrane surfaces
revealed that the motion was much more restricted on the inner
surface. Water soluble spin labels which penetrate the SR membrane
freely were also considerably more immobilized when located in the
trapped volume of the vesicle interior compared to the aqueous
medium. These data do not necessarily give any information about
the state of the water inside the vesicles. The data do allow
inferences to be made about the water based on the motion of the
spin label solutes dissolved in the aqueous interior. Polymer
projections protruding from the membrane surface may have effects
on the restrictions to spin label motion. Interactions that result
in hindering spin label motion may not have an equivalent effect
on the water molecules in the same environment.

The hindrance to molecular motion experienced by solutes
dissolved in the intracellular cytoplasm probably depends on several factors. Overall solute concentration, spacing between
membraneous structures, membrane interface charge composition and
distribution, and the presence of polymeric structures such as
microtubules and microfilaments are probably all important factors.

Since the initiation of the spin label method to measure the properties of the aqueous cytoplasm several investigations have been carried out. One system explored in this manner are auxotrophs of yeast deficient in the synthesis of one or more structural components required in biological membranes. Among these are the fatty acid auxotrophs of yeast. These are of two types. Auxotrophs which require supplementation of an unsaturated fatty acid (ole⁻) and another group of auxotrophs which require supplementation of saturated fatty acids (fas⁻) (13, 14).

In appropriate starvation studies it was determined that starvation for unsaturated fatty acids in the ole⁻ mutants stopped growth. Growth would resume after several hours of exposure to medium deficient in unsaturated fatty acid after adding an appropriate unsaturated fatty acid to the medium. Viability was high after several hours on deficient medium. The fas⁻ mutants on-the-other-hand behaved in a different manner. After from 6-8 hrs on medium deficient in a saturated fatty acid these requirers began to lose viability rapidly so that by the end of 8 hrs viability was usually about 10% (Fig. 4)(15). These mutants continue protein synthesis for the first few hours without cell division. The loss of protein synthesis is on the same general time scale as the loss of viability. Since these auxotrophs require a structural component for phospholipid biosynthesis it seemed reasonable that the physical properties of membrane and other lipids might be modified during the starvation process or at least by the time most of the cells were no longer viable. Several membrane spin labels were used on these auxotrophs at various times under conditions where the auxotrophs had developed on medium deficient in saturated fatty acid (16). The results of one membrane spin label are shown in Fig. 5A. No detectible modification in the restriction to rotational diffusion by membrane hydrocarbon zones was detected. Several other spin labels such as spin labeled fatty acids and spin label sterols were found to have no detectible modification in the spectra even after 24 hrs. However, when water soluble spin labels were used in conjunction with $NiCl_2$ it became clear that changes do take place in the aqueous component of the cytoplasm.

Figure 5B shows that the signal of TEMPONE residing in the cellular cytoplasm underwent a drastic increase in restriction to rotational diffusion. The general time scale of this increase, in general, paralleled the loss of viability. TEMPONE has many advantages as a spin label but also has some disadvantages for the present use in that it shows some partitioning between intracellular hydrocarbon zones and intracellular aqueous zones. The effects of signals originating from zones having different polarities could not be completely separated from the effects of increased cytoplasmic viscosity. Figure 5C shows the same experiments carried out on another water soluble spin label, PCA. PCA has a much more favorable partitioning into the aqueous phase and as a result, in the yeast cells, essentially contributes no hydrocarbon signal at all. The data in Fig. 5C show that the restriction to

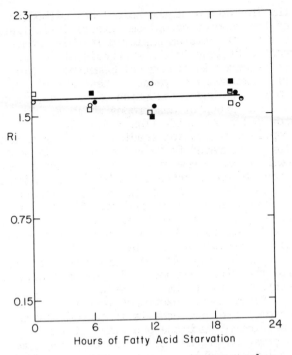

Figure 5. Spin label motion measurements made on fas⁻ cells. (A) The membrane hydrocarbon spin label, 5N10, was added to fas⁻ cells at different times after the initiation of saturated fatty acid depletion. R_i measurements are shown. $R_i = W_o[(h_o/h_{-1})^{\frac{1}{2}} - 1]$. W_o is the first derivative mid-line width. The midfield and high-field derivative line heights are h_o and h_{-1}. The expression, $R_i = \tau_c/constant$, is relevant for isotropic motion in the fast tumbling range. The constant is slightly different for each spin label and also depends on instrumental settings. The different notations shown are for fas⁻ cells supplemented and unsupplemented and both of these conditions in the presence and absence of cyclohexamide. Cyclohexamide is a protein synthesis inhibitor. (Figures 5(B) and (C) appear on opposite page.)

PCA motion behaves in the same manner as the data shown for TEMPONE.

Another membrane-affected auxotroph was analyzed in the same way (17). This auxotroph required inositol for growth and was observed to undergo loss of viability on about the same time scale as previously observed for the fas⁻ mutant (Fig. 6). This requirement is, however, different in several ways. There is no fatty acid involvement in the requirement. Inositol is required for the synthesis of phosphatidyl inositol involving polar zones of intact membrane structure and does not directly involve hydrocarbon zones

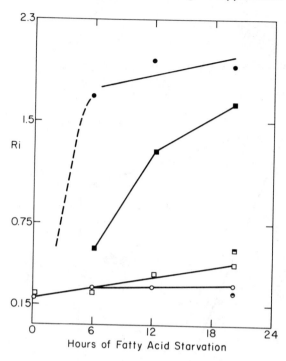

(B) The motion of TEMPONE in the presence of NiCl₂. Signals originate from inside the cells. Fas⁻ cells starved for a saturated fatty acid (closed circles and closed squares). Supplemented fas⁻ cells or unsupplemented with cyclohexamide added to medium (all other symbols).

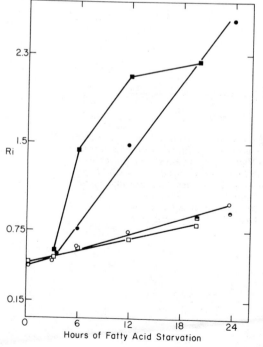

(C) Motion of PCA in fas⁻ cells. Fas⁻ cells starved for a saturated fatty acid (closed circles and closed squares). Fas⁻ cells supplemented with a saturated fatty acid or unsupplemented in the presence of cyclohexamide, a protein synthesis inhibitor (all other symbols).

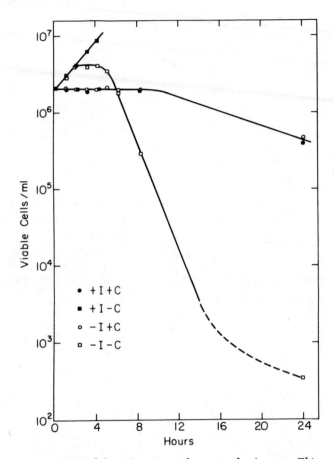

Figure 6. Viability of an inositol auxotroph of yeast. This plot shows the loss of viability as a function of time during starvation for inositol. Viability of an inositol requirer during inositol supplementation (closed squares). Viability of an inositol requirer during inositol starvation plus and minus the protein synthesis inhibitor, cyclohexamide (open and closed circles). Viability during inositol starvation (open squares). I, inositol. C, cyclohexamide.

of membranes. The two auxotrophic requirements are the same in that both are necessary for the synthesis of required membrane phospholipids.

This requirer also continues protein synthesis until the onset of viability loss. Fig. 7 shows that the motion of a spin label dissolved in membrane hydrocarbon zones is unaffected by the events which lead to cell death due to inositol starvation. A water soluble spin label, however, behaves in much the same way as with the fas⁻ requirer in that there is a large increase in the barriers that limit rotational diffusion in the aqueous cytoplasm.

These two studies on yeast show that the rotational diffusion of water soluble solutes are drastically modified during starvation for each of two nutritional requirements. The conclusion of this work was basically that diffusional processes were also drastically modified. The perturbation on the diffusion of metabolites was implicated as the cause of loss of viability.

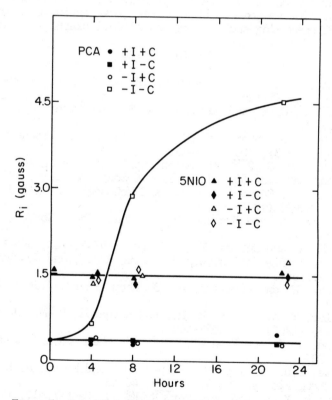

Figure 7. Spin label motion during inositol depletion. The molecular motion of 5N10 and PCA are shown as a function of starvation time. I, inositol. C, cyclohexamide.

ABSTRACT

The physical concept of and considerable experimental data dealing with the events that lead to close encounter dependent broadening of the N-hyperfine lines of nitroxides have been treated by several authors. This report presents the phenomena limiting line broadening due to spin-spin interactions and treats biological applications.

The concentration dependencies of the interactions between spin labels and other spin labels, paramagnetic ions, and paramagnetic chelates are presented. The line broadening efficiencies are compared. The effects of solvent viscosity and charge-charge interactions with respect to both interactions between spin labels and spin labels with other charged paramagnetic agents are treated. Collisional frequency in porous beads is treated in terms of diffusion barriers.

Biological applications are presented showing restriction of molecular motion of spin labels dissolved in the aqueous interior of cells. The phenomenon of increased restriction to spin label motion accompanying specific causes of cell death is treated.

LITERATURE CITED

1. McConnell, H. M. and McFarland, B. G. Q. Rev. Biophys., (1970), 3, 91-136.
2. Griffith, O. H. and Waggoner, A. S. Acc. Chem. Res., (1969), 2, 17-24.
3. Keith, A. D., Sharnoff, M., and Cohn, G. E. Biochem. Biophys. Acta (1973), 300, 379-419.
4. Smith, Ian C. P. "Biological Application of Electron Spin Resonance" 484-535, John Wiley & Sons, New York, 1972.
5. Devaux, P., Scandella, C.J., and McConnell, H. M. J. Amer. Chem. Soc., (1973), 9, 474-485.
6. Sackmann, E. and Trauble, H. J. Amer. Chem. Soc., (1972), 94, 4482-4497.
7. Keith, A. and Snipes, W. Science, (1974), 183, 666-668.
8. Keith, A. D., Snipes, W., Mehlhorn, R. J., and Gunter, T. (in manuscript).
9. Plachy, W. and Kivelson, D. J. Chem. Phys., (1967), 47, 3312-3318.
10. Miller, T. A., Adams, R. N., and Richards, P. M. J. Chem. Phys., (1966), 44, 4022-4023.
11. Freed, J. H. and Frankel, G. K. J. Chem. Phys., (1963), 39, 326-337.
12. Morse, P.D., Ruhlig, M., Snipes, W., and Keith, A. D. Arch. Biochem. Biophys., (1975), 168, 40-56.
13. Resnick, M. A. and Mortimer, R. K. J. Bacteriol., (1966), 92, 597-600.
14. Henry, S. and Fogel, S. Molec. Gen. Genetics (1971), 113,

1-19.
15. Henry, S. A.   J. Bacteriol., (1973), 116, 1293-1303.
16. Henry, S. A., Keith, A. D., and Snipes, W.   Biophys. J., (in press).
17. Keith, A. D., Henry, S. A., Snipes, W., and Pollard, E.   (in manuscript).

35

High Resolution NMR Studies of Micellar Solutions

JOHN R. HANSEN and ROY C. MAST

The Procter & Gamble Co., Miami Valley Laboratories, Cincinnati, Oh. 45247

The solubilization of water-insoluble compounds in aqueous solutions of surfactants is a process which has interested surface and colloid scientists for many years. By solubilization we mean the incorporation of water insoluble compounds into stable isotropic aqueous solutions of surface-active compounds. These "solutions" have been called micellar solutions and microemulsions. Of great interest and also of some practical importance is a more detailed knowledge of the solubilization phenomena. What are the "structures" of the aggregates formed? What interactions are responsible for solubilization? What molecular structural features promote the process?

Because of its sensitivity to relatively weak interactions, nuclear magnetic resonance (nmr) is ideally suited for answering such questions and has been used by a number of authors to gain information about the solubilization phenomenon. Proton nmr spectra of solutions containing benzene and sodium dodecyl sulfate or dodecyltrimethylammonium chloride (1) were interpreted in terms of incorporation of benzene into surfactant micelles above the cmc, with rapid (on the nmr time scale) exchange of benzene between aqueous solution and surfactant micelles. Solubilization of several aromatic alcohols and phenols by sodium dodecyl sulfate was studied by proton nmr (2) with different modes of solubilization being inferred from the different patterns of chemical shift changes observed. The initial and subsequent loci of solubilization of benzene, N,N-dimethylaniline, nitrobenzene isopropylbenzene and cyclohexane in micellar solutions of cetyltrimethylammonium bromide ($C_{16}TAB$) were determined from the proton chemical shift changes in both surfactant and solubilizate as a function of its concentration (3). Information about the structures and dynamics of micellar solutions of "reversed" phase (i.e. water-in-oil) has also been obtained using nmr (4,5).

The general purpose of this work was to obtain solution "structural" information on aqueous micellar solutions using high resolution nmr spectroscopy. Specifically, we wished to determine

440

the loci of maximum concentrations of xylene solubilized by
sodium dodecylbenzene sulfonate (NaDBS) with and without the
solubilization-promoting "hydrotrope" butyl-β-hydroxyethyl ether
(butyl cellosolve, C_4E_1) and to determine the dependence of the
nmr parameters on the nature of the phase. Although alkylbenzene
sulfonates are the most widely used anionic surfactants, few if
any investigations of micellar solubilized, water-continuous
systems have been reported with this important surfactant class.

Experimental

Materials. The sodium dodecylbenzene sulfonate used in
this work was a commercial material from Pilot Chemical Company:
Calsoft F-90. It contains a minimum of 90% surfactant, with most
of the remaining material being water and electrolyte. While
the chain length is stated to be dodecyl, the material is a
mixture of positional isomers of the sulfophenyl group on the
alkyl chain. We found the critical micelle concentration of
this material to be 0.055 g/100 ml water at 23°C by conductance
titration. The xylene was a mixture of isomers, containing 95%
meta isomer, based on proton nmr measurements. Xylene and C_4E_1
were the best grades available commercially and were used without
further purification.

Methods. The phase diagram was determined by mixing the
desired weights of NaDBS and xylene, then adding water dropwise
with thorough mixing until the liquid crystalline phase regions
were passed and the system was isotropic. The remaining
required weight of water was then added rapidly. The samples
prepared for nmr analysis were prepared by the same method,
using D_2O instead of H_2O. When C_4E_1 was included in a system
it was mixed with the NaDBS and xylene before water was added.

A Varian HA-100 nmr spectrometer was used. About 0.1%
deuterated TMSP (2,2,3,3-tetradeutero-3-(trimethylsilyl)-
propionic acid, sodium salt) was added to each sample to provide
an internal reference for measurement of chemical shifts. The
presence of TMSP in the samples did not affect the results
since the chemical shifts of various protons did not change and
there was no visual change in the micellar solutions on addition
of TMSP. A capillary of H_2SO_4 was used as the source of the
field-frequency stabilization signal. Chemical shifts were
measured in ppm downfield from internal TMSP and are accurate
to ±0.01 ppm. All measurements were made at 30°C.

Results and Discussion

The NaDBS/Xylene/Water System. Figure 1 shows the water-
rich portion of the phase diagram for the system NaDBS/water/
xylene at 25°C and the compositions of the samples prepared for
nmr analysis. Figure 2 is a plot of surfactant proton chemical

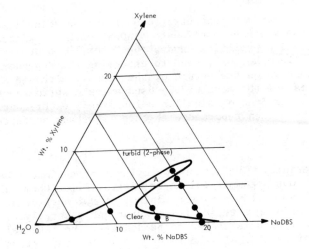

Figure 1. *Partial phase diagram of the NaDBS/xylene/*
H_2O system at 25°C

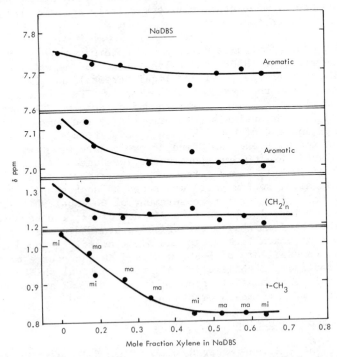

Figure 2. *NaDBS proton NMR chemical shifts vs. mole fraction*
xylene solubilized. Samples which are clear micellar solutions
or turbid macroemulsions are indicated by mi *and* ma, *respec-*
tively.

shifts <u>vs</u> concentration of added xylene. All samples contain at
least 5% surfactant which is well above the cmc. The xylene
concentration is expressed as mole fraction xylene = moles
xylene/(moles xylene + moles surfactant), since the relative
concentrations of xylene and surfactant turn out to be of
interest, rather than total concentration of xylene in solution.

The general trend of all surfactant resonances is an
upfield shift with increasing xylene concentration which suggests
that the solubilized xylene is uniformly distributed along the
surfactant molecules, rather than being restricted to the
micellar surface or interior only. This upfield shift is
presumably the result of the influence of the diamagnetic
anisotropy of the aromatic ring (6) on neighboring
surfactant protons. This interpretation is supported
by the absence of upfield shifts when benzene is replaced by
cyclohexane in these micellar solutions. The irregularity in
surfactant chemical shifts with the mole fraction of added
xylene is probably the result of these shifts also being a slight
function of the change in surfactant concentration. The
surfactant chemical shifts attain a constant value when the
xylene added reaches a mole fraction of about 0.3-0.4, before
the maximum level of solubilization is reached. This suggests
that above ∼0.3-0.4 mole fraction xylene, xylene tends to
accumulate in the micellar core, having "saturated" the
surfactant molecules in the outer part of the micelle.

This behavior can be contrasted to that found by Eriksson
and Gillberg (3) for the solubilization of benzene by the
cationic surfactant $C_{16}TAB$. They found benzene to initially
be absorbed close to the surfactant α-CH_2 groups (e.g. close
to the micelle-water interface) until at about one mole benzene
per mole surfactant a transition to solubilization in the micelle
interior occurred. These differences between the $C_{16}TAB$ and
NaDBS systems are probably the result of different surfactant
polar group charge and perhaps of the fact that the DBS molecule
itself possesses an aromatic moiety which can interact with the
aromatic solubilizate.

Figure 3 shows xylene proton chemical shifts as a function
of mole fraction xylene solubilized by NaDBS. As with the
surfactant chemical shifts, the xylene resonances also shift
upfield with increasing xylene concentration. They move further
from the aqueous solution chemical shift values for xylene but
do not become constant. These shifts are presumably also due to
the influence of the diamagnetic anisotropy of the aromatic
rings, acting intramolecularly to give rise to upfield shifts.
However the changes in xylene chemical shifts as a function of
concentration in <u>n</u>-octane-d_{18} (a model for a hydrocarbon solvent
also shown on Figure 3) are much smaller than those of xylene
solubilized by surfactant. That is, if one subtracts from the
xylene chemical shift in NaDBS solution the variation in
chemical shift caused by xylene-xylene interactions, there still

remains a sizable concentration dependence of the xylene shifts
in NaDBS. Interaction of xylene aromatic rings with DBS
aromatic rings in the surfactant micelles -- perhaps a
"stacking" -- is a possible explanation for this.

It is particularly interesting that the surfactant chemical
shifts are basically a function only of the xylene/surfactant
ratio; and not whether one or two isotropic phases are present.
That is, going across boundaries between single and two-phase
regions produces no discontinuities in chemical shifts. In
addition, single resonances are observed for each molecular
species in all samples. These results suggest that at the
molecular level there is no discoutinuous change in local
environment of either surfactant or xylene occurring at the
phase boundaries. Thus the macroemulsion (turbid) region may
be a mixture of micellar solution phase and another xylene-rich
phase, with rapid (< milliseconds residence times) exchange of
surfactant between phases. As with the surfactant chemical
shift data, no discontinuities in the xylene chemical shifts
occur in going from one phase region to another, in agreement
with continuity of molecular environment (of both surfactant
and xylene) in going from turbid to clear compositons.

The surfactant line widths are also a function of xylene
concentration as shown in Figure 4. There is little change in
the bulk viscosity of the solutions over this range of xylene
concentration, so that the line widths must reflect either
changes in overall aggregate reorientation rate due to changes
in aggregate size, or changes in surfactant molecular reorientation rate ("mobility"). The former explanation can be ruled
out since the aggregate size presumably changes greatly in
going from turbid emulsion to clear micellar solution phase,
while the line widths are dependent only on mole fraction xylene
(with a slight surfactant concentration dependence) and not on
phase. This leaves the hypothesis that the surfactant molecular
reorientation rate increases as xylene is solubilized. Since
the chemical shift data suggest that the xylene exists along
the entire hydrophobic length of the surfactant molecules, it
seems reasonable to attribute the line width variations to an
increase in mobility of the surfactant chains in the micelles
due to intercalation of xylene molecules among surfactant
molecules.

The NaDBS/Xylene/C_4E_1/Water System. "Hydrotropes" are
polar compounds whose water solubility is considerably greater
than that of typical surfactants and which often increase the
ability of surfactants to solubilize water-insoluble compounds.
The effect of the nonionic hydrotrope C_4E_1 on the micellar
"structure" was examined by measuring proton nmr chemical shifts
of surfactant, hydrotrope and solubilizate upon hydrotrope
addition.

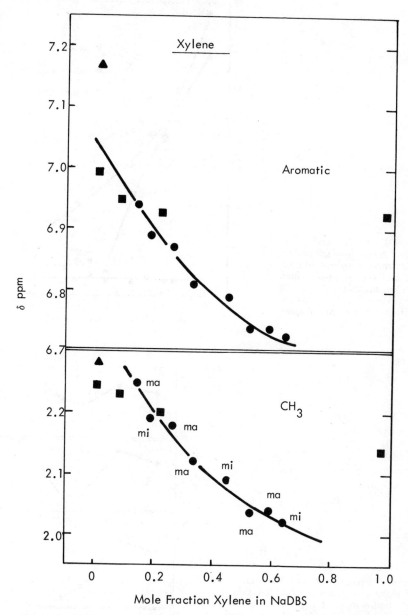

Figure 3.   Xylene proton NMR chemical shifts vs. mole fraction xylene in NaDBS (●). Filled squares (■) indicate chemical shift values for xylene vs. mole fraction in n-octane d_{18}, and ▲ indicates the xylene chemical shifts for a saturated solution of xylene in D_2O. Samples which are clear micellar solutions or turbid macroemulsions are indicated by mi & ma, respectively.

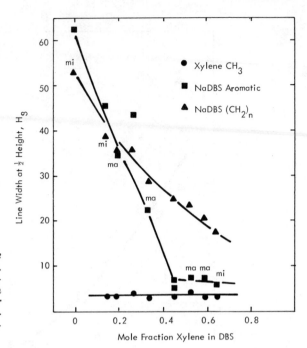

Figure 4. NaDBS and xylene proton NMR line widths vs. mole fraction xylene solubilized in DBS. Samples which are clear micellar solutions or turbid macroemulsions are indicated by mi & ma, respectively.

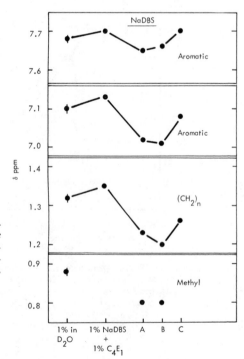

Figure 5. NaDBS proton NMR chemical shifts as a fraction of added xylene and "hydrotrope," C_4E_1. Samples indicated by A & B are turbid macroemulsions and C is a clear micellar solution of composition (wt/vol):

	Xylene	NaDBS	C_4E_1
A	0.5	1.0	0
B	0.5	0.5	0.5
C	0.5	1.0	1.0

Figure 5 shows NaDBS proton chemical shifts as a function of added hydrotrope (butyl-β-hydroxyethyl ether: C_4E_1) and solubilizate xylene). Addition of C_4E_1 to simple NaDBS solutions results in little variation in surfactant chemical shifts. Addition of xylene and C_4E_1 to DBS to form a macroemulsion gives chemical shifts similar to those in the system DBS plus xylene. Apparently in the macroemulsion system the surfactant environment is little affected by C_4E_1. On the other hand addition of enough DBS and C_4E_1 to xylene to form a micellar solution results in downfield shifts of all DBS resonances, implying (1) a more aqueous environment for DBS in the micellar solution, or (2) a lower concentration of xylene aromatic rings around the surfactant molecules, caused either by a "dilution" of xylene concentration by added C_4E_1 or by movement of the xylene to the micellar core, perhaps due to replacement of xylene in the interfacial region by C_4E_1.

Further information about the effect of hydrotrope addition on solubilization can be obtained from the hydrotrope chemical shifts in the four component NaDBS/C_4E_1/xylene/D_2O system shown on Figure 6. All C_4E_1 resonances are shifted somewhat upfield from their aqueous solution values in the presence of DBS or DBS plus xylene. There is, however, little change in C_4E_1 chemical shifts upon adding xylene or in going from macroemulsion to micellar solution phases, as was observed in the surfactant resonances. These results suggest that C_4E_1 is predominantly away from the xylene loci of solubilization.

Likewise the xylene proton chemical shifts for these systems shown on Figure 7 are not dependent upon addition of C_4E_1 (consistent with the chemical shift results for C_4E_1). The xylene resonances shift downfield toward their values in pure liquid xylene in going from C_4E_1-containing macroemulsion to micellar solution.

These results all point to minimum interaction of C_4E_1 with surfactant or xylene in the turbid phase and possible incorporation of C_4E_1 into the surfactant "film" with concomitant localization of xylene in the interior upon formation of clear micellar solution phase.

<u>Mathematical Treatments of Micellar Solution-nmr Data.</u> As xylene is added to surfactant in the NaDBS/xylene/water system, the surfactant resonances initially shift upfield and then approach a constant value. This behavior suggests "saturation" of an interfacial region with surfactant before maximum solubilization occurs, which in turn suggests the possibility of treating the nmr data in the manner of a Langmuir isotherm. If s is defined as $s = (\delta - \delta°)/(\delta^{sat} - \delta°)$ where δ and $\delta°$ are the surfactant proton chemical shifts in the presence and absence of xylene and δ^{sat} is the chemical shift at "saturation", i.e., the asymptotic value at high mole

fraction xylene, and the molar ratio of xylene to surfactant is called c, then in the usual Langmuir isotherm terminology (see e.g., Adamson, $\underline{7}$) where the "surface" is the interfacial region occupied by surfactant, we have

$$c/s = k_1 k_2 + k_1 c$$

where k_1 and k_2 are constants related to the "adsorption capactiy" of the micelle and the strength of the surfactant-solubilizate interaction, respectively. A plot of c/s $\underline{vs}$ c should thus be linear, if this treatment provides a reasonable description of the nmr data pertaining to the solubilization process. Figure 8 shows that such a plot for the surfactant terminal methyl proton resonance exhibits reasonably linear behavior. The number of sites available for xylene solubilization is limited and becomes filled as the mole fraction of xylene increases.

Another approach to considering this data is to use the equations derived to describe the variation in chemical shift when complex formation occurs between two molecular species. In the present case the relevant equation ($\underline{8}$) is:

$$(\delta - \delta^o)^{-1} = (\delta^c - \delta^o)^{-1}(1 + 1/KX_x^o)$$

(valid for large X_x^o) where δ and δ^o are as above, δ^c is the chemical shift of "complexed" surfactant, X_x^o mole fraction xylene added, and K is the equilibrium constant ($K = (X_c/X_x X_s)$) for the "reaction" $X + S \rightleftarrows C$, where X, S and C are xylene, surfactant and X-S "complex", respectively. Figure 9 is a plot of $(\delta - \delta^o)^{-1}$ $\underline{vs}$ $X_x^o{}^{-1}$, which is linear and fits the equation, except for very small X_x^o. From this plot K and δ^c are found to be 3.07 and 0.645 ppm respectively. K should be a measure of the "strength" of the surfactant-solubilizate interaction and δ^c an indication of the environment of solubilized material.

Agreement of the predictions of either of these approaches with the experimental data does not prove anything about the mechanism of solubilization, but these approaches could provide a framework for quantitative comparison of nmr data on solubilization of differing materials by various surfactants.

Conclusions

A. In the NaDBS/xylene/D$_2$O system:

1. Solubilized xylene is distributed uniformly around alkyl and aromatic portions of NaDBS surfactant molecules.

2. A "saturation" concentration of xylene in the outer part of the micelles is reached at xylene concentrations of ~0.3-0.4 mole fraction xylene in NaDBS. Xylene solubilized above this concentration is probably localized in the micellar interior.

3. There are no discontinuities in the local environments

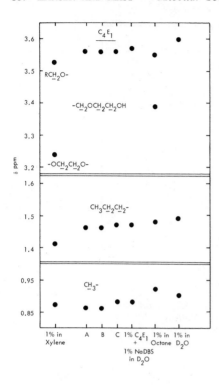

Figure 6. C_4E_1 ("hydrotrope") chemical shifts as a function of solvent and phase of NaDBS—xylene systems. Samples indicated by A & B are turbid macroemulsions and C is a clear micellar solution of composition (wt/vol):

	Xylene	NaDBS	C_4E_1
A	0.5	1.0	0
B	0.5	0.5	0.5
C	0.5	1.0	1.0

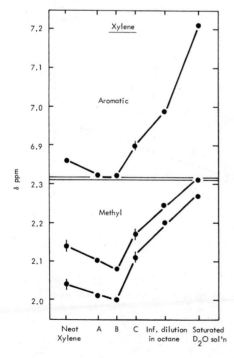

Figure 7. Xylene chemical shifts as a function of solvent and phase of NaDBS—xylene system. Samples indicated by A & B are turbid macroemulsions and C is a clear micellar solution of composition (wt/vol):

	Xylene	NaDBS	C_4E_1
A	0.5	1.0	0
B	0.5	0.5	0.5
C	0.5	1.0	1.0

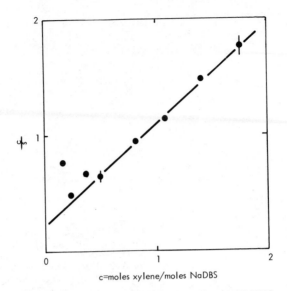

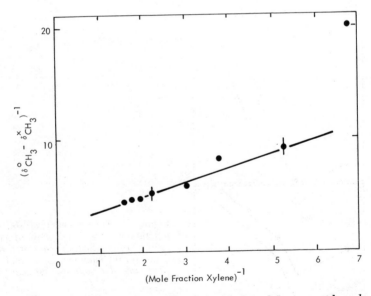

Figure 8. Langmuir isotherm-type plot for NaDBS terminal methyl proton NMR chemical shift vs. mole fraction xylene data. S is the normalized difference in surfactant chemical shifts with and without xylene.

Figure 9. NaDBS methyl proton chemical shift difference with and without added xylene as a function of mole fraction xylene, based on "complex" formation model

of surfactant or xylene molecules at the phase boundaries between turbid emulsions and clear micellar solutions. The turbid phase region may consist of a mixture of "clear" micellar solution phase and another phase, with rapid (msec) exchange of surfactant and xylene between the aggregates of these phases.

4. Surfactant molecular mobility increases upon solubilization of xylene. This appears to be the result of decreased surfactant-surfactant interactions (larger area per molecule) in the interfacial film, caused by incorporation of xylene molecules into this region.

B. The NaDBS/xylene/C_4E_1/D_2O Systems

1. In the turbid phase, addition of C_4E_1 has little effect on the surfactant or xylene environments, with C_4E_1 apparently existing mainly in the aqueous phase.

2. In the clear phase C_4E_1 appears to replace xylene in the outer part of the micelle with the xylene shifting into the micelle interior.

C. Nmr chemical shift data for solubilization of aromatic compounds by surfactants can be treated successfully by either Langmuir or "complex formation" models.

Acknowledgment

The authors would like to thank R. S. Treptow for determining the partial phase diagram for the NaDBS-xylene-water system.

Abstract

Information about the time-averaged "structures" of aqueous micellar solutions containing the surfactant sodium dodecyl-benzene sulfonate and xylene, with and without the hydrotrope butyl-β-hydroxyethyl ether was sought using high-resolution proton nuclear magnetic resonance (nmr) spectroscopy. The loci of solubilization of xylene by surfactant micelles was probed as a function of added xylene. The first xylene solubilized is intercalated among the surfactant molecules in the micelles, while above ≈ 0.3-0.4 mole fraction xylene in surfactant, xylene goes predominantly to the micellar interior. Continuity of local environments of surfactant and xylene was found across turbid (macroemulsion)-clear(micellar solution) phase boundaries. The hydrotrope appears to have little effect on the surfactant or xylene environments in the turbid phase, while replacing some of the xylene in the surfactant-containing interfacial film in the clear micellar phase.

Literature Cited

1. Nakagawa, T. & Tori, K., Kolloid-Z. & Z. Polym. (1964) 194, 143.

2. Tokiwa, F. & Aligami, K., Kolloid-Z. & Z. Polym. (1971) 246, 688.

3. Eriksson, J. C. & Gillberg, G., Acta Chem. Scand. (1966) 20, 2019.

4. Shah, D. O. & Hamlin, R. M., Jr., Science (1971) 171, 483.

5. Hansen, J. R., J. Phys. Chem. (1974) 78, 256.

6. Emsley, J. W., Feeney, J. & Sutcliffe, L. H., "High Resolution Nuclear Magnetic Resonance Spectroscopy", Vol 1, p. 94, Pergamon Press, Oxford, 1966.

7. Adamson, A. W., "Physical Chemistry of Surfaces", p. 399-401. Interscience, New York, 2nd Ed. 1967.

8. Williams, D. E., Thesis, University of Colorado, 1966.

Use of Nuclear Magnetic Resonance Techniques to Study Nonionic Surfactant Micelles and Mixed Micelles with Phospholipids

EDWARD A. DENNIS and ANTHONY A. RIBEIRO

Department of Chemistry, University of California at San Diego, La Jolla, Calif. 92093

Nonionic surfactants of the p,tert-octylphenylpolyoxyethyl-ene ether (OPE) class are widely used as detergents, solubilizers,

$$(CH_3)_3CCH_2C(CH_3)_2C \overset{\overset{\displaystyle H \quad H}{\displaystyle C=C}}{\underset{\underset{\displaystyle H \quad H}{\displaystyle C=C}}{}} C(OCH_2CH_2)_n OH \qquad (OPE)$$

and emulsifiers. Triton X-100 (Triton), a commercial product which has an average value of n = 9 to 10, forms micelles in aqueous solution and under certain conditions forms mixed micelles with phospholipids such as 1,2-diacyl-sn-glycero-3-phos-phorylcholine (phosphatidylcholine). These mixed micelles are of great current interest in biochemical studies as they are formed when Triton is used in the purification of membrane-bound pro-teins (1) and also serve as an excellent form for the phospho-lipid substrate (2) of various lipolytic enzymes such as phospho-lipase A$_2$ (3) and the membrane-bound enzyme phosphatidylserine decarboxylase (4).

Over the last few years, we have employed both ^{1}H and ^{13}C nuclear magnetic resonance (nmr) techniques to study the struc-tural aspects of OPE micelles, and particularly those of Triton X-100 (5-8). Furthermore, we developed a [1]H-nmr technique to follow the formation of mixed micelles with Triton and to deter-mine phase diagrams for the surfactant-phospholipid-water system (9,10). We have also utilized both [1]H- and ^{13}C-nmr relaxation techniques to gain information about the structure and mobility of the surfactant and phospholipid in the mixed micelles (6,7, 11). Nmr has been one of the major physical techniques utilized to characterize the physical state of phospholipid in multibi-layer (smectic mesophase) and single bilayer (sonicated vesicle)

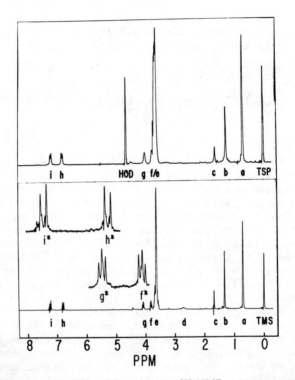

Figure 1. 220-MHz continuous wave ¹H-NMR spectra recorded
at 2500-Hz sweep width and 37° of 10mM Triton X-100 in D_2O
containing TSP (top spectrum) and in $CDCl_3$ containing TMS
(bottom spectrum). Inserts (*) were recorded at 500-MHz sweep
width. Peaks are identified above. Peak d is presumably the OH
of Triton in $CDCl_3$.

structures (12-15). However, until our work, little had been
done to characterize the physical state of phospholipid in mixed
micelles. In order to consider the evolution of our current
ideas about the structure of mixed micelles, we will first consider the structure of Triton micelles alone and the formation
of mixed micelles with phospholipid.

Nonionic Surfactant Micelles

Micelle Properties and Cloud Point. Triton X-100 is a polydisperse preparation of OPE's (16,17) with an average chain
length of 9.5 oxyethylene units and an average molecular weight
of 628; it may contain some heterogeneity in the hydrophobic portion (17). In aqueous solution, Triton X-100 forms micelles with
a critical micelle concentration (cmc) of about 0.3 mM (18) and
the cmc decreases slightly in high salt (19). Measurements of
the micelle size by light scattering suggest an average molecular
weight of 67,000 to 153,000 depending on experimental conditions
(20-22). Triton X-100 has the unusual property of exhibiting a
cloud point at about 65°C (23). Cloud points, which are characteristic of nonionic surfactants, are not well understood.
However, we have found that a fractionation of the polydisperse
surfactant occurs above the cloud point (5). In the temperature
region just below the cloud point, the [1]H-nmr spectrum loses
intensity and we have suggested that this is caused by the formation of very large structures by the Triton molecules and that
the structures are suspended in the aqueous phase (5). The cloud
point of Triton X-100 is known to be affected by the presence of
other substances (23) and we have found that phospholipids lower
the cloud point of this surfactant (5).

^{1}H-NMR Relaxation Studies. The [1]H-nmr spectrum of Triton
X-100 in $CDCl_3$ and as micelles in D_2O is shown in Figure 1 along
with the suggested assignment. We have found that the [1]H-nmr T_1
values for all peaks in Triton X-100 micelles increase with temperature and that for protons at the hydrophobic end of the
molecule, T_1/T_2^* is small and close to unity suggesting motion
in the extreme narrowing limit (6,7) and reflecting behavior of a
hydrocarbon liquid (24). Kushner and Hubbard (21) have suggested
on the basis of viscosity and turbidity studies that Triton
micelles contain a large amount of water bound to the polyoxyethylene chain. Studies on related surfactants which show the
T_1 values of the hydrophobic protons are independent of whether
the solvent is H_2O or D_2O, have led to the conclusion that water
does not penetrate the hydrophobic core of nonionic micelles to
any significant extent (25,26). Similar studies by us (6) and
Podo et al. (27) have confirmed that this is also the case for
Triton X-100. Thus, work on Triton and related surfactants suggests that the micellar core of Triton X-100 behaves as a hydrocarbon liquid and the polyoxyethylene group is heavily solvated

with water.

^{13}C-NMR Assignments and Relaxation Times. More recently, we
have analyzed the natural abundance ^{13}C-nmr spectra of p,tert-
octylphenol and various OPE's (Triton X-15, X-35, X-45, X-114,
and X-100) (8). The spectrum and a suggested assignment for
Triton X-100 is shown in Figure 2. In particular, ^{13}C-nmr allows
us to probe individual positions of both the hydrophobic and the
hydrophilic regions of these surfactants. The surfactants
studied all have the same hydrophobic p,tert-octyl (peaks a-e)
and phenyl (peaks m-p) groups, but have differing average oxyethylene chain lengths and the detailed spectral assignments
could be determined from chemical shifts, relative intensities,
non-proton decoupled spectra and one bond coupling constants.

Spin-lattice relaxation times (NT_1) of the protonated carbons were employed to assess the motion of these surfactants
(especially Triton X-100). Our results suggest that in organic
solvents, where the surfactant is very likely monomeric, free
tumbling occurs with segmental motion in the polyoxyethylene
chain. In aqueous media, where the nonionic surfactant forms
micelles, the surfactant molecules appear to be restricted from
free tumbling. The internal and segmental motions of the alkyl
and oxyethylene chains appear to be more dominant, and in particular, motion appears maximally restricted at the hydrophobic/
hydrophilic interface.

Formation of Mixed Micelles with Phospholipids

^{1}H-NMR Techniques. Several years ago, we showed that egg
phosphatidylcholine in mixed micelles with Triton X-100 gives
high-resolution ^{1}H-nmr spectra in which the phospholipid gives
rise to narrow lines and the Triton peaks, also narrow, are similar to those in pure Triton micelles (9). High-resolution nmr
spectra of unsonicated phospholipid preparations, i.e. multibilayers or smectic mesophases, result in low apparent intensities
because of the large line broadening, whereas mixed micelles show
full intensities. We have used this difference to follow the
transformation of phospholipid bilayers to mixed micelles upon
the addition of Triton X-100. We interpreted our studies to suggest that as Triton is added to phospholipid bilayers, three
specific changes occur (9): 1.) When Triton is added to a dispersion of phospholipid bilayers, the Triton is incorporated into
the bilayers until sufficient Triton is added that the bilayers
become saturated with it. 2.) Additional Triton solubilizes the
bilayers converting them into mixed micelles. Thus, at intermediate molar ratios, both bilayers (saturated with Triton) and
mixed micelles occur. 3.) Finally, at a molar ratio above about
2:1 Triton/phospholipid, all of the phospholipid is in mixed
micelles; further addition of Triton merely dilutes the phospholipid in the mixed micelles. This is illustrated schematically

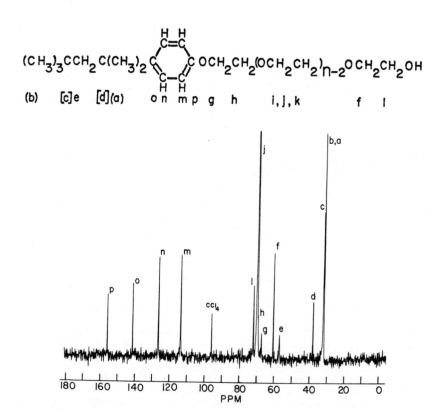

Figure 2. Natural abundance ^{13}C-NMR spectrum of 0.4M Triton X-100 in D$_2$O. The NMR tube was fitted with a coaxial capillary tube containing CCl$_4$. The probable assignments are indicated; carbons that could not be unambiguously distinguished from one another are indicated by parentheses.

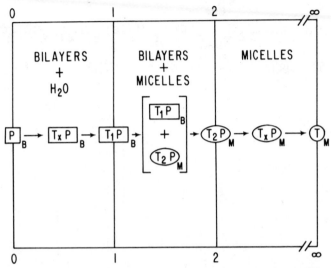

MOLAR RATIO (TRITON X-100 / PHOSPHATIDYLCHOLINE) ≡ X

Figure 3. Schematic of the average composition of the phases formed by Triton X-100 (T) and egg phosphatidylcholine (P) in the presence of an excess of water (at room temperature and certain experimental conditions). This is shown as a function of the molar ratio of Triton/phospholipid. For simplicity, the stoichiometry of the phospholipid bilayers (B) in the presence of an excess of Triton is assumed to be 1:1 and the stoichiometry of Triton micelles (M) in the presence of an excess of phospholipid is assumed to be 2:1. The monomer concentration of phospholipid and Triton is negligible under the experimental conditions employed and is not indicated.

in Figure 3.

Phase Diagram for Surfactant-Phospholipid-H$_2$O System.

Saturated phospholipids, such as dipalmitoyl phosphatidylcholine, undergo thermotropic phase transitions (28-30) and these phase transitions have been shown to affect membrane localized phenomenon (31,32). We have utilized nmr techniques to show that mixed micelle formation from dipalmitoyl phosphatidylcholine and Triton X-100 is affected by these phase transitions (10). Specifically, using the techniques developed with egg phosphatidylcholine, we have found that mixed micelle formation at 37°C and 49°C is similar to that from egg phosphatidylcholine, whereas at 20°C, mixed micelle formation with dipalmitoyl phosphatidylcholine is severely inhibited. These and other studies led to the formulation of the phase diagram for the dipalmitoyl phosphatidylcholine-Triton-H$_2$O system as shown in Figure 4.

Gel Chromatographic Confirmation.

We have also employed gel chromatography and centrifugation techniques to separate the mixed micelles and have found that under the experimental conditions employed (33), egg phosphatidylcholine bilayers can hold Triton up to a molar ratio of about 1:1 Triton/phospholipid and that mixed micelles are formed at a molar ratio of about 2:1 to 3:1 Triton/phospholipid. At molar ratios of Triton to phospholipid above about 10:1 Triton/phospholipid, the mixed micelles are about the same size as pure Triton micelles. Similar studies on dipalmitoyl phosphatidylcholine confirm that mixed micelles are only formed at very high molar ratios of Triton/phospholipid (greater than about 13:1) at room temperature. By contrast, dimyristoyl phosphatidylcholine which has its thermotropic phase transition at 23°C, readily forms mixed micelles at a molar ratio of 2:1 Triton/phospholipid at room temperature. These studies confirm and elaborate the phase diagrams shown above.

The chromatography results between a molar ratio of 3:1 and 10:1 Triton/phospholipid are similar for egg and dimyristoyl phosphatidylcholine and show a polydispersity of micelle size (33). Since the dimyristoyl phosphatidylcholine consists of a single molecular species of phospholipid, the polydispersity of micelle size does not arise from fractionation of the phospholipid. While the polydispersity may be simply that which occurs for mixtures of this polyoxyethylene surfactant and phospholipid under the experimental conditions employed, the molar ratios across the eluted peaks were found to not be constant, and this suggests that fractionation of the polydisperse Triton X-100 occurs. In order to determine if this is responsible for the polydispersity in micelle size, and to prepare mixed micelles at low molar ratios which are less polydisperse, it will be necessary to examine a homogeneous preparation of single-species Triton; these studies are underway.

Structure of Mixed Micelles

[1]H-Relaxation Times. The previous nmr and gel chromatographic studies show that at temperatures above the range of the thermotropic phase transition of the phospholipid and below the range of the cloud point of Triton, mixed micelles are formed at molar ratios above about 2:1 Triton/phospholipid. Under these conditions, the protons in both the Triton and phospholipid give rise to high-resolution nmr spectra with full intensities and narrow line widths as illustrated in Figure 5.

Proton T_1 and T_2^* (from line widths) relaxation times have been determined for pure Triton micelles and mixed micelles of Triton with egg, dipalmitoyl, and dimyristoyl phosphatidylcholines at a molar ratio of 3:1 Triton/phospholipid (6,32). Extensive proton relaxation measurements as a function of temperature and frequency (55-220 MHz) have also been conducted on these micelles and mixed micelles. In every case, we found that the T_1 and T_2^*, temperature dependence, frequency dependence, and activation energies of the Triton molecule are unchanged in the presence of phosphatidylcholine, suggesting that the mixed micelle has a structure not grossly different from that of pure Triton micelles. Some typical relaxation times for Triton X-100 micelles and mixed micelles with dimyristoyl phosphatidylcholine are given below (6). It is apparent that the T_1/T_2^* ratios and frequency dependence of the hydrophobic tert-butyl group (peak a) are very similar in micelles and mixed micelles. The data is consistent with nmr observations under extreme narrowing conditions and suggests motion reflecting a hydrocarbon liquid.

		Micelles		Mixed Micelles			
Peak (see Fig. 5)		a	b	a	x	y	z
100 MHz	T_1 (sec)	0.23	0.098	0.23	0.50	0.36	0.43
	$\Delta\nu_{1/2}$ (Hz)	2.1	6.6	2.1	4.7		1.5
	T_2^* (sec)	0.15	0.048	0.15	0.068		0.21
	T_1/T_2^*	1.5	2.0	1.5	7.3		1.9
220 MHz	T_1 (sec)	0.27	0.12	0.26	0.68	0.43	0.51
	$\Delta\nu_{1/2}$ (Hz)	2.1	4.4	2.3	5.3		1.9
	T_2^* (sec)	0.15	0.072	0.14	0.059		0.17
	T_1/T_2^*	1.9	1.7	1.9	11.4		3.0

The T_1 values of various protons of dimyristoyl phosphatidylcholine in mixed micelles (6) as well as for egg phosphatidylcholine and dipalmitoyl phosphatidylcholine above the phase transition (7) show that the T_1 relaxations are similar to those which occur in multibilayers and in sonicated vesicles which are used as membrane models. These relaxation times have been

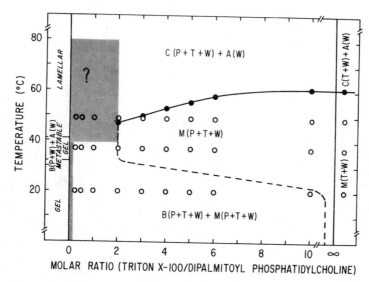

Figure 4.   Schematic of the Triton X-100 - dipalmitoyl phosphatidyl-choline–water system. On the left are shown the reported thermotropic phase transitions of unsonicated dispersions of dipalmitoyl phosphatidyl-choline (P) in the presence of excess water (W), which forms a bilayer phase (B) and an aqueous phase (A). On the right is indicated the re-ported thermotropic phase separation of concentrated solutions of Triton X-100 (T) in water from a micellar phase (M) to a cloud point phase (C), plus an aqueous phase. The contribution of free molecules (monomers) of phospholipid and Triton, which would be negligible for the concen-trated samples under consideration, is not indicated. The samples used to obtain the NMR results are indicated (○) in this diagram at the three temperatures employed for dipalmitoyl phosphatidylcholine in D_2O and the various molar ratios of Triton to phospholipid used as well as for a control sample of Triton X-100. The cloud point of these samples was determined visually and is indicated (●). The resulting diagram indicates the phases inferred to be present on the basis of the NMR studies and can be considered to be an approximate phase diagram for this system under defined experimental conditions. The situation in the shaded area is not clear at this time; no cloud point was observable for the mixtures at low molar ratios, and, presumably, there is a region containing B(P + T + W) + A(W).

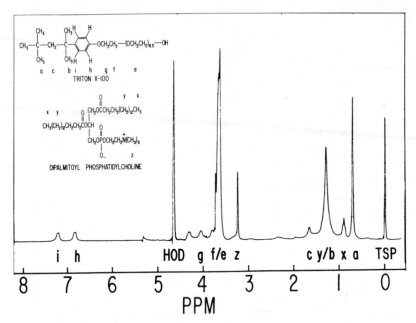

Figure 5.   220-MHz ¹H-NMR spectrum recorded at 37°C of a mixture of 200mM Triton X-100 and 100mM dipalmitoyl phosphatidylcholine in D_2O and containing TSP

attributed to coupled gauche–trans–gauche isomerizations (13,34–36) and this is presumably the mode of T_1 relaxation of the phospholipid in mixed micelles. Furthermore, T_1 measurements in H_2O/D_2O mixtures are consistent with the idea that water does not penetrate the hydrophobic core of the mixed micelles, while water does solvate the polar oxyethylene and choline methyl groups. Titration with Mn^{2+} confirms that the oxyethylene and choline methyl groups are on the exterior of the mixed micelle while the hydrophobic groups are located in the micellar interior.

The T_2^* values for the terminal methyl and choline methyl protons in the phospholipid were found to be much longer than those reported for these groups in multibilayers and even longer than in sonicated vesicles. In vesicles, there appears to also be a slower motion than the coupled motion as T_2 is significantly shorter than T_1 (13) and this is supported by the frequency dependence of T_1 (36). Although this slower component has been considered in terms of vesicle tumbling and lateral diffusion, arguments against these explanations have been advanced (35), and accepted by most workers in this area (15,34–36). Chan and co-workers (14,37) have postulated a theory to account for the narrower lines in sonicated vesicles compared with multibilayers which along with the results of Horwitz et al. (13) suggest that the increased T_2^* in vesicles results from less restricted motion of the phospho-

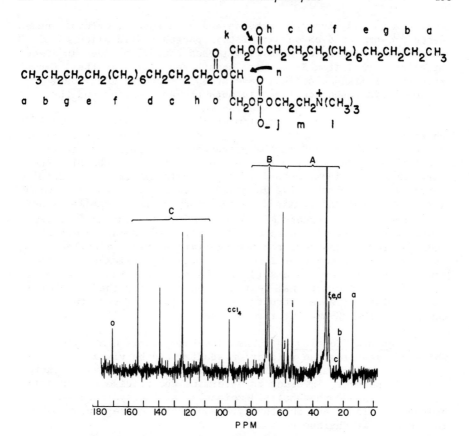

Figure 6. Natural abundance ^{13}C-NMR spectrum of 0.4M Triton X-100 plus 0.133M dimyristoyl phosphatidylcholine in D_2O. The NMR tube was fitted with a coaxial capillary tube containing CCl_4. Phospholipid carbons are assigned here. A, B, and C refer to the alkyl, oxyethylene, and phenyl carbon peaks of Triton shown in Figure 2.

lipid. Similar considerations suggest that the phospholipid in mixed micelles with Triton is in an even less restricted environment than in vesicles (6,7) and is, in fact, in a rather fluid environment. It should be noted, however, that the interpretation of relaxation times for phospholipid is complicated and recent deuterium nmr studies have reopened the question of whether vesicle tumbling is in fact responsible for line narrowing in vesicles (38).

^{13}C-NMR Relaxation Times. The ^{1}H-nmr studies on the surfactant/phospholipid mixed micelle system are unfortunately limited due to overlapping, envelope lines for the phospholipid acyl chains and glyceryl backbone and as a result, natural abundance ^{13}C-nmr studies on dimyristoyl phosphatidylcholine in CD$_3$OD and on dimyristoyl and dipalmitoyl phosphatidylcholines in mixed micelles with Triton X-100 were undertaken (11). In the ^{13}C-nmr spectrum of mixed micelles, resolution of several individual phospholipid resonance lines in both the acyl chains and the polar head group is possible as shown in Figure 6.

The ^{13}C-nmr studies complement the ^{1}H-nmr results. The ^{13}C chemical shifts, relaxation times, and line widths of the Triton carbons were unchanged in the presence of phospholipid, supporting the idea that the phospholipid does not affect the microenvironment or motional behavior of the surfactant micelle grossly. Moreover, the phospholipid in mixed micelles also exhibit similar T_1 values to the phospholipid in the unsonicated multibilayer and sonicated vesicle structures, corresponding to similar fast segmental motions of the phospholipid fatty acid chains in all of these systems. However, the phospholipid in mixed micelles very clearly exhibits narrower lines and longer T_2* relaxation times than in the bilayer systems (39). These ^{13}C-nmr results also suggest that less motional restrictions occur for the phospholipids in mixed micellar structures than in bilayer structures and complement the ^{1}H-nmr interpretations.

Structural Conclusion. The structure of mixed micelles which emerges from the relaxation studies can be visualized in a schematic fashion as shown in Figure 7. Further work is clearly required, however, in order to determine the precise conformation of the Triton and phospholipid molecules at the hydrophobic/hydrophilic interface and the type and degree of segmental motion in the polyoxyethylene chain.

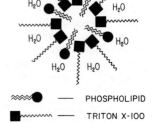

Figure 7. Schematic for the structure of mixed micelles of Triton X-100 and phosphatidylcholine at a molar ratio of 2:1 Triton/phospholipid

Acknowledgement

These studies were supported by grant NSF BMS 75-03560, A.A.R. was a NIH predoctoral fellow GM-1045, and the nmr spectrometers were operated by grant RR-00,708.

Literature Cited

1.   Helenius, A., and Simons, K., Biochim. Biophys. Acta (1975) 415, 29.
2.   Dennis, E. A., J. Supramol. Struct. (1974) 2, 682.
3.   Deems, R. A., Eaton, B. R., and Dennis, E. A., J. Biol. Chem. (1975) 250,9013.
4.   Warner, T. G., and Dennis, E. A., J. Biol. Chem. (1975) 250, 8004.
5.   Ribeiro, A. A., and Dennis, E. A., Chem. Phys. Lipids (1974) 12, 31.
6.   Ribeiro, A. A., and Dennis, E. A., Biochemistry (1975) 14, 3746.
7.   Ribeiro, A. A., and Dennis, E. A., Chem. Phys. Lipids (1975) 14, 193.
8.   Ribeiro, A. A., and Dennis, E. A., J. Phys. Chem. (1976) 80, in press.
9.   Dennis, E. A., and Owens, J. M., J. Supramol. Struct. (1973) 1, 165.
10. Ribeiro, A. A., and Dennis, E. A., Biochim. Biophys. Acta (Biomembranes) (1974) 332, 26.
11. Ribeiro, A. A., and Dennis, E. A., J. Colloid and Interface Sci. (1976) 55, 94.
12. Penkett, S. A., Flook, A. G., and Chapman, D., Chem. Phys. Lipids (1968) 2, 273.
13. Horwitz, A. F., Horsley, W. J., and Klein, M. P., Proc. Nat. Acad. Sci. USA (1972) 69, 590.
14. Seiter, C. H. A., and Chan, S. I., J. Amer. Chem. Soc. (1973) 95, 7541.
15. Metcalfe, J. C., Birdsall, N. J. M., and Lee, A. G., Annals New York Acad. Sci. (1973) 222, 460.
16. Becher, P., "Nonionic Surfactants, Surfactant Science Series", pp. 478-515, M. J. Schick, ed., Marcell Dekker, New York, 1967.
17. Enyeart, C. R., "Nonionic Surfactants, Surfactant Science Series", pp. 44-85, M. J. Schick, ed., Marcel Dekker, New York, 1967.
18. Crook, E. H., Fordyce, D. B., and Trebbi, G. F., J. Phys. Chem. (1963) 67, 1987.
19. Ray, A., and Némethy, G., J. Am. Chem. Soc. (1971) 93, 6787.
20. Mankowich, A. M., J. Phys. Chem. (1954) 58, 1027.
21. Kushner, L. M., and Hubbard, W. D., J. Phys. Chem. (1954) 58, 1163.
22. Kuriyama, K., Kolloid Z. (1962) 181, 144.

23. Maclay, W. N., J. Coll. Sci. (1956) 11, 272.

24. Hanson, J. R., and Lawson, K. D., Nature (1970) 225, 542.

25. Corkill, J. M., Goodman, J. G., and Wyer, J., Trans. Faraday Soc. (1969) 65, 9.

26. Clemett, C. J., J. Chem. Soc. (1970) (A), 2251.

27. Podo, F., Ray, A., and Némethy, G., J. Amer. Chem. Soc. (1973) 95, 6164.

28. Chapman, D., Williams, R. M., and Ladbrooke, B. E., Chem. Phys. Lipids (1967) 1, 445.

29. Melchior, D. L., and Morowitz, H. J., Biochemistry (1972) 11, 4558.

30. Hinz, H., and Sturtevant, J. M., J. Biol. Chem. (1972) 247, 6071.

31. Shimshick, E. J., and McConnell, H. M., Biochemistry (1973) 12, 2351.

32. Linden, C. D., Wright, K. L., McConnell, H. M., and Fox, C. F., Proc. Nat. Acad. Sci. Usa (1973) 70, 2271.

33. Dennis, E. A., Arch. Biochem. Biophys. (1974) 165, 764.

34. Horwitz, A. F., Klein, M. P., Michaelson, D. M., and Kohler, S. J., Annals New York Acad. Sci. (1973) 222, 468.

35. Chan, S. I., Sheetz, M. P., Seiter, C. H. A., Feigenson, G. W., Hsu, M., Lau, A., and Yau, A., Annals New York Acad. Sci. (1973) 222, 499.

36. Mclaughlin, A. C., Podo, F., and Blasié, J. K., Biochim. Biophys. Acta (1973) 330, 109.

37. Lichtenberg, D., Peterson, N. O., Girardet, J., Kainosho, M., Kroon, P. A., Seiter, C. H. A., Feigenson, G. W., and Chan, S. I., Biochim. Biophys. Acta (1975) 382, 10.

38. Stockton, G. W., Polnaszek, C. F., Tulloch, A. P., Hasan, F., and Smith, I. C. P., Biochemistry (1976) 15, 954.

39. Sears, B., J. Membrane Bio. (1975) 20, 59.

The Binding of Manganese to Photoreceptor Membranes

N. ZUMBULYADIS

Research Laboratories, Eastman Kodak Company, Rochester, N.Y. 14650

T. WYDRZYNSKI

Department of Physiology and Biophysics, University of Illinois, Urbana, Ill. 61801

P. G. SCHMIDT

Department of Chemistry, University of Illinois, Urbana, Ill. 61801

Recent work in the area of photosynthesis has demonstrated that manganese, through its effects on proton spin-spin and spin-lattice relaxation can serve as a probe of light induced chemical and structural changes in photoreceptor membranes. The purpose of the present paper is to set the framework for an analysis of proton relaxation results in photo-receptor membrane systems containing manganese. We report the water proton spin-lattice relaxation rates of aqueous dipalmitoylphosphatidylcholine vesicle and liposome suspensions in the presence of manganese as a function of resonance frequency.

In the presence of paramagnetic ions, the fluctuation of electronic magnetic fields, ordinarily dominates proton relaxation. For nuclei at a distance r from a paramagnetic center one obtains the Solomon–Bloembergen equations:

$$1/T_{1M} = (2/5)(\gamma_I^2 g^2 s(s+1)\beta^2/r^6)(3\tau_c/1 + \omega_I^2\tau_c^2), \quad (1)$$

where $1/T_M$ is the spin-lattice relaxation rate, ω_I the resonance frequence, β the Bohr magneton. The correlation time is given by

$$1/\tau_c = 1/\tau_S + 1/\tau_R + 1/\tau_M, \quad (2)$$

where

τ_S = the electronic spin-lattice relaxation rate,
τ_R = the rotational correlation time of the water manganese complex,
$1/\tau_M$ = the ligand exchange rate.

Three features of the above equations deserve attention:

1. The shortest correlation time will dominate.
2. The relaxation rate has a maximum when $\omega_I \tau_c = 1$.
3. τ_S and hence τ_c can be frequency dependent.

τ_M and τ_R relate to molecular dynamics and cannot be affected by the magnetic field strength. τ_S, however, is itself a relaxation time and contains an explicit frequency dependence, given in the simplest case by the Bloembergen-Morgan equations:

$$1/\tau_S = B\left[(\tau_v/1 + \omega_S^2\tau_v^2) + (4\tau_v/1 + 4\omega_S^2\tau_v^2)\right]. \quad (3)$$

τ_v is a correlation time related to the modulation of electric crystal field gradient by molecular vibrations.

The Solomon-Bloembergen equation describes the relaxation of nuclei bound to the paramagnetic centers. It is related to the observed overall relaxation rate by

$$1/T_1(obs.) = M_b q^*/N(1/T_{1M})_b + M_f q/N(1/T_{1M})_f + (1/T_{1A}). \quad (4)$$

$1/T_{1A}$ is the proton relaxation in the absence of manganese; q^* and q are the water coordination numbers of membrane bound and free manganese, M_b and M_f are the concentrations of bound and free manganese, respectively. N is the molarity of water.

$(1/T_{1M})_f$ has a weak frequency dependence. It decreases monotonically with increasing resonance frequency. If τ_S is the dominating correlation time, however, $(1/T_{1M})_b$ and thus $1/T_1(obs.)$ can display a strong, characteristic frequency dependence, with a maximum around $\omega_I \tau_S = 1$.

We have therefore undertaken a water proton relaxation study of manganese chloride-phospholipid dispersions to determine the relaxation mechanism and compare molecular dynamics in vesicles and large liposomes.

Materials and Methods

Spin-lattice relaxation measurements were made on a home built spectrometer designed in our laboratory. The inversion-recovery method was used. Details of the spectrometer will be published elsewhere.

Dipalmitoyllecithin dispersions were prepared in a sonicating bath; a probe was used for the sonication of vesicles. Both samples were prepared by sonicating the phospholipid in $5 \times 10^{-5}\underline{M}$ MnCl$_2$ solution in the absence of any buffer. Sample composition and relaxation rates are summarized in Table I.

Table I

The Dependance of Observed Water Proton Relaxation on Resonance Frequency.

Resonance Frequency MHz	MnCl$_2$ (aq.)[1] $1/T_1$ (sec^{-1})	Liposomes[2] $1/T_1$ (sec^{-1})	Vesicles[3] $1/T_1$ (sec^{-1})
8	0.92[4]	3.68	5.20
12	0.91	4.67	7.04
24	0.74	4.90	7.15
64	0.82	3.15	5.40
90	0.63	2.20	3.45

[1] MnCl$_2$ = $5 \times 10^{-5}\underline{M}$.

[2] Lecithin = $0.8 \times 10^{-1}\underline{M}$.

[3] Lecithin = $0.6 \times 10^{-1}\underline{M}$.
Phospholipid concentrations were determined by spectrophotometric phosphorus assay.

[4] Estimated experimental error ± 5%.

Results and Discussion

The functional dependance of $1/T_1$(obs.) on the resonance frequency strongly suggests that the electron spin-lattice relaxation time is the appropriate correlation time for the water proton spin-lattice relaxation (2). The functional form of $1/T_1(\omega_I)$ furthermore gives an estimate of $0.7 - 0.9 \times 10^{-8}$ sec for the correlation time of the fastest dynamic process (i.e., τ_S) in both vesicles and liposomes, suggesting that molecular reorientation in these systems takes place at a slower time scale.

The higher relaxation rate obtained for vesicles indicates their enhanced affinity for manganese over liposomes. Let K be the apparent binding constant of manganese to a phospholipid molecule, then

$$K = M_b/M_f L_t, \tag{5}$$

where L_t is the total lipid concentration. A value of $53 \underline{M}^{-1}$ has been reported for K in the literature (3). Around the relaxation rate maximum,

$$(1/T_{1M})_b K L_t \gg (1/T_{1M})_f \tag{6}$$

and assuming q* = q, Eq. (4) can be rearranged to give

$$\frac{\left[1/T_1(\text{obs.}) - 1/T_{1A} \right]_{\text{vesicle}}}{\left[1/T_1(\text{obs.}) - 1/T_{1A} \right]_{\text{liposome}}} \quad \frac{(M_b)_{\text{vesicle}}}{(M_b)_{\text{liposome}}} \, .$$

The value of $1/T_{1A}$ (0.36 sec^{-1}) was taken from Reference 3.

The ratio 1.6 indicates that the apparent binding constant of liposomes is about one-fifth of the binding constant for the vesicles. A possible explanation is that manganese binds preferentially to lipids with a head group conformation favored in vesicles more than in the tightly packed liposomes.

Literature Cited

1. Wydrzynski, T., Zumbulyadis, N., Schmidt, P. G. and Govindjee, Biochim. Biophys. Acta, (1975) 408, 349.
2. Dwek, R. A., "Nuclear Magnetic Resonance (NMR) in Biochemistry: Application to Enzyme Systems" Clarendon Press, Oxford (1973).
3. Nolden, P. W. and Ackermann, T., Biophysical Chem., (1975) 3, 183.

NMR Studies on Chloroplast Membranes

T. WYDRZYNSKI and GOVINDJEE

Department of Physiology and Biophysics, University of Illinois, Urbana, Ill. 61801

N. ZUMBULYADIS,[†] P. G. SCHMIDT, and H. S. GUTOWSKY

Department of Chemistry, University of Illinois, Urbana, Ill. 61801

In green plant photosynthesis the mechanism by which water is photo-oxidized and oxygen is produced still remains largely unsolved (see review [1]). However, it is known that manganese is directly involved ([2]). Inasmuch as the unpaired electron spin of Mn(II) can lead to large increases in magnetic relaxation rates of nuclei bound near the ion, it appeared to us that manganese would be a natural paramagnetic probe and that proton magnetic relaxation could be used to study the oxygen evolving mechanism. In this communication we present our initial findings, some of which have been reported earlier ([3], [4]). The results indicate that a significant contribution to proton relaxation rates of chloroplast membrane suspensions does arise from interactions with membrane-bound manganese. Furthermore light-induced changes in the relaxation rates suggest that proton relaxation is monitoring the oxygen evolving system.

Materials and Methods

Chloroplast Preparation. Chloroplast thylakoid membranes were isolated either from commercial spinach (Spinacea oleracea) or green house grown peas (Pisum sativa) in a medium consisting of 50 mM N-2-hydroxyethylpiperazine-N'-2-ethanesulfonic acid (HEPES) buffer adjusted to pH 7.5 with NaOH, 400 mM sucrose and 10 mM NaCl. The chloroplasts were given an osmotic shock in a similar medium containing 100 mM sucrose and finally resuspended in the original isolation medium. Chlorophyll concentration was adjusted to 3 mg Chl/ml in all samples.

Nuclear Relaxation Measurements. The inversion recovery method ($180°$ - τ - $90°$ sequence) was used to determine the spin-lattice relaxation rate (T_1^{-1}). The spin-spin relaxation rate (T_2^{-1}) was measured from the exponential decay of the echo amplitudes in a Carr-Purcell (Meiboom-Gill modification) (CPMG) train of rf pulses. The experimental uncertainties in T_1^{-1} and T_2^{-1} data are within $\pm 5\%$.

†Present Address: Research Laboratories, Eastman-Kodak Company, Rochester, New York 14650

In order to measure light-induced changes the nmr probe was
designed to provide the best optical geometry while still main-
taining a good signal-to-noise ratio. A tight fitting Plexiglas
plug was inserted into the bottom of a 12 mm nmr tube to support
a thin layer of sample ($\sim$100 μl total volume) in the region of
the nmr coils. Illumination was from the top. This arrangement
allowed for a large surface area and hence maximum absorption of
light by the whole sample.

In the flashing light experiments T_2^{-1} was measured after a
sequence of light flashes. The CPMG train was initiated simultaneously with the last light flash of the sequence. The time
interval between successive flashes in a sequence was 2 sec. A
dark adaptation period of 7 min was allowed between each sequence
of light flashes. Although this procedure is somewhat modified
from the one usually employed to measure oxygen, we found that it
did not affect the oxygen yield pattern.

Light flashes were obtained from a strobe light (Strobotac
type 1538-A, General Radio Co.) and were of short duration (2.4
μsec at half height with an extended tail up to 10 μsec).

Results and Discussion

Paramagnetic Contributions to the Water Proton Relaxation
Rates of Chloroplast Membrane Suspensions. Suspensions of dark-
adapted chloroplast membranes have a large effect on the water
proton relaxation rates. Upon washing the membranes twice in
buffer medium T_1^{-1} decreases in general by about 50%. Simple
washing usually has little effect on chloroplast activity but
undoubtedly serves to remove loosely bound paramagnetic ions.
However, not all ions are removed; for example, it has been reported that about 35% of the manganese is lost upon repeated
washings (5). Washed chloroplasts represent the control in the
following experiments.

In washed chloroplasts any paramagnetic contribution to
water proton relaxation will depend on the accessibility of water
to the tightly bound metal ions in the membrane. When EDTA is
added to washed chloroplasts the relaxation rates decrease. For
example, at 26 MHz and 26°C 1 mM EDTA reduces T_1^{-1} to about one
fourth of the control value (TABLE I). As shown in TABLE I the
magnitude of the effect of EDTA depends to a large degree on
temperature and nmr frequency. It appears from these results
that the tightly bound paramagnetic ions do have a major influence on the relaxation rates in washed chloroplasts.

For a system such as chloroplast membranes the measured
relaxation rate, $T_1^{-1})_{obs}$ (or $T_2^{-1})_{obs}$) can be considered as the
sum of contributions from all sites in the membrane accessible to
the solvent water, plus the relaxation rate of free water:

$$T_1^{-1})_{obs} = \sum_i \frac{P_i}{T_{1,i}} + T_1^{-1})_{free} \qquad (1)$$

TABLE I. Effect of EDTA and Tris-Acetone Washing on Water Proton Relaxation Rates of Pea Chloroplasts

Conditions	$a_{T_1}^{-1}$ (sec^{-1})					$a_{T_2}^{-1}$ (sec^{-1})				
	16 MHz		26 MHz			16 MHz		26 MHz		
	26°	8°	38°	25°	8°	26°	8°	38°	24°	9.5°
[b] Washed Chloroplasts	0.50	0.44	0.56	0.75	0.78	2.80	2.94	2.69	3.02	3.34
Washed Chloroplasts + 1 mM EDTA	0.14	0.22	0.16	0.15	0.40	2.45	2.89	2.34	2.42	2.64
[c] Tris-Acetone Washed Chloroplasts	0.20	0.18	0.18	0.19	0.21	2.30	2.24	2.29	2.27	2.39

[a] Relaxation rates corrected by subtracting rates of buffer medium from observed rates of chloroplast suspensions.

[b] After isolation chloroplasts were washed twice in buffer medium and resuspended to a concentration of 3 mg chlorophyll/ml.

[c] Chloroplasts were treated with tris-acetone medium according to procedure of Yamashita and Tomita (9) and then washed once with buffer media.

where P_i is the fraction of water in site i. Most water is free ($P_{free} \cong 1$) and $T_1^{-1})_{free}$ is taken as the relaxation rate in the buffer medium without chloroplasts. The quantity $T_1^{-1})_{obs} - T_1^{-1})_{free}$ is therefore the relaxation contribution due to the membranes and is denoted simply T_1^{-1} (or T_2^{-1}) in this communication.

In macromolecular systems T_1^{-1} of H_2O is usually influenced most strongly by paramagnetic sites. The relaxation rate T_{1m}^{-1} of water at such a site is usually dominated by electron nuclear dipole-dipole interactions:

$$\frac{1}{T_{1m}} = \frac{2}{15} \frac{\gamma_I^2 g^2 S(S+1)\beta^2}{r^6} \left[\frac{3\tau_c}{1+\omega_I^2\tau_c^2} + \frac{7\tau_c}{1+\omega_s^2\tau_c^2}\right] \qquad (2)$$

where γ_I is the nuclear magnetogyric ratio, S is the total electron spin, g is the electronic g-factor, β is the Bohr magnetron, r is the distance between the nucleus and the paramagnetic ion, ω_I and ω_s are the nuclear and electronic Larmor frequencies respectively, and τ_c is the correlation time.

The dipole-dipole interaction may be modulated by any of several time-dependent processes such that:

$$\frac{1}{\tau_c} = \frac{1}{\tau_s} + \frac{1}{\tau_R} + \frac{1}{\tau_M}$$

where τ_s is the electronic relaxation time, τ_R is the rotational correlation time and τ_M is the exchange lifetime. The shortest of these correlation times dominates.

An expression similar to (2) can be obtained for the spin-spin relaxation rate, T_{2m}^{-1}, but contains additional terms associated with scalar coupling not usually important for T_1^{-1}.

$$\frac{1}{T_{2,m}} = \frac{1}{15} \frac{\gamma_I^2 g^2 S(S+1)\beta^2}{r^6}\left(4\tau_c + \frac{3\tau_c}{1+\omega_I^2\tau_c^2} + \frac{13\tau_c}{1+\omega_s^2\tau_c^2}\right) +$$

$$\frac{1}{3} S(S+1)\left(\frac{A}{\hbar}\right)^2 \left(\frac{\tau_e}{1+\omega_s^2\tau_e^2} + \tau_e\right) \qquad (3)$$

where A is the electron-nuclear hyperfine coupling constant and τ_e is the correlation time for the scalar interaction.

Chloroplast Manganese and Water Proton Relaxation. Several treatments are known to affect chloroplast managnese (e.g. see ref. 6). For example, washing chloroplasts with 0.8 M tris (hydroxymethyl) aminomethane (tris) buffer at pH >8 alters the environment of manganese such that a Mn(II) esr signal appears (7). The current hypothesis is that some of the manganese is released to the inside of the membrane vesicle (6), but is not removed from the chloroplasts. However, the amount of manganese affected is still a matter of controversy (see ref. 6). Nevertheless, tris washing inactivates the oxygen evolving mechanism, but leaves the rest of the electron transport chain intact and functional to the extent that photoreduction of NADP+ can be restored by adding exogenous electron donors (8). This suggests that tris washing does not affect other paramagnetic centers in the electron transport chain up to the primary acceptor of Photosystem I. We find that tris washing generally changes T_1^{-1} of chloroplasts, but that the magnitude and direction of the change vary with the source of the plant material. Some examples are shown in TABLE II.

TABLE II. Effect of Tetraphenylboron and Tris-Washing on Water Proton T_1^{-1} of Spinach Chloroplasts

	$^a T_1^{-1}$ (sec)			
Conditions	Sample No.			
	1	2	3	4
[b]Washed Chloroplasts	0.86	0.90	1.04	1.03
Washed Chloroplasts + 5 mM TPB⁻	1.64	1.70	-	-
[c]Tris Washed Chloroplasts	0.82	-	1.36	0.41
Tris Washed Chloroplasts + 5 mM TPB⁻	0.88	-	-	-

[a]Rates corrected by subtracting rates of buffer medium from observed rates of chloroplast suspensions. Measurements were made at 26 MHz, 24°C.

[b]As in TABLE I.

[c]Tris washing according to procedure of Yamashita and Butler (8).

Recently Yamashita and Tomita (9) have found that a more complete extraction of manganese from the membrane is obtained when 20% acetone is included during tris washing. Again photoreduction of NADP+ can be restored with added electron donors. When chloroplasts are treated in this way T_1^{-1} and T_2^{-1} is considerably reduced (TABLE I). Interestingly, the rates do not show either a marked frequency or temperature dependency.

Although these results indicate that bound manganese does influence the proton relaxation, contributions from other paramagnetic centers cannot be ruled out; however, they probably do not have a dominating effect. For example, the copper bound in plastocyanin, a component of the electron transport chain, is not accessible to the bulk water and has little effect on observed water proton relaxation rates (10). With respect to iron, high spin Fe(II) and high and low spin Fe(III) have much faster electronic relaxation rates than Mn(II) and are less efficient in relaxation by comparison (11).

Water Proton Relaxation as a Monitor of Manganese Oxidation
States. It is not known what oxidation states of manganese exist
in chloroplast membranes. The electronic relaxation rate, however, is strongly dependent on the oxidation state. For example,
the values of τ_s for Mn(II) are generally 10^{-8} - 10^{-9} sec, depending on the nmr frequency and chemical environment (11). Mn(III),
on the other hand, has a much shorter electron spin relaxation
time. A recent study by Villafranca et. al. (12) yielded a value
of $\tau_s \simeq 3 \times 10^{-11}$ sec for Mn(III) bound to a superoxide dimutase
from E. coli. This difference in τ_s is sufficient to account
for a much greater relaxation effect by Mn(II) than Mn(III). If
the electronic relaxation of metal ions is dominating the proton
relaxation in chloroplast membranes, then changes in oxidation
state will be reflected in the relaxation rates.
    Oxidation states of bound ions can be shifted by adding
redox reagents. But many redox reagents upon oxidation or reduction give rise to free radical intermediates which could interfere with the proton relaxation rates. One reductant which
does not appear to form free radical intermediates is the tetraphenylboron anion (TPB$^-$). The oxidation of TPB$^-$ is a two electron transfer (13):

$$B(C_6H_5)_4^- \xrightarrow{0.7v} (C_6H_5)_2 + B(C_6H_5)_2^+ + 2e^-$$

$$B(C_6H_5)_2^+ + HOH \longrightarrow (C_6H_5)_2BOH + H^+$$

    TPB$^-$ is known to act as a reductant in the oxygen evolving
system of chloroplasts (14, 15). When TPB$^-$ is added to the
chloroplast suspension, T_1^{-1} increases (TABLE II). Figure 1
shows T_1^{-1} as a function of TPB$^-$ concentration in unwashed
chloroplasts. The titration curve shows several plateaus which
may be indicative of several fractions of ions being successively
reduced by TPB$^-$. TPB$^-$ itself has no effect on the buffer medium.
Interestingly, TPB$^-$ also has no effect in tris-washed chloroplasts (TABLE II) suggesting that it is acting on manganese involved in O_2 evolution.
    In a number of cases it has been found (11) that τ_s of Mn(II)
and other paramagnetic ions depends on the strength of the
applied magnetic field. The value of τ_s is determined by crystal
lattice field fluctuations having a correlation time, τ_v, such
that:

$$\frac{1}{\tau_s} = B\left[\frac{\tau_v}{1+\omega_s^2\tau_v^2} + \frac{4\tau_v}{1+4\omega_s^2\tau_v^2}\right] \tag{4}$$

where B is a constant containing the value of the resultant
electronic spin and the zero field splitting parameters.
    At low magnetid fields τ_s is often the shortest correlation

time and dominates the relaxation of nuclei bound to the paramagnetic sites (Eq. 3 and 4). At higher field strengths τ_s can
increase and τ_R or τ_M may then become the dominant correlation
time. T_1^{-1} reaches a maximum as the field increases and then
declines. On the other hand, T_2^{-1} has terms depending directly
on τ_c (Eq. 3). For the case of a field dependent τ_s, T_2^{-1} is
found to increase to a plateau as the magnetic field increases.

Figure 2 shows the frequency dependence for T_1^{-1} and T_2^{-1}
for a normal chloroplast suspension and for one containing 5 mM
TPB⁻. The T_1^{-1} for normal chloroplasts shows a broad maximum
and then a slow decline as the nmr frequency is increased. However, the T_2^{-1} increases significantly at the higher frequencies.
This behavior does suggest that electronic relaxation dominates
the proton relaxation in normal chloroplasts. However, the lack
of a distinct peak in T_1^{-1} is peculiar. This may indicate the
existence of a distribution of correlation times. On the other
hand when TPB⁻ is added to the chloroplasts, T_1^{-1} and T_2^{-1} show
a frequency dependence distinctly characteristic of electronic
domination of proton relaxation. The correlation time calculated
at the peak in T_1^{-1} with TPB⁻ is approximately 6×10^{-9} sec at
24 MHz, which is within the expected range of τ_s for Mn(II). This
result is consistent with the idea that TPB⁻ reduces a fraction of
manganese in a higher oxidation state to a lower oxidation state
which is more efficient in proton relaxation.

<u>Light Effects on Water Proton Relaxation Rates of Chloro-
plast Membranes: Relationship to the Oxygen Evolving Mechanism.</u>
In a series of microsecond light flashes the yield of oxygen
evolved from isolated chloroplasts or whole algal cells shows a
damped oscillatory pattern, having a period of four with peaks
after the 3rd, 7th, and 11th flashes ([1]). Based on this unique
pattern Kok and co-workers ([16]) have proposed a four step model
in which some chemical intermediate accumulated up to four oxidizing equivalents upon successive photoactivations of the oxygen
evolving centers:

$$S_0 \xrightarrow{h\nu} S_1 \xrightarrow{h\nu} S_2 \xrightarrow{h\nu} S_3 \xrightarrow{h\nu} S_4 \qquad (5)$$

$$4H^+ + O_2 \longleftarrow \longleftarrow 2\ HOH$$

Here S indicates the oxidation state of the intermediate; S_4
represents the most oxidized state. The primary photoreaction
of the oxygen evolving system is the excitation of the reaction
center chlorophyll molecule P_{680}, which is oxidized upon reduction of the primary electron acceptor Q; P_{680}^+ then receives an
electron from the S intermediate, perhaps via another intermediate labeled Z (for details see review, [17]). When four oxidizing equivalents have accumulated and the S_4 state is formed,
two water molecules react to produce oxygen and the original S_0

state.

The identity of the charge accumulating intermediate is
unknown, although it has been suggested to involve manganese
(2, 18-20). However, there has been no direct experimental
evidence to show that chloroplast manganese undergoes changes
in oxidation state during photosynthesis. Data from previous
sections indicate that proton relaxation monitors membrane-bound
manganese and suggest that the relaxation rates are sensitive to
changes in oxidation states. To determine whether proton relaxation could be related to the oxygen evolving mechanism we
measured the spin-spin relaxation rate in brief flashes of light.

Figure 3 shows T_2^{-1} of chloroplast membranes as a function
of flash number (4). Similar data have been obtained from seven
other preparations of spinach and lettuce chloroplasts. The
oscillatory pattern for T_2^{-1} shows some striking similarities to
the oxygen yield pattern. As in oxygen measurements, maxima
occur after the 3rd, 7th, 11th and 15th flashes. Also, the T_2^{-1}
oscillations damp out after the 17th flash, corresponding to a
similar damping of the oscillations in the oxygen yield. These
important parallels in the two types of data strongly imply that
proton relaxation is monitoring the oxygen-evolving mechanism.

However, there are some significant differences. After the
first flash where no oxygen is evolved, the relaxation rate shows
a large decrease which has no subsequent counterpart. Minima in
the relaxation rates then occur after the 4th, 8th, 12th, etc.
flashes. Minima in the oxygen yield, on the other hand, occur
after the 6th, 10th, 14th, etc. flashes. The relaxation rates
steadily increase from the 4th to the 7th flash, from the 8th to
the 11th flash and so on, while the trend is the opposite for
oxygen evolution, the yield steadily dropping from the 3rd to
the 6th flash and from the 7th to the 10th flash. These differences in T_2^{-1} and oxygen yield patterns may be explained on the
basis that the relaxation rates differ for each of the S states
whereas oxygen evolution only takes place during the S_4 to S_0
transition.

The time scales for the formation ($t_{1/2} \sim 600$ μsec) and life-
times ($t_{1/2} \sim 10$-30 sec) of the individual S states are sufficiently different (1) from the spin-spin relaxation times ($T_2 \sim$
100 msec) so as not to introduce a complex behavior in the T_2
data. The spin echo amplitudes always yield a single exponential
decay.

The changes in T_2^{-1} are not caused by the oxygen produced in
photosynthesis. We estimated that the amount of oxygen produced
after the third flash is less than 4% of the total oxygen present
in the sample when equilibrated as it is with the air. The
amount was calculated to have less than 1% effect on T_2^{-1} whereas
the maximum light-induced changes are about 20%.

Figure 4 shows T_2^{-1} as a function of flash number for tris
washed chloroplasts. There is an initial light-induced decrease
in T_2^{-1}, but the oscillations are absent. As pointed out earlier,

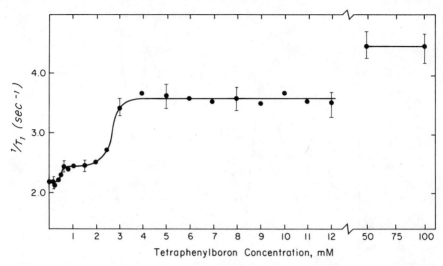

Figure 1. T_1^{-1} *measured as a function of tetraphenylboron (TPB⁻) concentration in unwashed spinach chloroplasts. Measurements were made at 26.9 MHz, 24°C.*

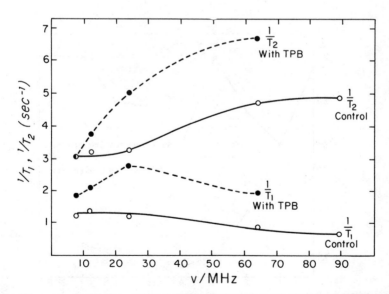

Figure 2.   Frequency dependence of T_1^{-1} and T_2^{-1} for a spinach chloroplast suspension and for one containing 5mM tetraphenylboron (TPB⁻); 26°C.

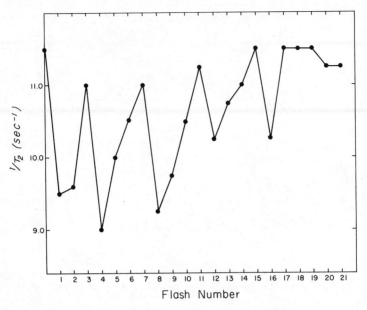

Biochimica Biophysica Acta

Figure 3. T₂⁻¹ measured as a function of number of light flashes in unwashed spinach chloroplasts. The procedure is given in "Materials and Methods." Measurements were made at 26.9 MHz, 24°C (5).

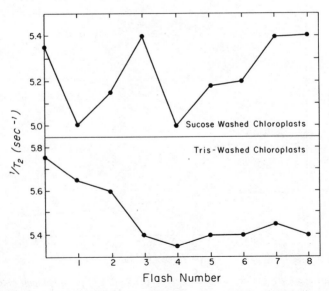

Figure 4. T₂⁻¹ measured as a function of number of light flashes in washed and tris-washed lettuce chloroplasts. Measurements were made at 26.9 MHz, 24°C.

tris washing inactivates the oxygen-evolving apparatus, but leaves the rest of the electron transport chain intact and functional.

Although stepwise changes in manganese oxidation states and consequent changes in electronic relaxation time could provide a simple qualitative explanation of the flashing light results, other mechanisms could lead to oscillations in T_2^{-1}. Such possibilities include differences in the access of water to the bound paramagnetic ions and modifications in chemical exchange rates as an indirect result of change accumulation. We hope that further experiments will clarify the mechanism involved.

Concluding Remarks

Chloroplast membranes represent an extremely complex system for physical chemical studies. Unfortunately, experiments on the oxygen evolving mechanism are confined to the use of intact membranes as the oxygen evolving capacity is rapidly lost in attempts to isolate submembrane protein fragments.

NMR relaxation measurements of water in chloroplast suspensions in part reflect the system's complexity. On the other hand, some simplification is achieved in that the major contribution to relaxation enhancement appears to be membrane bound manganese ions. Most importantly the pattern of relaxation rate (T_2^{-1}) in flashing light bears close similarities to the oxygen yield demonstrating that nmr can serve as a probe of the oxygen evolving mechanism. The details underlying this observation are the subject of our current work.

Acknowledgements

We thank the National Science Foundation for financial support to G (GM 36751) and to HSG (MPS 73-04984), the National Institutes of Health to PGS (GM 18038) and the Office of Naval Research to HSG (NR 056-547). TW was supported by HEW PHS GM 7283-1 (Sub Proj. 604) training grant in Cellular and Molecular Biology.

Literature Cited

1. Joliot, P. and Kok, B. in "Bioenergetics of Photosynthesis" (Govindjee, ed.) pp. 387-412, Academic Press, New York (1975).
2. Cheniae, G. Ann. Rev. Plant Physiol. (1970) 21, 467-498.
3. Wydrzynski, T., Zumbulyadis, N., Schmidt, P. G. and Govindjee Biochim. Biophys. Acta (1975) 408, 349-354.
4. Wydrzynski, T., Zumbulyadis, N., Schmidt, P. G., Gutowsky, H. S. and Govindjee Proc. Natl. Acad. Sci., U.S.A. (1976) 73, 1196-1198.
5. Blankenship, R. E., Babcock, G. T. and Sauer, K. Biochim.

Biophys. Acta (1975) 387, 165-175.

6. Blankenship, R. E. and Sauer, K. Biochim. Biophys. Acta (1974) 357, 252-266.

7. Lozier, R., Baginsky, M. and Butler, W. L. Photochem. Photobiol. (1971) 14, 323-328.

8. Yamashita, T. and Butler, W. L. Plant Physiol. (1968) 43, 1978-1986.

9. Yamashita, T. and Tomita, G. Plant Cell Physiol. (1974) 15, 252-266.

10. Blumberg, W. E. and Peisach, J. Biochim. Biophys. Acta (1966) 126, 269-273.

11. Dwek, R. A. "Nuclear Magnetic Resonance (NMR) in Biochemistry: Application to Enzyme Systems" Claredon Press, Oxford (1973).

12. Villafranca, J. J., Yost, F. J. and Fridovich, I. J. Biol. Chem. (1974) 249, 3532-3536.

13. Geske, D. H. J. Chem. Phys. (1959) 63, 1062-1070.

14. Homann, P. Biochim. Biophys. Acta (1972) 256, 336-344.

15. Erixon, K. and Renger, G. Biochim. Biophys. Acta (1974) 333, 95-106.

16. Kok, B., Forbush, B. and McGloin, M. Photochem. Photobiol. (1970) 11, 457-475.

17. Govindjee and Govindjee, R. in "Bioenergetics of Photosynthesis" (Govindjee, ed.) pp. 1-50, Academic Press, New York (1975).

18. Olson, J. M. Science (1970) 168, 438-446.

19. Renger, G. Z. Naturforsch. (1970) 25b, 966-971.

20. Earley, J. E. Inorg. Nucl. Chem. Lett. (1973) 9, 487-490.

Hyperfine Induced Splitting of Free Solute Nuclear Magnetic Resonances in Small Phospholipid Vesicle Preparations

ADELA CHRZESZCZYK, ARNOLD WISHNIA, and CHARLES S. SPRINGER, JR.

Department of Chemistry, State University of New York, Stony Brook, N. Y. 11794

In recent years there have been a number of reports of elegant studies of the ^{1}H ($\underline{1-9}$, $\underline{17a}$), ^{31}P ($\underline{10-16}$), and ^{13}C ($\underline{17b,c}$) nuclear magnetic resonance (NMR) spectra of small phospholipid vesicles in the presence of paramagnetic lanthanide ions. These have focused on the splitting of phospholipid headgroup resonances induced by hyperfine interactions with the paramagnetic metal ions which are generally introduced on one side of the bilayer membrane only. This study will show that the spectrum of a marker species present as free solute in the aqueous solution outside and/or inside the vesicles can provide important, often decisive, and sometimes unique information in four areas:

1. Transport of the solute, of course.
2. Size and asymmetry of the vesicle bilayer membrane.
3. The mechanism of vesicle formation by sonication.
4. Quantitative description of the (usually unappreciated) electrostatic interaction between the charged groups of the dipolar phospholipids, and its effect on cation binding and lipophilic anion incorporation.

Some of our data are preliminary and incomplete and thus some of our conclusions must, of necessity, be provisional. These will be indicated.

Figure 1 depicts the ^{31}P NMR spectrum of a solution of small L-dipalmitoyl-α-lecithin (DPL) vesicles prepared in the presence of sodium dimethylphosphate (DMP$^-$). In the absence of paramagnetic metal ions (Fig. 1a), only two features are observed: one sharp, spin-coupled (in the absence of {^{1}H}-noise) septet, clearly attributable to DMP$^-$, and one broad resonance assigned to DPL.

The addition of a small amount of Pr^{3+} produces the spectrum shown in Fig. 1b. The Pr^{3+} ion is known not to penetrate into the vesicular inner cavity for at least a period of days ($\underline{6,12,13}$). The unshifted DPL and (small) DMP$^-$ resonances clearly arise from ^{31}P nuclei in the bilayer inner surface and cavity, respectively,

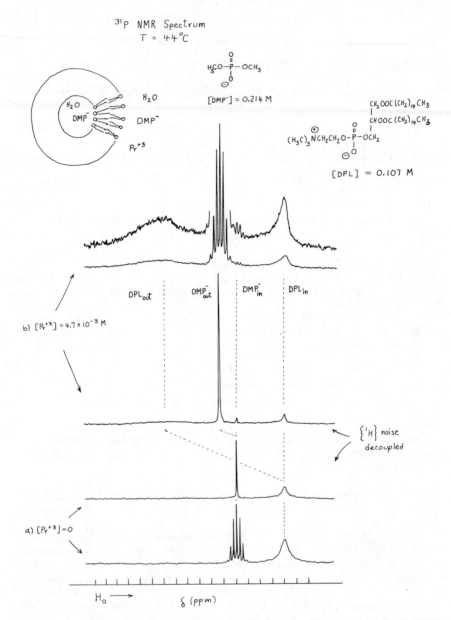

Figure 1. ^{13}P NMR spectra of a DPL (0.107M) small bilayer vesticle and DMP$^-$ (0.214M) solute dispersion in D_2O/H_2O at 44°C. (a) Undecoupled and ^1H-noise decoupled spectra of the system in the absence of any Pr^{+3} ion. (b) Undecoupled and ^1H-noise decoupled spectra of the same solution after addition of Pr^{+3} ion (0.005M) to the external volume. Small spinning side-bands have been deleted for clarity.

protected from interaction with Pr^{3+}. The downfield DPL and DMP$^-$ resonances, whose positions vary with the amount of Pr^{3+}, equally clearly represent the weighted average of the chemical shifts of free and Pr^{3+}-complexed external ^{31}P nuclei, arising from the rapid exchange of Pr^{3+}. (Notation: $\Delta^o \equiv [\delta(Pr^{3+}$ complex) $- \delta$(free)], $\Delta \equiv [\delta$(obs) $- \delta$(free)]; shifts downfield counted positive; δ, in ppm from the carrier frequency.) The splitting of the DMP$^-$ resonance reported in Fig. 1b is, to our knowledge, the first example of the splitting of the resonance of a free solute partitioned between outside solvent and vesicular inner cavity. The observed splittings for DMP$^-$ anions are smaller than for DPL (e.g., $\Delta_{DMP^-} = 1.55$ ppm, $\Delta_{DPL} = 9.63$ ppm in Fig. 1b, uncorrected for possible small bulk susceptibility effects (2,6)), reflecting differences both in limiting shift, Δ^o, and Pr^{3+} affinity (vide infra).

It is important to note that the longitudinal relaxation time, T_1, of ^{31}P in DMP$^-$ is long: 25.2 sec in deaerated 3/2(v/v) H_2O/D_2O at 0.23 M, 30.2^{o}C (Reported values for the neutral ester $OP(OCH_3)_3$, as neat liquid, are 11.6 sec (25°C) and 14.1 sec (55°C) (22).) Addition of Pr^{3+} reduces T_1 sharply: 5.2 sec at [DMP$^-$] = 0.23 M, [Pr^{3+}] = 0.005 M, 30.2^{o}C, no vesicles. Experimental design for quantitative interpretation of spectra must take these facts into account.

1. DMP$^-$ Transport

Figure 2 emphasizes the DMP$^-$ splitting. The fact that inner and outer peaks are observed shows that DMP$^-$ transport across the bilayer is slow on the NMR time scale. At lower values of ρ_{DMP^-} ($\equiv$[Pr^{3+}]/[DMP$^-$], stoichiometric), where the NMR time scale is as long as possible, the DMP$^-$ resonances remain sharp: no evidence of exchange broadening is observed. The average residence time of a DMP$^-$ anion on either side of the bilayer is thus greater than milliseconds (cf. ref. 20). Exploratory saturation transfer experiments (cf. ref. 21) show that the residence time is greater than the inner T_1, 25 sec. In fact, the exchange may be observed classically: when NaDMP and Pr^{3+} are introduced into the outer solution together, only after vesicles have already been formed, the small upfield resonance of inner DMP$^-$ is first observed after about 10 minutes, reaching its equilibrium intensity only after an hour.

2. Size and Asymmetry of the Bilayer

The spherical annulus of the bilayer portion of a small phospholipid vesicle is fundamentally asymmetric (This truism is worth repeating, because its implications are often neglected). If we idealize the time-average volume per DPL molecule (probably less variable than other parameters) into a bilayer composed of concentric outer and inner honeycombs, characterized by three radii

$R_C > R_B > R_A$ (where R_A is the radius of the inner interface; R_C, the radius of the outer interface; and R_B, the radius of the effective boundary between monolayers), we make each DPL cell a truncated hexagonal pyramid. The details are uncritical; what is important is that the area per DPL is necessarily greater at R_C than at R_B for outer cells, and greater at R_B than at R_A for inner cells, and considerably so, given the likely values of R_C and R_A. Since the outer headgroup of DPL occupies part of the broad base of its cell, whereas the tail end of the inner DPL must cover its broad base, the packing of inner and outer layers is intrinsically different. Values of area per head group (23) or bilayer thickness (17b) appropriate to essentially flat bilayers (and hexagonal prisms) may be used only to contrast them with the vesicle values, lest one beg the question. If, for example, one chooses to consider the areas per headgroup at R_C and R_A as equal, the area per outer tail at R_B is smaller, whereas the area per inner tail is larger. To preserve constant volume per DPL, the inner cell is necessarily shorter and squatter, the packing interactions in the two layers are then intrinsically different, and the a priori justification for the choice vanishes.

   If the volume, V_o, per DPL molecule is the same in both layers, and thus calculable from the partial specific volume of DPL ($\bar{v}$ = 0.99 ml/gm (23)), the vesicle is then characterized by the three parameters $\bar{R}_C$, R_B, and R_A. Three equivalent parameters are: (where N ≡ the number of DPL molecules, V is volume, and M

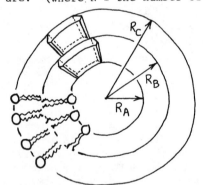

$$x \equiv \frac{N_{DPL,OUT}}{N_{DPL,IN}} = \frac{(R_C^3 - R_B^3)}{(R_B^3 - R_A^3)}$$

$$y^3 \equiv \frac{(V_{DPL} + V_{IN})}{V_{IN}} = \frac{R_C^3}{R_A^3}$$

$$M = N_{Avog} \frac{4}{3}\pi \frac{(R_C^3 - R_A^3)}{\bar{v}}$$

is molecular weight). These have the considerable advantage of being proportional to (length)^3, and thus likely to give smaller errors in the important volumes than parameters linear in length.

   Two of these parameters are obtainable from NMR studies, if suitable precautions regarding T_1 (i.e., saturation effects) (25) and Nuclear Overhauser Effects (NOE) (26), are taken, so that areas under peaks can be translated into molecular ratios. The first is the well known x. Our result, measured from the inner and outer DPL ^{31}P resonances (and corrected for NOE and T_1 effects), is x_{DPL} = 2.12 ± .20. This is in agreement with carefully measured literature values, from ^1H (2.2(2), 2.12(4), 2.80 (5), 1.85-1.90(7), 1.9(8), and 2.31(17a)), ^{31}P (2(10) and 2.2

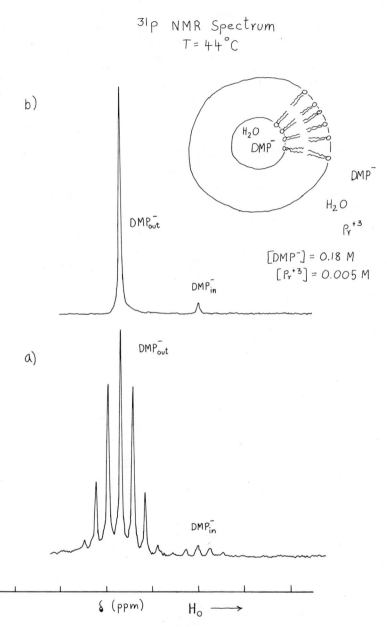

^{31}P NMR Spectrum
$T = 44°C$

Figure 2. ^{31}P NMR spectra of a DPL small bilayer vesicle and DMP$^-$ (0.18M) solute dispersion in D_2O/H_2O at 44°C with Pr^{+3} (0.005M) added to the external solution. (a) Undecoupled. (b) ^1H-noise decoupled. This figure is similar to Figure 1 but has been expanded to emphasize the DMP$^-$ splitting.

(12)), and ^{13}C (2.1(17b)) data.

The special contribution of the inclusion of DMP^- is the value of y^3. For a given solution, we know (either from chemical analysis, or by construction) the ratio of actual DPL volume (= mass x $\bar{v}$) to total volume of solution, say V_{DPL}/V_{TOT}. We can measure the ratio of the volume accessible to DMP^- (but not to Pr^{+3}) to the volume accessible to both, say $R_{I/O}[\equiv V_{IN}/V_{OUT} =$ (area, upfield DMP^- resonance)/(area, downfield DMP^- resonance)]. The value of V_{IN} must be closely representative of the entire vesicle inner cavity, since H_2O and DMP^- ought both to be able to make contact with the $-N^+(CH_3)_3$ end of DPL presented to the solution. In that case, $V_{TOT} = V_{OUT} + V_{IN} + V_{DPL}$, and $y^3 = (V_{DPL} + V_{IN})/V_{IN}$ is easily calculated. For example, a solution containing 0.270g DPL and 3.0ml H_2O (DMP^-, 0.19M; Pr^{+3} 0.005M; $44°C$) gives $R_{I/O}$ = 1:14.7 for the ratio of the central peaks of the undecoupled DMP^- septets (this ratio is 1:17.1 for the $\{^1H\}$-noise decoupled spectrum, demonstrating a slight differential NOE). This ratio becomes 1:8.0 when corrected for the difference in T_1 values (27) between the inner and outer DMP^- resonances, but this will be considered to be only an upper limit here.* The values of $R_{I/O}$ of 1:14.7 and 1:8.0 lead to values of y^3 equal to 2.40 and 1.80, respectively.

For the third parameter, we must use literature values: I)M = 2.0×10^6 (aggregation number, $\bar{n}$, 2667)(ref. 23), II)R_C = 125 Å (23), or III)R_C = 105 Å (17b, 29). The derived values of R_A, the thickness, $d[\equiv R_C-R_A]$, and other parameters, for these three cases are listed in Table I for values of $R_{I/O}$ equal to 1:25, 1:14.7, and 1:8. These employ only y^3 from the NMR data plus the literature datum and serve to characterize the average size of the vesicles.

The magnitudes of the parameters are somewhat surprising, perhaps even startling. For example, the bilayer thickness, perhaps between 27 and 32 Å, as obtained from the NOE corrected ratio, would be considered quite small when compared with values typical of planar bilayers, 42 Å (ref. 28a) to 50 Å (ref 28b), or other values, 53 Å, recently suggested for small vesicles. (29) We consider it quite important to remeasure $R_{I/O}$ as precisely and accurately as possible.

---

*In this particular case, it is impossible to make this correction very accurately for at least two reasons. First, the vesicle solution was not deaerated, while the T_1 value reported above for the ^{31}P resonance of DMP^- in the absence of Pr^{+3} was obtained for a deaerated solution. Second, the value of ρ_{DMP^-} for the T_1 study in the presence of Pr^{+3} was different from that of the present solution. Thus, the value $R_{I/O}$, equal to 1:8.0 should be considered only as a probable upper limit. A lower limit is probably given by another solution, 0.250g DPL and 2.50ml H_2O, where the observed $R_{I/O}$, uncorrected for either NOE or T_1 effects, is 1:25 (correction for T_1 gives 1:6.8). We are presently repeating these experiments with NMR pulse parameters such that T_1 corrections will be unnecessary.

Table I. Average Size Parameters for Small Phospholipid Vesicles

Case	I	II	III
$R_{I/O}$=1:25			
R_A (A^o)	67.8	81.8	68.7
R_C (A^o)	103.6	125*	105*
d (A^o)	67.8	43	36
M (daltons)	2.0×10^6*	3.6×10^6	2.1×10^6
$\bar{n}$**	2667	4684	2777
$R_{I/O}$=1:14.7			
R_A (A^o)	82.8	93.3	78.4
R_C (A^o)	111	125*	105*
d (A^o)	28	32	27
M (daltons)	2.0×10^6*	2.9×10^6	1.7×10^6
$\bar{n}$**	2667	3867	2267
$R_{I/O}$=1:8			
R_A (A^o)	99.7	102.5	86.1
R_C (A^o)	121.6	125*	105*
d (A^o)	21.9	23	19
M (daltons)	2.0×10^6*	2.2×10^6	1.3×10^6

*assumed from the literature (see text).

**a molecular weight of 750 daltons was used for the phospholipid monomer in calculating the value of $\bar{n}$.

In order to evaluate the molecular packing, one must combine the other NMR parameter, x, with those used to obtain the results exhibited in Table I.   Table II lists packing parameters, expressed in terms of the average area per head group, found for the three cases of Table I.

Table II. Average Packing Parameter (Area/Headgroup) for Small Phospholipid Vesicles

Case	I	II	III
$R_{I/O}$=1:25			
inner surface	67.6 $(A^o)^2$	56.0 $(A^o)^2$	66.6 $(A^o)^2$
outer surface	74.4	61.7	73.4
$R_{I/O}$=1:14.7			
inner surface	100.6	88.1	106.1
outer surface	85.5	74.8	90.0
$R_{I/O}$=1:8.0			
inner surface	145.9	140.3	167.6
outer surface	102.6	98.6	117.8

Again, we find the somewhat startling result that, for the most probable value of $R_{I/O}$ (i.e. between 1:15 and 1:8), the average area per headgroup is greater on the inner surface than on the outer surface! The values can also be compared with the value of 60 $(A^o)^2$/headgroup, estimated for planar lipid films. (23)

3. Vesicle Formation

The contribution of this study to an understanding of the mechanism of sonication is at the following level: suppose one resonicates the solution of Fig. 1b (DMP$^-$ inside the vesicles, DMP$^-$ and Pr^{3+} outside). Do the vesicles break in a one-shot process, the inner surface and cavity equilibrating completely with the external Pr^{3+} concentration, or are there processes (leakage, preliminary fusion, fragmentation, etc.) that permit partial equilibration of a given vesicle?

Partial equilibration processes would cause both the upfield (inner) and the downfield (outer) resonances to move toward each other (with no change in intensities, for smooth leakage, or with more complex behavior for the more complex processes).

What is observed experimentally is a clean, first-order decrease in the intensities of the inner (upfield) DPL and DMP$^-$ lines with resonication time (the apparent half-time for the DPL "line" at the power setting used is 23 min.), with essentially no change in their positions. This is the classic expression of a random all-or-nothing process.

The outer (downfield) DPL and DMP$^-$ lines shift asymptotically toward the equilibrium positions with resonication; after one hour, when the upfield resonances have almost disappeared, Δ_{DMP} has changed by 48%, from 1.55 ppm to 0.81 ppm, and Δ_{DPL} by 18%, from 10.0 ppm to 8.2 ppm. Both the magnitudes and the differential responses of DPL and DMP$^-$ may be qualitatively understood from our data on Pr^{3+}-DMP$^-$ and Pr^{3+}-DPL binding equilibria (vide infra), but there are two processes involved, and we cannot at this time make the definitive quantitative apportionment to each.

Consider a related, static experiment: Pr^{3+} is added to a mixed solution (0.23 DMP$^-$, 0.075M DPL) to total concentrations of 0.0036M, 0.007M, and 0.010M. Binding to DPL and DMP$^-$ increases, of course, as does the concentration of free Pr^{3+} (which remains small compared to the amount bound). But the apportionment changes: DPL-Pr^{3+} complexes make up 60-odd, 40-odd, and 30-odd % of the total bound Pr^{3+}, respectively. At low [Pr^{3+}], DPL complexes are preferred. In the resonication experiment, (0.21M DMP$^-$, 0.11M DPL, 0.0045M Pr^{3+}) increasing amounts of inner DPL (30% of the total DPL) are exposed to the outer solution, and bind significant amounts of Pr^{3+}. As a consequence, not only does the free Pr^{3+} concentration decrease, lowering the percent of DMP$^-$ and DPL-Pr^{3+} complexes, and thus Δ_{DMP^-} and Δ_{DPL}, but the percent of DMP$^-$-Pr^{3+} complexes is reduced disproportionately (from, say, 50% of bound Pr^{3+} initially to perhaps 30% at the end) decreasing Δ_{DMP^-} still further.

There is another effect which concerns DPL only. The contribution of inner DMP$^-$-Pr^{3+} complexes to the downfield DMP$^-$ line is always small, because the cavity volume is small compared to the outer volume, so that the Δ_{DMP^-} reflects the true equilibrium value for the current external [Pr^{3+}]. But the contribution of inner-surface DPL-Pr^{3+}, from reformed vesicles, is considerable. Moreover, the probability of breaking reformed vesicles and virgin vesicles is the same, so that, for example, when 50% of the virgin vesicles have been broken, more than 50% of the reformed vesicles will not have been reopened and will have inner DPL-Pr^{3+} concentrations, and thus values of Δ_{DPL}, which reflect earlier, higher external [Pr^{3+}] values. The observed downfield DPL "line" is thus skewed and broadened, and the apparent peak will not have shifted upfield as far as the true position required by the current outer DPL-Pr^{3+} equilibrium. The shifts of the downfield lines are thus also compatible with a "reverse pandora" all-or-nothing process.

4. DMP$^-$-Pr^{3+} and DPL-Pr^{3+} Complexes

By obtaining a quantitative description of DMP$^-$-Pr^{3+} complex formation we hope to establish the observed Δ_{DMP^-} as an indicator of free Pr^{3+} concentration, much as indicator absorbance is used classically to determine, say pH or pMg (Williams (24) has already extended this idea to shift reagents, but in a form useful only for 1:1 complexes, and in a way which obscures its utility in determining the <u>stoichiometry</u> of binding of the lanthanide ion to other ligands). The parameters are such that, although we have achieved a degree of internal consistency, more data are needed to establish uniqueness.

One can imagine that DMP^- (L) and Pr^{3+} (M) can form at least three complexes, ML^{2+}, ML_2^+, and ML_3, each with its own formation constant, β_i, and intrinsic change in chemical shift, Δ_i^o, i=1,2,3. Our binding isotherms cannot, for solubility reasons, approach asymptotic values of Δ, nor are the indicated values of β_i such that each ML_i contributes overwhelmingly in its own portion of the isotherm. Since, experimentally, there is no need to ascribe major differences in Δ_i^o to the complexes, we have fit the isotherms using only one NMR parameter: we assume that Δ^o, the limiting shift <u>per ligand</u>, is identical in ML, ML_2, and ML_3.

Our non-linear least-squares fitting algorithm (30) was based on the following formulation.

$$\beta_1 = [ML]/([M][L]) \cdot \equiv \cdot F_1 K_0$$

$$\beta_2 = [ML_2]/([M][L]^2) \cdot \equiv \cdot F_2 K_0^2$$

$$\beta_3 = [ML_3]/([M][L]^3) \cdot \equiv \cdot F_3 K_0^3$$

$$\Delta_{DMP^-} = ([ML] + 2[ML_2] + 3[ML_3]) \, \Delta^o/[L]_{TOTAL}$$

However, computing experience showed the formation constant parameter space to have large, flat valleys rather than sharp, distinct minima, so that quite small variations in observed Δ_{DMP^-} in replicate experiments result in very different β_i/β_1 ratios, when the β_i are permitted to vary independently. The fitting parameters were therefore restricted to Δ^o and K_0. The F_i were fixed at specific values: a) $F_1 = F_2 = 0$, $F_3 = 1$ (only ML_3 exists); b) $F_1 = F_2 = F_3 = 1$ (progressive facilitation of ML_i formation); c) $F_1 = 9$, $F_2 = 27$, $F_3 = 27$ (independent, equivalent "sites", statistical ratios); d) $F_1 = 28$, $F_2 = 150$, $F_3 = 150$ (an electrostatic free energy model, reflecting the values of G^o_{el} of ML_3, ML_2^+, ML^{2+}, M^{3+}).

The fits to case d are given in Fig. 3. The fits are excellent; the only problem is that they are not unique - case c does essentially as well. We need data over a wider range of DMP^- concentrations to make sharper distinctions. The following trends are clear: case a gives poor fits; case b gives an acceptable fit only for the 28°C data; case c is marginally better than case d at 28°C, marginally worse at 44°C, and distinctly worse at 52°C

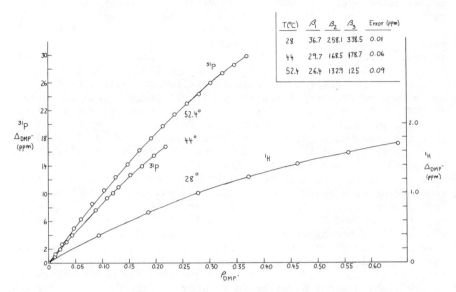

Figure 3.   The isotropic hyperfine shift of DMP⁻ as a function of the ratio of stoichio-metric Pr⁺³ to stoichiometric DMP⁻, ρ. The resonances studied and temperatures are indicated. For the various curves, the [DMP⁻] continuously varies as ρ increases as fol-lows: 52.4°, from 0.297M to 0.265M; 44°, from 0.235M to 0.203M; and 28°, from 0.164M to 0.127M. The points represent experimental data while the solid curves represent the theoretical dependence required by the equilibrium quotients noted and the Δ° values discussed in the text. The fitting algorithm employed the stoichiometric concentrations and not simply the value of ρ.

(which may reflect a real change).

Studies of DPL-Pr^{3+} interactions have proceeded in parallel. Although complete interpretation requires more data, and a clarification of the DMP^--Pr^{3+} situation, some conclusions may still be drawn. The evidence for preferential binding of Pr^{3+} to DPL vesicles over free DMP^- is compelling: in every case, the ratio $\Delta_{DPL}/\Delta_{DMP^-}$ falls steadily with increasing $[Pr^{3+}]$, indicating a belated catchup of DMP^- after an initial DPL spurt (for the series sketched in section 3, comparing ^1H shifts of DPL N-methyl protons to DMP^- O-methyl protons, the ratio goes through 3.7, 1.8, 1.39, 1.23, 1.05 as added Pr^{3+} goes from 0.0036M to 0.0205M).

We can go further, and extract preliminary values of Δ^o_{DPL} and values of DPL-Pr^{3+} formation constants proportional to those for DMP^-. For a given choice (say case c, statistical ratios), one calculates concentrations of free Pr^{3+} and DMP^--bound Pr^{3+} from the observed Δ_{DMP}; the difference between the total added Pr^{3+} and the sum of free and DMP^--bound Pr^{3+} is, necessarily, the DPL-bound Pr^{3+}, independent of any choice of Δ^o_{DPL}. It turns out that although the calculated free Pr^{3+} concentration changes by ten-fold between cases c and d, for a given solution, DPL-bound Pr^{3+} is more or less the same. Moreover, the ratio of observed Δ_{DPL} to DPL-bound Pr^{3+} is approximately constant over the range studied (Pr^{3+} bound/outer DPL of 0.1). The values of Δ^o_{DPL} thus calculated are surprising. Assuming that Pr^{3+} is coordinated to three DPL molecules, one obtains Δ^o_{DPL} ($N(CH_3)_3$) = $2.3\pm.3$ ppm, compared to 4.0 or 2.8 ppm (cases c or d) for the O-methyl protons of DMP^-, which might be expected to approach the Pr^{3+} more closely. The corresponding result for ^{31}P, where the groups are locally identical (($RCH_2O)_2PO_2^-$), is Δ^o_{DPL} = 200 ± 40 ppm, compared to 37 or 56 ppm for $\Delta^o_{DMP^-}$ (cases c or d).

For the free Pr^{3+} concentrations calculated for case c, the ^1H results quoted above yield values of DPL-Pr^{3+} formation constants, calculated on a Langmuir-isotherm type basis,

$$K_{DPL} \cdot \equiv \cdot [DPL\text{-bound } Pr^{3+}]/\{[Free\ Pr^{3+}][Remaining\ outer\ DPL] / 3\}$$

that fall from 1×10^4 M^{-1} to 3×10^3 M^{-1} with increasing $[Pr^{3+}]$. The corresponding value of β_3 for ML_3 is 7×10^3 M^{-3}. (The values for case d are lower by an order of magnitude). At this point it seems premature to extract intrinsic formation constants from these values, but one should note the following: β_3 includes a cratic term for entropy of dilution of reactants arising from the arbitrary standard state of 1M (β_3 should be multiplied by $(55.5M)^2$ for direct comparison with K_{DPL}); both constants contain complex electrostatic contributions.

As has been observed elsewhere (2,6,15,17b), the affinity of M^{3+} for DPL vesicles falls off well before saturation of the possible "sites". (Our methods will permit us to establish absolute stoichiometries of binding by NMR methods, free of any hypothesis regarding Δ^o_{DPL}, and obtaining [Free Pr^{3+}] and

[DPL-bound Pr^{3+}] from the Δ_{DMP}- of the sharp DMP^- resonance. Equilibrium dialysis is the only other effective way to obtain such values.) But we note that when the Pr^{+3} bound/outer DPL ratio is 0.1, the cationic charge added to the DPL vesicle is 500. We feel that the initial fall off in K_{DPL} can be accounted for by electrostatic considerations, and suspect that later stabilization arises from parallel condensation of counterions; we see no need for postulating structurally-different strong and weak sites.

Finally, there are indications that Pr^{3+} binding to the <u>inner</u> surface of DPL vesicles is stronger than to the outer surface (the alternative is that Δ^o is larger). DPL vesicles were prepared by sonication in the presence of both DMP^- and Pr^{3+}, and then dialyzed against a DMP^- solution to remove Pr^{3+}; ^{31}P spectra were taken at intervals. (Fig. 4) The smaller broad peak (at $\Delta=8.69$ ppm from hypothetical unshifted DPL), presumably representing inner surface DPL, did not move upfield during dialysis. The larger broad peak ($\Delta=7.65$ ppm), attributable to outer DPL, as well as the sharp DMP^- peak (originally at $\Delta=1.11$ ppm), moved upfield as Pr^{3+} was removed from the external solution. If our assignments are correct, the larger initial value of Δ for the inner DPL implies either a higher affinity for Pr^{3+}, or a modified geometry or bonding. The effect, important for an understanding of headgroup packing, ion binding, and ion-dipolar ion interactions, is being studied.

EXPERIMENTAL

The vesicle solutions were prepared by a 20-30 minute sonication, under N_2, (Branson W185) of the DPL (GIBCO) and DMP^- dispersions in H_2O/D_2O. These were then centrifuged at 12000g (10000 rpm) (Sorvall RC-2B) for 10 minutes at $\sim30^oC$. The Pr^{+3} ion was added, where indicated, in the text as an aliquot of $Pr(NO_3)_3$ stock solution (0.32 M). The concentration of the Pr^{+3} stock solution was determined by cation exchange on a Dowex 50W-X8 (BioRad) column. For dialysis, samples were injected into ¼" dialysis tubing (A. H. Thomas) and suspended in a vigorously stirred DMP^- solution. The ^{31}P NMR spectra were obtained on a Varian XL-100-12a NMR spectrometer in the FT mode, with pulse widths on the order of 35 μsec and acquisition times on the order of 2.5-3 seconds, in general. Programs supplied by Varian with the 6201L computer were used for the T_1 experiments on DMP^-, both in the absence and presence of Pr^{3+}. The appropriate precautions were taken for the T_1 measurements, namely: samples were N_2 purged and pulse delays were on the order of $5T_1$. Computing for fitting of the binding equilibria was performed on a Univac 1110.

ACKNOWLEDGEMENTS

We would like to thank the National Institutes of Health for Biomedical Sciences Support Grant #31-H056C and Biomedical

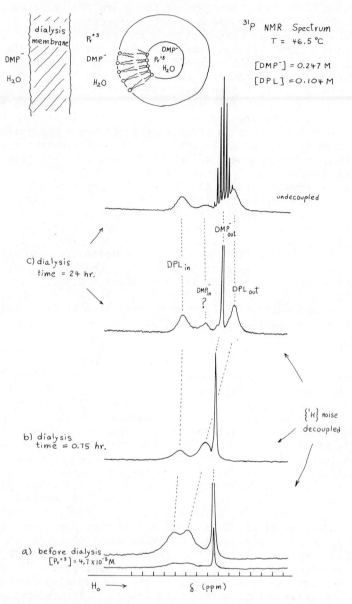

Figure 4. ³¹P NMR spectra of a DPL (0.247M) vesicle and DMP⁻ (0.104M) solute dispersion in D₂O/H₂O prepared by sonication in the presence of Pr⁺³ (0.005M) and subsequently dialyzed against DMP⁻. (a) ¹H-noise decoupled spectrum of solution before dialysis. (b) ¹H-noise decoupled spectrum after 45 min of dialysis. (c) ¹H-noise decoupled and undecoupled spectrum after 24 hr of dialysis.

Research Support Grant #31-HO71E (old #31-HO71D), and the National Science Foundation for Grant #PCM75-14788A01 (old #BMS75-14788).

LITERATURE CITED

1. Lau, A.L.Y. and Chan, S.I., Proc. Nat. Acad. Sci. (USA), (1975), 72, 2170.
2. Huang, C.-H., Sipe, J.P., Chow, S.T., and Martin, R.B., ibid, (1974), 71, 359.
3. Fernandez, M.S., Celis, H., and Montal, M., Biochim. Biophys. Acta, (1973), 323, 600.
4. Andrews, S.B., Faller, J.W., Gilliam, J.M., and Barrnett, R.J., Proc. Nat. Acad. Sci. (USA), (1973), 70, 1814.
5. Fernandez, M.S. and Cerbon, J., Biochim. Biophys. Acta, (1973), 298, 8.
6. Levine, Y.K., Lee, A.G., Birdsall, N.J.M., Metcalfe, J.C., and Robinson, J.D., ibid, (1973), 291, 592.
7. Kostelnik, R.J. and Castellano, S.M., Jour. Mag. Res., (1972), 7, 219.
8. Bystrov, V.F., Dubrovina, N.I., Barsukov, L.I., and Bergelson, L.D., Chem. Phys. Lipids, (1971), 6, 343.
9. Bergelson, L.D., Barsukov, L.I., Dubrovina, N.I., and Bystrov, V.F., Dokl. Akad. Nauk (SSSR), (1970), 194, 222.
10. Yeagle, P.L., Hutton, W.C., Huang, C.-H, and Martin, R.B., Proc. Nat. Acad. Sci. (USA), (1975), 72, 3477.
11. McLaughlin, A.C., Cullis, P.R., Berden, J.A., and Richards, R.E., Jour. Mag. Res., (1975), 20, 146.
12. DeKruijff, B., Cullis, P.R., and Radda, G.K., Biochim. Biophys. Acta, (1975), 406, 6.
13. Barsukov, L.I., Shapiro, Yu.E., Viktorov, A.V., Volkava, V.I., Bystrov, V.F., and Bergelson, L.D., Biochem. Biophys. Res. Commun., (1974), 60, 196.
14. Berden, J.A., Barker, R.W., and Radda, G.K., Biochim. Biophys. Res. Commun., (1975), 375, 186.
15. Michaelson, D.M., Horwitz, A.F., and Klein, M.P., Biochem., (1973), 12, 2637.
16. Bystrov, V.F., Shapiro, Yu.E., Viktorov, A.V., Barsukov, L.I., and Bergelson, L.D., FEBS Letters, (1972), 25, 337.
17. a) Hauser, H. and Barratt, M.D., Biochem. Biophys. Res. Commun., (1973), 53, 399.
    b) Sears, B., Hutton, W.C., and Thompson, T.E., Biochem., (1976), 15, 1635.
    c) Yeagle, P.L., Hutton, W.C., Martin, R.B., Sears, B., and Huang, C.-H., Jour. Biol. Chem., (1976), 251, 2110.
18. Morgan, W.E. and Van Wazer, J.R., Jour. Amer. Chem. Soc., (1975), 97, 6347.
19. Yeagle, P.L., Hutton, W.C., and Martin, R.B., ibid, (1975), 97, 7175.
20. Tanny, S.R., Pickering, M., and Springer, C.S., ibid, (1973), 95, 6227.

21. Hoffman, R.A. and Forsen, S., Prog. in NMR, (1966), 1, 15.
22. Dale, S.W. and Hobbs, M.E., Jour. Phys. Chem., (1971), 75, 3537.
23. Huang, C.-H., Biochem, (1969), 8, 344.
24. Williams, D.E., Tetrahedron Letters, (1972), 1345.
25. Farrar, T.C. and Becker, E.D., "Pulse and Fourier Transform NMR", Academic Press, New York (1971).
26. Noggle, J.H. and Shirmer, R.E., "The Nuclear Overhauser Effect", Academic Press, New York, (1971).
27. Waugh, J.S., Jour. Mol. Spectroscopy, (1970), 35, 298.
28. a) Lee, A.G., Prog. Biophys. Molec. Biol., (1975), 29, 1.
    b) Eisenberg, M. and McLaughlin, S., Bioscience, in press (1976).
29. Newman, G.C. and Huang, C.-H., Biochem., (1975), 14, 3363.
30. Margenau, H. and Murphy, G.M., "Mathematics of Physics and Chemistry", pp. 517-519, D. Van Nostrand Co., New York (1956).

Motions of Water in Biological Systems

I. D. KUNTZ, JR., A. ZIPP, and T. L. JAMES

Department of Pharmaceutical Chemistry, University of California,
San Francisco, Calif. 94143

While much has been written about the hydration of biomacro-
molecules, most measurements cannot be reduced to simple molecular
pictures (1). The most successful approach to (static) structural
questions has been diffraction experiments (e.g. x-ray and neutron
studies of protein single crystals,2,3). Kinetic information is increasingly available.The most informative kinetic experiments seem
to be the various dispersive techniques - especially dielectric
relaxation and NMR relaxation investigations (4,5).

To summarize a number of experiments for a simple system
such as a dilute aqueous solution of a globular protein (1,6):
1) most of the water molecules, at any instant, tumble and diffuse
as freely as in a nonproteinaceous solution of equivalent viscosity
2) The motions of a small number of water molecules, at any instant
are coupled to the relatively slow rotation of the protein molecules (5). 3) A small number of water molecules have intermediate
rotational correlation times between the value of about 10 pico-
seconds for "pure" water and the values of .1 to 1 microsecond for
typical proteins. All water motions appear to be sampled by any sin
gle water molecule on a time scale of milliseconds or longer. A
number of models based on such a description can account for the
dielectric and NMR dispersion experiments referred to above, as
well as the proton and deuteron spin-lattice and spin-spin relaxation times commonly measured at megahertz frequencies (6).

We have for some time used subfreezing temperatures to examine molecular motion in frozen protein solutions. We report,here,
some proton NMR data including spin-lattice, spin-spin, and rotating frame spin-lattice relaxation times. The samples were frozen
hemoglobin solutions and frozen packed red blood cells.

Experimental.

Whole blood and cell free hemoglobin samples were prepared as
described earlier (7) The NMR equipment and pulse techniques are
standard and have been discussed in detail elsewhere (8). We measured relaxation times at two spectrometer frequencies - 44 MHz
and 17 MHz, and we could obtain $T_{1\rho}$ data at frequencies up to

40 KHz for the temperature range of -60 to -20°C. The short T_2 values were obtained directly from the free induction decay.

Results.

1. The amplitude of the NMR signal decreases dramatically near the freezing point of water as most of the water changes into ice. Some of the water does not freeze (.4 g H_2O/ g Hb). The NMR properties of this "nonfreezing" water are described below.

2. In a previous communication (8), we reported T_1 values for hemoglobin solutions and for packed red cells (Figure 1). The solutions and cells exhibit the same behavior, going through a shallow, frequency-dependent, minimum as the temperature is varied. The minimum T_1 value is ca. 45 millisec at 44 MHz and 25 millisec at 17 MHz.

3. Arrehenius plots for T_2 and $T_{1\rho}$ (extrapolated to zero H_1 field) are shown in Figure 2. The uncertainty in these values is approximately 20%. T_2 is always much shorter than T_1 in these systems. The (small) differences between T_2 and "zero field" $T_{1\rho}$ may be due to experimental error.

4. The $T_{1\rho}$ results - obtained at a Larmor frequency of 44 MHz are shown in Figure 3A. At "high" temperatures (above -30°C), $T_{1\rho}$ is approximately equal to T_2 at all H_1 fields we could reach. At lower temperatures, the spin-lattice relaxation time in the rotating frame decreases as H_1 increases, giving a dispersion.

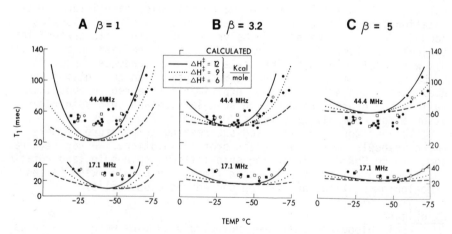

Figure 1. Experimental (data points) and calculated (curves) temperature dependence for the water proton spin-lattice relaxation time for 30 gm % hemoglobin solutions (44.4 MHz) and packed red cells (17.1 MHz). Curves calculated assuming a bimodal log-normal distribution of correlation times with the distribution width, β, and the activation enthalpies as given in figure.

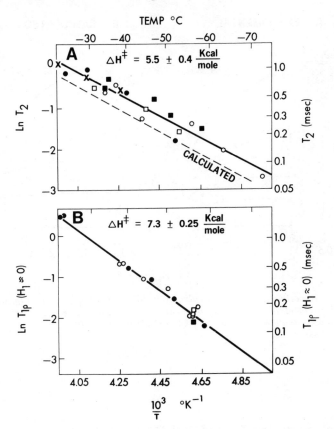

Figure 2.   Arrhenius plots of the water proton spin-spin and rotating frame spin lattice ($H_1 = 0$) relaxation times for frozen packed red cells. The Larmor frequency is 44.4 MHz.

Discussion.

We have developed the following model to fit these results for frozen solutions. First, the T_1 minimum can be used to characterize a rotational correlation time from simple Solomon-Bloembergen theory ($\underline{9}$). We find a correlation time of 2 nanoseconds at -35°C. Second, we can use the conventional approach of a "log normal" distribution of correlation times about the 2 nanosecond value to reproduce the finding that T_1 is considerably greater than T_2 (see, for example, ref. 10).   Such a distribution <u>does not</u> reproduce the $T_{1\rho}$ data. Thus we assume <u>two</u> correlation times for proton motion; the first being about 2 nanoseconds, the second being much longer, ca. a few microseconds. Approximately 3/4 of the nonfreezing water molecules are associated with the faster correlation time. Using narrow distributions about each of these times, and the "rigid lattice limits" as described by Resing ($\underline{11}$),

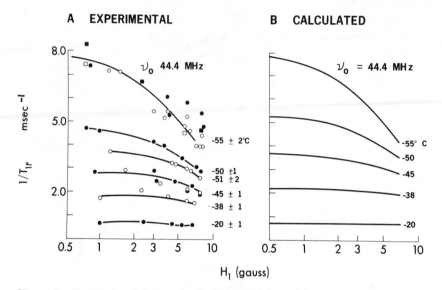

Figure 3. Experimental (3A) and calculated (3B) dependence of the rotating frame spin-lattice relaxation rate on H_1 for packed red cells at low temperatures. Curves calculated assuming a bimodal log-normal distribution of correlation times.

we can reproduce the $T_{1\rho}$ results (Figure 3B) as well as the T_2 and T_1 data (see dotted curve of Figure 2A and solid curves of Figure 1). The slower motion makes the dominant contribution to T_2 and appears to have an activation enthalpy of 5-7 kcals/mole. The faster motion largely determines the 17 and 44 MHz T_1 results and seems to have a somewhat larger activation enthalpy, although the curve fitting involved makes this conclusion somewhat uncertain.

At this point we cannot assign these motions to specific physical relaxation mechanisms. The use of D_2O should permit us to decide the importance of rotational and translational contributions. We can only speculate on the possible significance of the two different water motions. It may be that the lower frequency term comes from water coupled to the slow sidechain motions (12), while the higher frequency term comes from "loosely held" water molecules in a "hydration" shell about the protein. Other interpretations are certainly possible.

Literature Cited.

1. I.D.Kuntz and W. Kauzmann. Adv. Protein Chem. 28, 239 (1974).
2. K.D.Watenpaugh, L.C.Sieker, J.R.Herriott, and L.H. Jensen, Cold Spring Harbor Symposia on Quantitative Biology, 36, 359 (1972).

3. B.P.Schoenborn, ibid, 36, 569 (1972).
4. S.Harvey and P.Hoekstra, J. Phys. Chem. 76, 2987 (1972).
5. K. Hallenga and S.H. Koenig, Biochem., in press.
6. R. Cooke and I.D.Kuntz, Ann. Rev. of Biophys. and Bioeng. 3, 95 (1974).
7. A. Zipp, I.D.Kuntz, S.J. Rehfeld, and S.B.Shohet, FEBS Lett. 43, 9 (1974).
8. A. Zipp, T.L. James, and I.D.Kuntz, Biochim. Biophys. Acta, in press.
9. N.Bloembergen, E.M.Purcell, and R.V.Pound, Phys. Rev. 73, 679 (1948).
10.H.A. Resing, Adv. Molec. Relax. Processes 1, 109 (1967-8).
11. H.A.Resing and R.A.Neihot, J. Coll. and Int. Sci. 34, 480 (1970).
12.G.T.Koide and E.L. Carstensen, J.Phys. Chem. 80, 55 (1976).

41

Cytoplasmic Viscosity

FREDERICK SACHS

Pharmacology Department, State University of New York, Buffalo, N.Y. 14214

What are the physical properties of a cell's interior? The transport of substances across cell membranes is an area of intense activity at the present, but what happens to these substances after they enter the cell. And what of the movement of substances synthesized within the cell? Is the cell interior organized into an ion exchanger by long range ordering forces of cellular macromolecules? These are the kind of questions that prompt the study of cytoplasmic viscosity.

Anyone that has looked at an electron micrograph of a cell will appreciate that the cytoplasm is anything but homogeneous. There are numerous types of membrane bound bodies, filaments of various sorts and extracellular passageways. It is obvious that no one measurement, or type of measurement, can give a true picture of cellular activity. The forces that act on the movement of vesicles or even macromolecules down an axon may not be the same as those that limit the diffusion of calcium out of sarcoplasmic reticulum into the myofibrils of a muscle. We need many kinds of measurements to fully understand the movement of substances within cells.

Since the turn of the century people have measured what has been known as "protoplasmic viscosity" by a variety of clever, and not so clever means. To quote L. V. Heilbrunn, "Muscle physiologists have at times made measurements of the 'viscosity' of an entire muscle fiber, outer membrane, cortex and interior, all considered as one material, or they have even measured the viscosity of an entire muscle. One might as well make viscosity measurements of a solution and include in the value obtained the viscosity of the bottle which contained it." (1). Nonetheless, if we divide the cytoplasm into small enough and homogeneous regions to make precise measurements, we have in a way begged the question of how the system functions as a whole.

Heilbrunn and others, until quite recently, concerned themselves with the viscosity of cytoplasm as measured by the movement of micron sized particles through the interior of cells under the influence of centrifugal, gravitational or magnetic fields (see reference 1 for a rather complete review of the experiments up to 1958). These experiments measured a particular property of the cell interior on a particular time scale. The results of such experiments, if taken at face value, lead to 'viscosities' of a few centipoise (cp) to thousands of cp, depending upon the system and the method. Not surprisingly non-Newtonian behavior was reported along with viscoelastic properties. Crick and Hughes (2), after studying the cytoplasm of fibroblasts containing small magnetic particles, concluded that the cytoplasm appears like "Mother's Work Basket--a jumble of beads and buttons of all shapes and sizes with pins and threads for good measure, all jostling about and held together by 'colloidal forces'". These conclusions are appropriate perhaps for the movement of large particles within cells, but what of the movement of water and metabolites?

Translational Diffusion of Molecular Species

Lehman and Pollard (3) compared the viscosity of the extruded cytoplasm of E. coli as measured by an Ostwald viscometer with that measured by the diffusion of tracer molecules. The macroscopic viscosity measured by the viscometer indicated a value of 100 poise whereas the viscosity measured by following the diffusion of radioactive sucrose or dextran was only about 3 cp. Galactosidase, however, diffused in an environment with an apparent viscosity of some 500 cp. Thus, not unexpectedly, the size of the probe significantly affects its translational mobility. A similar result was obtained by Hodgkin and Keynes (4, 5) for the diffusion of potassium and dyes in the cytoplasm of squid giant axons. Potassium had a diffusion coefficient very close to that of solution, while that of the dyes was depressed by a factor of 2 to 5.

This kind of tracer diffusion experiment can be done in vivo, and the almost universal conclusion for a wide variety of atomic and molecular species is that the effective viscosity seen by a diffusing molecule is between 2 and 3 times that of bulk water (3, 4, 5, 6, 7, 8). The measurement is restricted however to relatively large cells of essentially cylindrical geometry where the tracer can be introduced locally and the preparation is long enough so that a tracer profile free of edge effects can be observed. Certain exceptions are to be expected. The diffusion of Ca^{++} down a muscle fiber is severely depressed relative to bulk solution since there are sequestering sites (the sarcoplasmic reticulum) distributed throughout the muscle (6). Secondly, any molecule which may be incorporated into proteins transported on the fast axoplasmic transport system (9) will show anomolously fast translation, although this is clearly an energy requiring process. The passive viscosity, however, will, in part, determine how much energy the transport system has to expend in order to function.

What about translational diffusion in non-cylindrical cells? The best measurements to date have been the field gradient nuclear magnetic resonance (NMR) measurements of water diffusion (10, 11). A magnetic field gradient is imposed upon the preparation in a spin echo NMR spectrometer. As spins diffuse from areas of one magnetic field to another they become dephased (T_2 decreases). The faster the molecules diffuse, the greater the magnetic field differences they experience in a given time, and the faster they become dephased. The change in T_2 with changes in the gradient permit calculation of the average diffusion constant. The method seems insensitive to artifacts, and diffusion constants may be measured on a time scale of milliseconds. The results give a diffusion constant for water inside cells 1.5 to 3 times less than that observed in bulk water. If the translational diffusion of small molecules in cells is quite rapid, then we would expect that rotational diffusion should also be rapid. In fact the same viscosity appears in the Stokes-Einstein formulation of rotational and translational diffusion. Rotational diffusion measurements however, may be carried out on a much shorter time scale than the translational measurements and thus may reveal different properties.

Rational Diffusion Measurements in Cells

NMR Probe Experiments. The rotational motion of cell water itself is a complicated problem since water in cells is 40 to 50 molar, and consequently is involved in a variety of sites. Water hydrates proteins (12), phospholipids (34) and other polar molecules within cells. Cell water is also segregated to some degree within the variety of membrane bound compartments of the cell mitochondria, Golgi apparatus, endoplasmic reticulum, nuclei and extracellular spaces penetrating into the cell, such as the transverse tubular system of muscle. Inhomogeneity and exchange between these different sites may produce complicated measures of rotational correlation time of water within cells (13). Despite the complications, it appears that less than 2% of the cell water has a correlation time longer than 1 nsec, and 4% has a correlation time longer than 100 psec. The basic view is one of water of hydration, less than 20% of the cell water, rapidly exchanging with what is probably free water, but the resolution of the system for short correlation times is not good.

Using an entirely different kind of NMR probe London, Gregg and Matwiyoff (14) found the mobility of hemoglobin inside a red cell to be only 25% lower than in solution. They fed mice a diet enriched in ^{13}C histidine which was incorporated into hemoglobin. Then, using pulsed NMR techniques, they obtained the rotational time for hemoglobin in, and out of red cells (a sort of giant spin probe). The correlation time in the cells was 0.41 nsec. Upon lysing the cells, the correlation time fell by 25%. Using the Stokes-Einstein equations (taking the diameter of hemoglobin to be 55 Å), they obtained a value of 1.94 cp inside the cell, and 1.56 cp outside.

Fluorescent Probe Experiments. The basis for inference of rotational rates from polarization is that if a molecule absorbs light with its transition dipole in a particular direction, diffusion of the molecule will cause the dipole to be at a different angle at the time of emission. The further the molecule can rotate before emitting the fluorescent radiation, the lower the polarization of the fluorescence. Polarization is related to hydrodynamic viscosity by the Perrin equation:

$$1/p = 1/p_o + (1/p_o - 1/3) \; RT\tau/\eta V$$

where p is the polarization of the emitted light (assuming polarized excitation), p_o is the extrapolated maximum possible polarization (which includes the effects of very rapid relaxation processes as well as any angular differences between the absorption and emission dipoles), R is the gas constant, T is absolute temperature, τ is the lifetime of the excited state producing the fluorescence, η is the viscosity and V is the molecular volume.

The first quantitation of the polarization of cytoplasmic fluorescent probes was made by Burns (15) in yeast and Euglena. He measured the fluorescence polarization of fluorescein in the cells and compared it to the polarization in standards prepared from sucrose-water mixtures. He estimated the viscosity of the cytoplasm at 15°C to be 14 cp and 6.3 cp for yeast and Euglena respectively. The rotational mobility increased with temperature. Between 15 and 27°C viscosity decreased from 14 to 10 cp and from 6.3 to 5.2 cp in yeast and Euglena respectively. Later work (16) substantiated that the fluorescence lifetime in the cells was the same as that in aqueous solution, adding further support to the validity of the rotational times and suggesting that there was probably no binding of the probe since binding usually alters the lifetime of the excited state.

A fluorescent probe study by Gamaley and Kaulin (7) in frog muscle came to much the same experimental conclusions. Using the polarization of fluorescein, they calculated a viscosity of about 6 cp, which increased with dehydration of the muscle. An Arrhenius plot of the apparent viscosity was not monotonic indicating a variety of processes contributing to the observed polarization. (See the section on fast exchange under paramagnetic probes.) A very interesting finding was that the translational diffusion of the dye down a fiber was only 1.5 times slower than that observed in bulk solution.

Electron Paramagnetic Resonance (EPR) Probe Experiments. The determination of rotational times from EPR spectra is well developed, particularly for the nitroxides, and will not be

treated here. The interested reader is referred to any of the
excellent available reviews such as (17, 18).

The first use of spin probes to measure cytoplasmic viscosity seems to have been made by Cook and Morales (19) incidental to
a study on spin labelled muscle proteins of glycerinated muscle.
They noticed that the rotational mobility of free (non-covalently
linked) nitroxide probes in muscle was not very different from
that observed in solution, although the results were not quantitated. Since the muscles in the glycerinated state are capable
of contraction in response to normal biochemical stimuli, it appeared that long range structured water was not necessary to the
contractile function of the cell, and in fact was not produced by
the ordered array of muscle proteins.

Dr. Ramon Latorre and I decided to see if we could find some
evidence of structured water in living muscle using the spin
probe technique (20, 21). We used the giant muscle fibers of the
barnacle, Balanus nubilus, which are often two to three milli-
meters in diameter, so that we could inject probes without causing massive damage to the cells. We initially attempted to use
Fremy's salt (nitrosyl di-sulfonate) as the probe since it is
extremely water soluble and would have little tendency to bind to
membranes in the cell. Unfortunately, the probe was so rapidly
reduced by the cell that the experiments proved impossible. We
then switched to another nitroxide probe, TEMPOL, shown below in
Figure 1. This probe is amphiphilic with an octanol-water partition coefficient of 2.8, but is stable in the muscle at temperatures below 25°C and decays sufficiently slowly at higher temperatures to permit reliable measurements.

Our initial experiments were designed to test whether the
cytoplasm is ordered on the order of the probe dimensions (approximately 6 Å diameter) and the probe relaxation time, which for
the nitroxides is given by the anisotropy of the hyperfine split-
ing (approximately 70 Mhz, or $\sim 10^{-8}$ seconds). We reasoned that
if the cell water was ordered, perhaps by cellular proteins, the
probe should tumble significantly slower than in bulk water.
Further, if that ordering was manifest only close to the proteins,
then by dehydrating the cell and removing the "freer" water we
should be able to see the probe become increasingly immobilized
as it was forced into the more viscous, or ordered cytoplasm.

Spectra of the probe in a normal, and in a severely dehydrated (5% of normal water content) muscle are shown in Figure 1.
In the normal muscle the spectral lines are sharp and of almost
equal height characteristic of the probe tumbling rapidly compared to the hyperfine anisotropy. The dehydrated specimen
showed at least two distinct spectral components, a component
with a rotational correlation time on the order of 1 nsec, and a
strongly immobilized fraction, with a correlation time longer
than 10 nsec. In a normally hydrated muscle the immobilized
component was not visible even under maximum amplification.
When the results were quantitated using Kivelson's (1960) theory

relating linewidths and rotational correlation time for rapid
isotropic motion, the results shown in Figure 2 were obtained.
For reference, the rotational correlation time, τ_c, for TEMPOL in
water is 10 to 20 psec. (The uncertainty arises because the
probe is tumbling so fast that averaging of the anisotropy is almost complete and the spectra are rather insensitive.)

In the normally hydrated muscle, the probe was apparently
moving 5 to 10 times slower than in pure water, i.e. the local
viscosity appeared to be 5 to 10 cp. A most interesting feature
of the curves is the small slope of the viscosity versus water
content. Removal of half the water increased the viscosity by
only a factor of two. This suggests that the observed viscosity
is not due to a high solute concentration. For comparison, dou-
bling the solute content of a 5 cp solution of glycerol results
in a 100 fold increase in viscosity. As water was removed from
the muscle, the filament lattice contracted, and yet, with 75%
dehydration, the spectrum appeared homogeneous. Firmly immobilized probe did not appear until the water content fell below 25%
of normal. If the cellular proteins were capable of ordering
water into an ice-like array, they could not exert such an effect
over more than 25% of the cell's water.

In as much as the experiments were being performed on whole
cells, the question arose as to how much of the cell volume was
sampled by the probe. We first assumed that the probe was
passively distributed, and by extracting label from an equilibrated muscle we compared the concentration of the probe in the
muscle to that in the labelling solution. We found that, under
the passive distribution assumption, the probe was distributed
through about 90% of the cell water. Since spin-spin interactions produce spectral broadening with probe concentrations in the
mM range, we tested for local concentration of the label by equilibrating the muscle with concentrations of the probe that
showed broadening in solution and looked for broadening in the
muscle. We found that concentration broadening did not occur in
the muscle until it was observed in the labelling solution, i.e.
the label was not concentrated by the cell. Thus the signals we
observed represented an average property of most of the cell.

We had established that the probes were rapidly moving in
the cytoplasm, but the spectra did indicate that they were moving
slower than observed in pure water and in fact slower than expected from the diffusion experiments mentioned earlier (6, 11).

There are several possible sources of these apparent discrepancies. We had attributed all of the broadening to incomplete
isotropic rotational averaging, however, other forms of spectral
broadening are possible. The probe may be moving somewhat anisotropically (23) or there may in fact be several populations of
probes with different mobilities, all in the fast rotational
regieme such that the line heights are dominated by the fastest
probes and the central linewidth by the slower probes. (For all

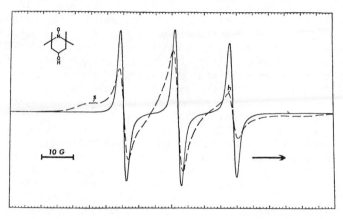

Figure 1.   The ESR spectrum of TEMPOL (structure at upper left) in single barnacle muscle fibers. The solid line is the spectrum at normal hydration. The dotted line is the spectra at 5% of muscle water (note the spectrometer gain at 5% is 4.7 times that at 100%). The symbols s and h indicate, respectively, the low field peak of the strongly immobilized component and the high field peak of the rapidly tumbling component. The arrow indicates the direction of increasing field. Muscles were dehydrated with N_2 and loaded with TEMPOL by diffusion (21).

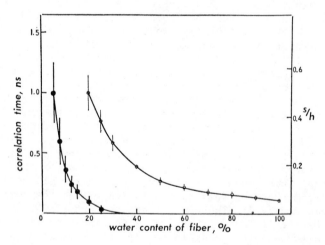

Figure 2.   The correlation time (open circles), and s/h (filled circles) of TEMPOL in single barnacle muscle fibers as a function of hydration. s/h represents the relative proportion of strongly immobilized (SI) to liquid-like probe. The bars represent standard errors of the mean. Note that since the SI spectrum is much broader than the liquid-like one, the ratio of peak heights underemphasizes the proportion of SI probes (21).

$\tau_c < 10^{-9}$ sec, the absorption peaks with the same nuclear spin, overlap and thus are not visible as distinct populations of probes). Another possibility, one that is often used in the NMR analysis of cell water, is a fast exchange process where the probe spends part of its time in one environment, and part in another, moving between the two faster than the fastest relaxation rate so that the spectral properties in the different environments are averaged. The nitroxide probes have g factors and hyperfine splittings which are solvent sensitive (24). Thus one may average not only anistropies of the g and hyperfine tensors, but the solvent shifts (25).

The requirement for some sort of exchange process in at least some systems was vividly pointed out by Stryukov, Stunzhas and Kirillov (26) in a study of diffusion of TEMPO (2,2,4,4 tetramethyl pyrrolidine-N-oxyl) in polymers (vaseline and polyethylene). Using ^{15}N and ^{14}N derivatives, they showed that anisotropic motions could not account for the different correlation times observed with the otherwise identical probes. Furthermore, the saturation of the probes was not characteristic of a homogeneous population, although the spectra were not visibly inhomogeneous. Using a computer program, they were able to fit the spectra of both isotopes properly using an exchange model.

In order for the probes not to show up as two distinct populations, they must move between one environment and the other rapidly compared to the differences to be averaged. In the case of the hyperfine anisotropies which are large (70 Mhz), the probe cannot remain in its "slow" environment for more than about 10^{-9} seconds without appearing as a separate spectral component. In 10^{-9} sec, TEMPOL, with an average radius of 3.5 Å, would diffuse a distance of about 10 Å. It is hard to imagine that all the probes in the cell are within 5 or 10 Å of similar immobilizing substrates, so perhaps we were observing in fact an inhomogeneous spectrum, where the inhomogeneities are too small to make the separate populations visible.

Another possibility for sample inhomogenity causing misinterpretation of the spectra, comes from the fact that solvent shifts of the g and hyperfine parameters are such as to make the low and midfield lines stay together, while the high field lines separate (17). If the probes are distributed between regions of somewhat different polarity, whether in fast exchange or not, they may appear to be somewhat immobilized.

The net effect of all of these potential sources of broadening is to emphasize that some probes must be tumbling faster than the average since there are many ways to broaden the spectrum, and only one, rotational averaging, to narrow it.

To test whether the apparent viscosity that we observed was in fact a simple viscosity, or an overlap of multiple processes, Dr. Latorre and I looked at the temperature dependence of the correlation time in aqueous model systems and in the muscle (21).

In model systems such as glycerol–water, guanidine HCl–water sucrose–water, we found simple Arrhenius plots with an enthalpy of activation which increased with bulk viscosity (from about 8.6 kcal/mole in water to 11.1 kcal/mole in 96% glycerol). The entropies of activation were all about 20 cal $mol^{-1}deg^{-1}$. In the muscle, however, the Arrhenius plots were nonlinear showing enthalpies of –0.3 to +2 kcal/mole and apparent entropies of activation of –5 to –12 cal $mol^{-1}deg^{-1}$. (Nonlinear Arrhenius plots were also seen in muscle by Gamaley and Kaulin (7) using a fluorescent probe.) The anomalous behavior of apparent slow tumbling with low activation energies was not confined to the muscle, however. In a model system of 20% bovine serum albumin (BSA), TEMPOL showed similar behavior. Note that 20% gelatin showed quite normal behavior, so that the effect is not characteristic of just any protein solution. Our interpretation of the data was that TEMPOL in the muscle was probably undergoing some exchange process, and that the observed average correlation time was representative of an upper limit for the local viscosity of water in the barnacle muscle.

In contrast to the relatively short correlation times observed in barnacle (21), frog muscle (13) and nerve (9). Keith and his co-workers (27, 28) have observed some very long correlation times, indicative of some 30 to 50 cp in a number of cells and organelles using nitroxide spin probes. There are some methodological differences however. In order to eliminate the signal arising from extracellular probe, they treated the preparations with 0.5 to 2 M $NiCl_2$ which broadens the extracellular nitroxide signal to such a degree that it is effectively invisible. $NiCl_2$ is essentially impermeant for the duration of the experiment and thus does not directly affect the intracellular or intramembranous probes although it may have hyperosmotic effects. Keith and Snipes (27) found a correlation between the isotropic hyperfine splittings and the apparent correlation time, τ_c: probes in cells with high apparent viscosity had smaller splittings indicative of a less polar environment; an effect which they attribute to a lower dielectric constant for water which has been ordered by nearby membranes. One suspicious indication of an inhomogeneous population is that there was no relation between the correlation time and the temperature dependence of the correlation time. Increased viscosity should be correlated with increased temperature dependence in a homogeneous system (29).

In the paper by Morse et al. (28), the same techniques were used to probe the viscosity "profile" of sarcoplasmic reticulum vesicles. They found that their amphiphilic or water soluble labels had apparent correlation times between 30 and 50 times as great as observed in bulk water. In those spectra in which the isotropic hyperfine splitting was observable, the value was between the values for water and hydrocarbon. This is indicative of an exchange process at the membrane interface. It is easy to imagine probes bobbing in and out of the membrane–cytoplasm interface.

If membranes had the ability to organize water for hundreds of angstoms, then those effects should show up in mitochondria which have extensive membrane systems. Using Mn^{++} as an EPR probe Pushkin and Gunter (30) permitted mitochondria to accumulate manganese (as a calcium substitute). After broadening the external Mn^{++} signal with EDTA, a narrow sextet characteristic of highly mobile Mn^{++} was visible as a fraction of the total intramitochondrial manganese. By comparing the linewidths of this rapidly moving fraction with those observed in standard solutions, they arrived at an upper limit of 1.5 cp for the intramitochondrial viscosity.

Further evidence against long range ordering of water by membranes is provided by the work of Stilbs and Lindemann (31). They studied the rotational mobility of the paramagnetic ion VO^{++} in micellar solutions of sodium decyl sulfate. Below the critical micelle concentration (CMC) they observed a correlation time of 50 psec. This increased to 70 psec at the CMC and remained at that value independent of concentration of the surfactant. These correlation times are in accord with those measured by Chasteen and Hanna (32) for the vanadyl ion in 0.9 mol % ethanol in water.

One additional piece of information obtained with spin probes suggests that membranes don't significantly order water over long distances. Menger et al. (33) introduced nitroxide spin probes into small pools of water (10 to 100 Å diameter) formed in di-2-ethylhexyl sodium sulfosuccinate. They observed that in general as the water pool got smaller, the isotropic hyperfine splitting became smaller, (characteristic of more apolar environments) and the rotational correlation time increased. These are the kinds of correlations noted by Morse et al. (28) in the sarcoplasmic reticular vesicles. However, Menger et al (33) noted that when they used an ionized probe, the high field line split into two lines characteristic of the probe in two distinct environments. The concluded that although the uncharged probes could exchange across the interface fast enough to average the hydrophilic and interfacial environments, the charged label could not and consequently showed up in both environments.

Concluding this section, the work on the muscles and nerves seems to show a reasonable agreement between the rotational and translational viscosities, at least within the limits of the Stokes-Einstein formulation of viscosity. The work on other cells and systems seems to be more seriously complicated by inhomogeneity, perhaps because of the larger surface to volume ratios involved.

Summary

The probe techniques using EPR, NMR and fluorescence have in general indicated that small molecules may freely move about in the many compartments of the cell that compose the cytoplasm.

Certain exceptions are found, particularly for molecules that are
specifically stored, transported, or metabolized.

Literature Cited

1. Heilbrunn, L.V., "The viscosity of protoplasm",
    Wein, Springer-Verlag, Vienna, 1958.
2. Crick, F.H.C. and Hughes, A.F.W., Exptl. Cell. Res. (1950)
    1, 37-80.
3. Lehman, R.C. and Pollard, E., Biophys. J.   (1965) 5, 109-119
4. Hodgkin, A.L. and Keynes, R.D., J. Physiol. (1953) 119,
    513-528.
5. Hodgkin. A.L. and Keynes, R.D., J. Physiol. (1956) 131,
    592-616.
6. Kushnevick, M.J. and Podolsky, R.J., Science (1969) 166,
    1297-1298.
7. Gamaley, I.A. and Kaulin, A.B., Physiol. Chem. Phys. (1974)
    6 (5), 445-456.
8. Edzes, H.T. and Berendsen, H.J.C., Ann. Rev. Biophys. Bioeng.
    (1975) 4, 256-285.
9. Ochs, S. and Smith, C., J. Neurobiol. (1975) 6, 85-102.
10. Tanner, J.E. and Stejskal, E.O., J. Chem. Phys. (1968)
    49, 1768-1777.
11. Finch, E.D. Harmon, J.F. and Muller, B.H., Arch. Bioch.
    Biophys. (1971) 147, 299-310.
12. Kuntz, I.D., Brassfield, T.S., Gaw, G.D. and Purcell, G.V.,
    Science (1969) 163, 1329-1331.
13. Cooke, R. and Kuntz, I.D., Ann. Rev. Biophys. Bioeng. (1974)
    3, 95-126.
14. London, R.E., Gregg, C.T. and Matwiyoff, N.A., Science (1975)
    188, 266-268.
15. Burns, V.W., Bioch. Biophys. Res. Comm. (1969) 37, 1008-1014.
16. Burns, V.W., Exptl. Cell Res. (1971) 64, 35-40.
17. McConnell, H.M. and McFarland, B.G., Quart. Rev. Biophys.
    (1970) 3 (1), 91-136.
18. Keith, A.D., Sharnoff, M. and Cohn, G.E., Bioch. Biophys.
    Acta. (1973) 300, 379-419.
19. Cook, R. and Morales, M.F., Biochem. (1969) 8, 3188-3194.
20. Sachs, F. and Latorre, R., Biophys. J. (1973) 13, 263a.
21. Sachs, F. and Latorre, R. Biophys. J. (1974) 14, 316-326.
22. Kivelson, D., J. Chem. Phys. (1960) 33, 1094-1106.
23. Goldman, S.A., Bruno, G.V., Polnaszek, C.F. and Freed, J.F.,
    J. Chem. Phys. (1972) 56 (2), 716-735.
24. Griffith, O.H., Dehlinger, P. J. and Van, S.P., J. Membrane
    Biol. (1974) 15, 159-192.
25. Dye, J.L. and Dalton. L.R., J. Phys. Chem. (1967) 71 (1),
    184-191.
26. Stryukov, V.B., Stunzhas, P.A. and Kirillov, S.T., Chem.
    Phys. Lett. (1974) 25 (3), 453-456.
27. Keith, A.D. and Snipes, W., Science (1974) 183, 666-668.

28. Morse, P.D., Ruhlig, M., Snipes, W., and Keith, A.D., Arch.
    Bioch. (1975) 168, 40-56.
29. Glasstone, S., Laidler, K.J. and Eyring, H., "The theory of
    rate processes", p. 666, McGraw-Hill, N.Y., 1941.
30. Pushkin, J.S. and Gunter, T.E., Bioch. Biophys. Acta. (1972)
    275, 302-307.
31. Stilbs, P. and Lindman, B., J. Colloid Interface Sci. (1974)
    46 (1), 177-179.
32. Chasteen, M.D. and Hanna, M.W., J. Phys. Chem. (1972)
    76 (26), 3951-3958.
33. Menger, F.M., Saito, G., Sanzero, G.V. and Dodd, J.R., J. Am.
    Chem. Soc. (1975) 97, 909-911.
34. Zaccai, G., Blasie, J.K. and Shoenborn, B.P., Proc. Natl.
    Acad. Sci. U.S.A. (1975) 72 (1), 376-380.

42

Bounds on Bound Water: Transverse and Rotating Frame NMR Relaxation in Muscle Tissue

H. A. RESING and A. N. GARROWAY

Naval Research Laboratory, Washington, D. C. 20375

K. R. FOSTER

The University of Pennsylvania, Philadelphia, Pa. 19104

Most of the water in muscle tissue has an NMR transverse relaxation time (T_2) about one-fiftieth that of the pure liquid, an effect known for twenty years (1) without an interpretation of concensus (2,3), despite proposed use of NMR relaxation measurements for cancer diagnosis (4,5). NMR data in rat (3) and frog (6) skeletal muscle suggest that a small observed proton fraction with a T_2 of about one msec exchanges at an "intermediate" rate (7,8) with the major portion of cell water, and thus lowers the relaxation time of this major fraction below that of pure water. We reported earlier (9) that single muscle cells from the giant barnacle, Balanus nubilus, also exhibit a small proton fraction with a millisecond T_2. However, within experimental limits, these protons do not exchange with D_2O in times as long as 24 hours, and cannot be the cause of the proton relaxation in the major portion of tissue water. In contrast the non-freezing water protons which we observed (9) at -34°C (attributed to "bound water" in studies (10) of protein solutions and tissues) do exchange with D_2O and are clearly different from the non-exchanging protons which we observe in these cells. Finally, we showed (9) that in the barnacle proton and deuteron transverse relaxation times for the major portion of the muscle water do fit an "intermediate exchange rate" model which requires only 0.1% of the water molecules to be in an "irrotationally bound" state at room temperature. The reader is referred to the earlier work (9) for experimental details and the more global picture.

The purpose of this paper is to describe more fully the "intermediate exchange rate" model proposed earlier (9) and to examine its consequences for rotating frame relaxation time ($T_{1\rho}$) measurements in muscle systems. The model predicts a $T_{1\rho}$ dispersion in "H_1 space" centered about a value of H_1 equal to the r.m.s. local field of the irrotationally bound water molecule, ca. 3 gauss. Available data for frog (11) and mouse (12) muscle systems are in semiquantitative agreement with the model.

In muscle tissue, proton transverse relaxation can generally be described as the sum of four exponential decays; the proton fractions (i.e. fractional amplitudes corresponding to each of the exponential terms) are labeled (3,9) by the mnemonics "protein", "msec", "major" and "extracellular" in order of increasing respective T_2 values. Fractions and T_2 values for the rat (3) and barnacle systems are assembled in Table I. In this paper we concentrate on the "major" proton fraction, which accounts for over 90% of the water in the barnacle fiber and has a T_2 about 1% that of bulk water. Other important features found for the barnacle fiber (9) are these: a) the ratio of the deuterium transverse relaxation time (in a deuterated barnacle) to that of protons (in undeuterated barnacle) is much smaller than for the bulk liquids; b) on deuteration the proton relaxation time in the barnacle does not rise as much as would be expected for a similar deuteration level in the bulk liquid; and c) on lowering the temperature to 5°C there is a small but significant rise in T_2. All of these facts are consistent with an exchange model in which exchange rates are just at the borderline between "intermediate" and fast", as we discuss below. The simplistic interpretation for the small T_2, that the intracellular water is one hundred times more viscous, is, of course, rejected, mainly because diffusion in intracellular water is not much less than in bulk water (13).

Model for Transverse Relaxation

The low value of T_2 is believed to arise as an "averaging" effect by exchange of water of normal viscosity with bound water molecules or other proton species which are relatively less free to move and hence have smaller T_2 values. If the fraction P_b of bound protons is relatively small then equation (1) is applicable for the observed transverse relaxation time T_{2a}' of the major fraction (8):

$$(T_{2a}')^{-1} = T_{2a}^{-1} + P_b \, [(1-P_b) \, \tau_b + T_{2b}]^{-1}. \qquad (1)$$

where T_{2a} is the intrinsic relaxation time of the "free" water (about that of bulk water), τ_b (τ_a) is the lifetime of a proton in the bound (free) state, and T_{2b} is the intrinsic relaxation time in the bound state. By detailed balance

$$\tau_a/(1-P_b) = \tau_b/P_b \qquad (2)$$

Each "term" of equation (1) dominates in a given temperature (i.e. exchange rate) range, as is shown schematically in Figure 1 for a hypothetical system in which $P_b = 0.01$. These ranges are labeled "slow", "intermediate", and "fast" accordingly as the first (intrinsic relaxation of the "free" state), second (rate of exchange into the bound state) or third (weighted average relaxation in the bound state) "term" of equation (1) predominates respectively. In this paper the intrinsic relaxation rate in the free state is neglected because the observed value of $T_{2a}{}'$ is so small. The complete range of behavior illustrated in Figure 1 has been observed, for example, in a recent study (8) of a hydrolyzed zeolite system. A set of experiments involving deuteration was carried out by Woessner, et al. (14) and T_2 minima were observed for aqueous agarose gels; interpretation, along the lines given in this paper, required strange temperature dependence for P_b however; that very complete set of data are not subject to the simple interpretation we give here for the barnacle fiber.

The working hypothesis is that the bound state consists of water molecules irrotationally bound to a rigid substrate (15). The only motion such a molecule executes is its jump out of the bound state; this means that there is no dipolar (or quadrupolar, in the case of deuterons) motional averaging in the bound state, and that the correlation time for the bound molecule is its life-time in the bound state, τ_b. Thus, in the motional narrowing region the intrinsic relaxation time for the bound state is given as (16)

$$T_{2bo}{}^{-1} = \sigma_o{}^2 \tau_b \qquad (3)$$

Here $\sigma_o{}^2$ is the Van Vleck (17) second moment of a proton in a bound molecule due both to its intramolecular partner and to protons in the substrate to which it is bound. The subscript zero on T_{2b} indicates a muscle of natural isotopic composition. A consequence of the equality of exchange and correlation times is that the P_b/T_{2b} and the $P_b/[(1-P_b) \tau_b]$ contributions are governed by the same activation energy, giving rise to slopes of T_2 vs T^{-1} equal in magnitude but opposite in sign for the fast and intermediate exchange regimes. It is the small positive slope observed (9) for T_2, referred to earlier, which leads us to suppose that the intermediate exchange rate regime may represent the relaxation for the major proton fraction in the barnacle muscle.

Isotopic Effects

Of the parameters necessary for further development P_b, the fraction of protons (and/or deuterons) in the bound state, and the bound state lifetime τ_b do not depend significantly on the deuteration level. There may indeed be a mass effect on τ_b, but if whole water molecules traverse the exchange reaction coordinate, this effect should be minimal. Only T_{2b}, through the second moment, is significantly affected by isotopic composition. The second moment, a mean square interaction strength, is of dipolar origin for protons and of quadrupolar origin for deuterons. For

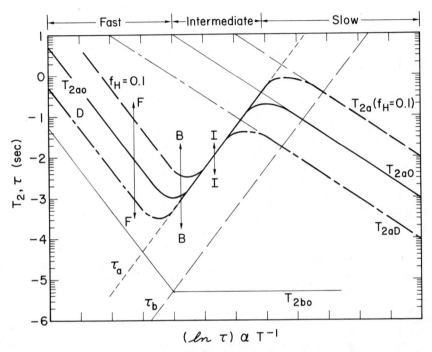

Figure 1.　Theoretical transverse relaxation times vs. inverse temperature for the model of irrotationally bound water molecules ($P_b = 0.01$) in exchange with free water, according to Equation 1 of text. The inverse temperature scale is uncalibrated, but the section B–B corresponds roughly to room temperature for barnacle fibers. The three exchange rate regimes are indicated above. The curves correspond (from top to bottom) to proton relaxation in a partially deuterated system, proton relaxation in a system of normal isotopic composition (T_{2ao}'), and deuteron relaxation (D). The intrinsic relaxation time T_{2bo} is also indicated.

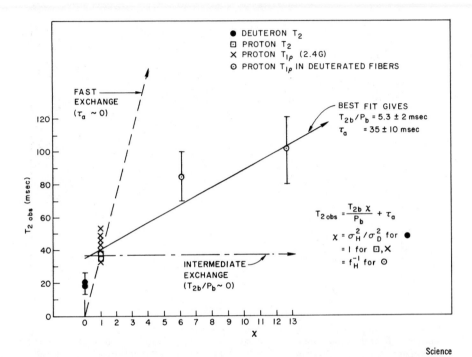

Figure 2.   Plot of T_{2a}' vs. x for the "major" water fraction in barnacle muscle fiber. Line labeled "fast exchange" corresponds to section F–F of Figure 1; likewise "intermediate" to section I–I, and "best fit" to B–B.   Note: $T_{2obs} = T_{2a}'$ of text (9).

the deuterons it is predominantly intramolecular in origin.   For
the protons, if the intramolecular part were dominant, or to the
extent that the surroundings are also deuterated, the second
moment in the partially deuterated system may be written as
$\sigma_d^2 = f_H \sigma_o^2$, where f_H is the fraction of protons remaining after
deuteration.   (As a preliminary estimate σ_o^2 is taken to be
1.6×10^{10} $rad^2 s^{-2}$, appropriate to a single water molecule.)   Thus
the intrinsic relaxation time for bound protons in a partially
deuterated system T_{2bd}, becomes (18)

$$T_{2bd}^{-1} = f_H \sigma_o^2 \tau_b. \tag{4}$$

Here the effects on σ_o^2 of all non-proton magnetic fields are
excluded.   For the deuterons, the mean square interaction strength
of the quadrupole moment with its local electric field gradient
in a deuteroxyl group is about ten times stronger than the proton-
proton nuclear magnetic dipolar second moment in a water molecule (8, 19); for a deuteron the intrinsic relaxation time in
the bound state may then be written as

$$T_{2bD}^{-1} \simeq 10 \, \sigma_o^2 \, \tau_b. \tag{5}$$

In (5) all relaxation mechanisms other than quadrupolar are
neglected.   The consequences of equations (4) and (5) on
equation (1) are also illustrated in Figure 1, under the assumption that T_{2a} and T_{2b} are similarly affected by deuteration or
nuclear species.

Transverse Relaxation in Barnacle Muscle Fiber

Our experiment consists in making an isothermal section
through Figure 1.   We desire to find out in what exchange rate
regime the barnacle muscle system is by comparison of the
observed $T_{2a}{}'$ section to those illustrated in Figure 1, namely:
a) I-I, intermediate, no dependence of $T_{2a}{}'$ on deuteration level
or isotopic species; b) F-F, fast exchange, full dependence;
c) B-B, borderline, moderated dependence.

A simple summary of the above, very restrictive, model is
given by the expression:

$$T_{2a}{}' = \tau_a + x \, T_{2bo}/P_b. \tag{6}$$

This conveniently linear equation allows the sections through
Figure 1 to be plotted vs the parameter x, which is f_H^{-1} for
protons in the deuterated muscle, or $\sigma_{oD}^2/\sigma_o^2 \sim 0.1$ for the
deuterons.   The data for the barnacle muscle are plotted vs x
in Figure 2, along with the limiting sections for fast and intermediate exchange according to equation (6).   Figure 2 suggests
that a borderline secion such as B-B is most nearly applicable;

the best fitting parameters are indicated. From the slope
(T_{2bo}/P_b) and intercept (τ_a) and equations (2) and (3) above, the
freely adjustable parameters are deduced as $P_b = (0.08 \pm 0.02)\%$,
and $T_{2b} = 3.5 \pm 1.5$ µs (or equivalently $\tau_b = 20 \pm 10$ µsec). Thus
about one molecule per thousand, irrotationally bound for only
tens of microseconds, can account for the transverse relaxation
times of the great majority of the water in barnacle muscle cells.
Such a small fraction is unlikely to be found by searching for
its contribution to the free induction decay, even though it is
in principle observable. We make no attempt here to explain
the value of P_b in terms of biochemical structure; the only
implication of this model is that there is some substrate which
can bind a water molecule for 20 microseconds and not move
itself during this time.

On the other hand the values of T_{2bo} or τ_b which emerge are
eminently reasonable in terms of the n.m.r. model proposed. The
model of an exchanging, irrotationally bound water molecule, as
sketched above, requires that the T_{2a}' minimum be at the same
temperature as the motional narrowing onset (knee) for T_{2bo} of
the bound molecule (see Figure 1). This knee is marked by the
condition that

$$\tau_b \sim T_{2bo} \sim T_{2RL} \qquad (7)$$

where T_{2RL} is the intrinsic, rigid-lattice relaxation time in
the bound state, about 10 µsec. Clearly T_{2bo} cannot become
smaller than T_{2RL}. The plot of experimental values of T_{2a}'
vs x (Figure 2) implies a temperature slightly lower than that
of the T_{2a}' minimum (Figures 1 and 2). The values of T_{2bo} and
τ_b derived from the data by application of the model are of the
order of 10 µsec demanded by (7); this is a crucial test; the
hypothesis lives! Note however that equations (3), (4) and (5)
apply only in the motional narrowing region, and begin to break
down (not catastrophically though) as the rigid lattice region,
i.e. intermediate exchange region is approached.

Relaxation in the Rotating Frame

The "irrotationally bound water molecule" model may be
tested further, for it has certain, not superficially obvious,
implications for the rotating frame relaxation experiment.
And though such $T_{1\rho}$ data do exist in the literature for other
muscle systems (11, 12), the interpretations given assume the
fast exchange condition, under which $T_{1\rho}$ observed for the major
fraction $(T_{1\rho a}')$ is simply $P_b/T_{1\rho b}$, and any $T_{1\rho}$ dispersion which
occurs in the bound state reflects its weighted contribution in
the major fraction (look ahead to equation (9') for $T_{1\rho b}$). The
correlation time deduced from the $T_{1\rho a}'$ dispersion is then
characteristic of some motion in the bound state which may or may
not be the exchange reaction. (Equation (1) for T_{2a}' is assumed

to hold as well for $T_{1\rho a}{}'$, mutatis mutandis.)

In the intermediate exchange regime the situation is not so simple, as the following example will make clear. Recall, from equations (1) and (2), that intermediate exchange is characterized for $T_{2a}{}'$ by the inequalities

$$T_{2a} > \tau_a > \tau_b > T_{2b} \qquad (8)$$

where the center inequality holds because we are interested in a system where $P_a > P_b$. Retreat for a moment from the specific model of irrotationally bound water molecules and allow the bound state to have a correlation time τ_{cb} which is much less than the bound state lifetime τ_b. Suppose that the bound state is in the motional narrowing (16), i.e. weak collision (20), region of τ_{cb}, so that an equation like (3) applies for T_{2b}. Under these conditions the intrinsic rotating frame relaxation time for the bound state is given (21) as

$$T_{1\rho b} = (1 + 4\gamma^2 H_1{}^2 \tau_{cb}{}^2)/(\sigma_o{}^2 \tau_{cb}) \qquad (9)$$

$$= T_{2bo} (1 + 4\gamma^2 H_1{}^2 \tau_{cb}{}^2) \qquad (9')$$

where γ is the nuclear gyromagnetic ratio, and H_1 is the applied radio frequency field. For small H_1, $T_{1\rho} \sim T_2 < \tau_b$, and the intermediate exchange rate condition must hold for $T_{1\rho}$ as well as for T_2. However, when H_1 becomes so large that $T_{1\rho b} \sim \tau_b$ two things occur: a) the system passes into the fast exchange regime for $T_{1\rho}$, and b) a dispersion in the observed rotating frame relaxation time for the "major" fraction, $T_{1\rho a}$ commences which depends on both the correlation time and the lifetime in the bound state. That is, under the assumption that $\gamma^2 H_1{}^2 \tau_{cb}{}^2 > 1$,

$$T_{1\rho b} = 4 \gamma^2 H_1{}^2 \tau_{cb}{}^2/(\sigma_o{}^2 \tau_{cb}) \sim \tau_b \qquad (10)$$

and, since the local field in the bound state (H_{Lb}) is $\gamma^{-1}\sqrt{\sigma_o{}^2/3}$ (22), we have at this point

$$H_1 = H_{Lb} (3/4 \; \tau_b/\tau_{cb})^{1/2} \qquad (11)$$

Under such conditions the dispersion region observed for the major fraction does not truly reflect the dispersion going on for $T_{1\rho b}$ and does not furnish an unambiguous measurement of some correlation or exchange time. Indeed, if one considers the possible values of the ratio $\tau_b/\tau_{cb} > 1$ and of the local field (~ 1 gauss), the $T_{1\rho b}$ dispersion may be completely hidden for lack of large enough H_1.

There is a strong temptation to recover the model of irrotationally bound water molecules by setting $\tau_{cb} = \tau_b$ (as discussed earlier). Then the content of equation (11) is that there should be a $T_{1\rho a}{'}$ dispersion when the applied r.f. field is about the same as the local nuclear magnetic dipolar field. This limiting procedure is not strictly allowed, because the intermediate exchange regime for the irrotationally bound water molecule implies a rigid-lattice state, i.e. a strong-collision (20) regime for $T_{1\rho b}$, contrary to the assumption in the previous paragraph. But the validity of using the dispersion of $T_{1\rho a}{'}$ in H_1 space as a technique to measure the local field for the bound state can indeed be shown. The result obtained when $\tau_{cb} = \tau_b$ is (23)

$$T_{1\rho a}{'} = T_{2a}{'} \, (H_1{}^2 + H_{Lb}{}^2)/H_{Lb}{}^2 \qquad (12)$$

The consequences of this equation are illustrated in Figure 3. There it can be seen, in contrast to Figure 1 for T_2, that "variability" of $T_{1\rho}$ occurs in the intermediate exchange regime and "constancy" in the fast exchange regime; and for all practical purposes, the borderline regime is equivalent to the intermediate. Based on the second moment for a single water molecule, H_L for an irrotationally bound water molecule is 2.7 G. In Figure 4 the $T_{1\rho a}$ data for mouse (12) and frog (11) muscle are plotted along with theoretical plots of equation (12) in which H_L was taken as 3 G (the excess over 2.7 G being arbitrary allowance for an intermolecular field). The agreement, with essentially no adjustable parameters, is encouraging. The model of irrotationally bound water molecules exchanging with "free water" at (borderline) intermediate rate predicts a $T_{1\rho a}{'}$ dispersion exactly where it does occur.

Now Figure 3 shows quite clearly the temperature dependence expected for $T_{1\rho a}{'}$; as the temperature decreases $T_{1\rho a}{'}$ (as well as T_{2a}) should rise. We did see this behavior for barnacle muscle fiber (Table I) (9). On the other hand, Homer and Finch (11) in their study of frog muscle saw a small temperature dependence for $T_{1\rho a}$ opposite in sign to that we would expect. At the moment we are reluctant to allow this contrary sign of the temperature dependence for frog muscle to discredit the model completely. Rather we await a more complete set of measurements on many systems so that the model may be tested further.

Summary

The model of irrotationally bound water molecules has been proposed to account for certain NMR relaxation effects, in muscle systems namely (a) the small value of T_2 generally observed for intracellular water, b) the effect of dilution by deuterium on the transverse relaxation time of protons in the

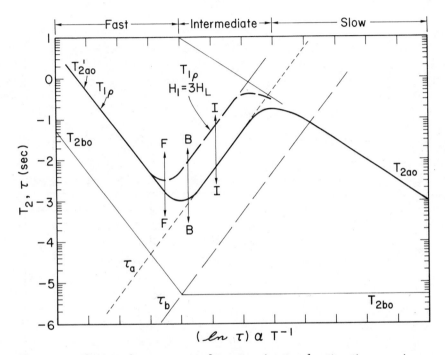

Figure 3. *Theoretical transverse and rotating frame relaxation times vs. inverse temperature according to Equations 1 and 12 of text. The* $T_{1\rho}$ *plot corresponds to* $T_{1\rho a}$ *for a rotating magnetic field* H_1 *which is about three times the intramolecular field for an irrotationally bound water molecule.*

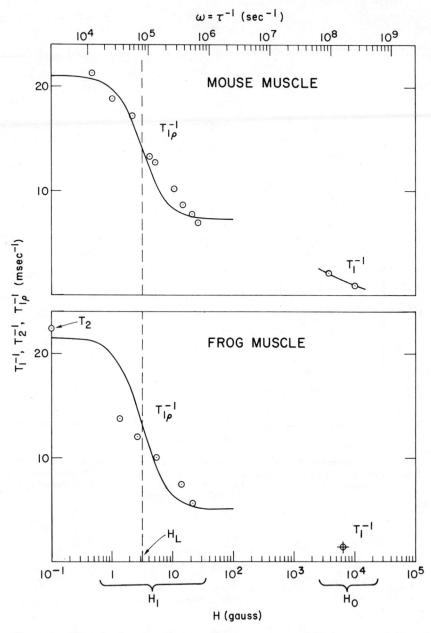

Figure 4.   Plots of relaxation rates vs. effective magnetic field for mouse (12) and frog (11) muscle tissue.   Theoretical curves for "T_{1p} dispersion in H_1 space" according to Equation 12 are also given; the center of the dispersion region necessitated by irrotationally bound water molecules in intermediate exchange is indicated as H_L.   The conventional inverse correlation time scale is given above.   Note: H_L is H_{Lb} of text.

intracellular water of barnacle muscle fibers, c) the ratio of the deuterium transverse relaxation time to that of protons in the same system, and d) the dispersion of the rotating frame relaxation time for the protons in the intracellular water of mouse and frog muscle. And in each case it has been successful. The fraction of the total number of protons in the bound state, according to the model, is about 0.1%, and the local field in the bound state is found to be that expected for a water molecule.

Table I

Summary of Room Temperature Transverse Relaxation Data
in Single Barnacle Muscle Cells and in Rat
Gastrocnemius Muscle Tissue

Proton fraction Designation	T_2, (msec)	Relative Amplitude	Tentative Proton Source
Rat[a]			
Protein[b]	~ 0.02	~ 0.2	tissue proteins
Msec	0.4	0.08	"bound water"
Major	45	0.82	intracellular water
Extracellular	200	0.10	extracellular water
Barnacle			
Protein[b]	~ 0.02	~ 0.2	tissue proteins
Msec	$0.75 \pm .25$	$0.033 \pm .006$	tissue proteins, lipids, water?
Major	35	0.92	intracellular water
	$T_{1\rho} = 45$ at $25°C$[c]		
	$T_{1\rho} = 55$ at $5°C$[c]		
	$T_{2D} = 20$ at $25°C$[d]		
Extracellular	400	0.03	extracellular water

a) Data from reference 3
b) Not counted in the water: other fractions sum to unity.
c) Rotating frame relaxation time $T_{1\rho}$ at $H_1 = 2.4$ G, from CPMG sequence at $2\tau = 50$ μs.
d) Deuteron transverse relaxation time.

Literature Cited

1. E. Odeblad, B. Bhar, and G. Lindstrom, Arch. Biochem. Biophys. 63, 221 (1956).
2. R. Cooke and I. D. Kuntz, Ann. Rev. Biophys. and Bioeng. 3, 95 (1974).
3. C. F. Hazlewood, D. C. Chang, B. L. Nichols, and D. E. Woessner, Biophys. J. 14, 583 (1974).
4. R. Damadian, Science 171, 1151 (1971).
5. R. R. Knispel, R. T. Thompson, and M. M. Pintar, J. Mag. Res. 14, 44 (1974).
6. E. D. Finch and L. D. Homer, Biophys. J. 14, 907 (1974).
7. The "intermediate exchange rate" regime is that in which the observable relaxation rate for the major fraction is determined by its lifetime against exchange into the bound state rather than by the relaxation time of the bound state. For fundamental details see J. R. Zimmerman and W. E. Brittin, J. Phys. Chem. 61, 1328 (1957) and H. A. Resing, J. Phys. Chem. 78, 1279 (1974); for a particularly complete example see ref. 8.
8. J. S. Murday, R. L. Patterson, H. A. Resing, J. K. Thompson and N. H. Turner, J. Phys. Chem. 79, 2674 (1975); H. A. Resing, ibid, 80, 186 (1976).
9. K. R. Foster, H. A. Resing and A. N. Garroway, Science, in press.
10. I. D. Kuntz, Jr., T. S. Brassfield, G. D. Law, and G. V. Purcell, Science 163, 1329 (1969); P. S. Belton, K. J. Packer and T. C. Sellwood, Biochim. Biophys. Acta 304, 56 (1973); D. R. Woodhouse, D. W. Derbyshire and P. Lillford, J. Mag. Res. 19, 267 (1975).
11. E. D. Finch and L. D. Homer, Biophys. J. 14, 907 (1974).
12. R. R. Knispel, R. T. Thompson and M. M. Pintar, J. Mag. Res. 14, 44 (1974).
13. I. L. Reisiu and G. N. Ling, Physiol. Chem. 5, 183 (1973).
14. D. E. Woessner and B. S. Snowden, Jr., J. Colloid and Interface Sci. 34, 290 (1970).
15. The concept of "irrotationally bound" water was introduced to explain NMR relaxation in protein solutions (H.A. Resing and J. J. Krebs, Abstr. 140th Meeting, Am. Chem. Soc. 13C (1961)) but such a model pushed to its logical conclusion is evidently not supported by recent data and interpretation in protein solutions (see S. H. Koenig, K. Halenga, and M. Shporer, Proc. Nat. Acad. Sci. USA 72, 2667 (1975)). See also (2). In contrast to the work in protein solutions, we assume the substrate to which the bound molecule is attached to be stationary.
16. N. Bloembergen, E. M. Purcell, and R. V. Pound, Phys. Rev. 73, 679 (1948).
17. J. H. Van Vleck, Phys. Rev. 74, 1168 (1948).
18. W. A. Anderson and J. T. Arnold, Phys. Rev. 101, 511 (1956).

19. D. E. Woessner, J. Chem. Phys. 40, 2341 (1964); J. G. Powles and M. Rhodes, Proc. Colloq. AMPERE 14, 757 (1967).

20. D. Wolf and P. Jung, Phys. Rev. B12, 3596 (1975).

21. D. C. Look and I. J. Lowe, J. Chem. Phys. 44, 2995 (1966); ibid 44, 3437 (1966).

22. C. P. Slichter and D. C. Ailion, Phys. Rev. A135, 1099 (1964).

23. J. S. Murday and H. A. Resing, to be published. See also H. A. Resing, Adv. Mol. Relax. Proc. 3, 199 (1972).

INDEX